Maynooth Library
D1321081

Detection Theory

Applications and Digital Signal Processing

Detection Theory

Applications and Digital Signal Processing

Ralph D. Hippenstiel

CRC PRESS

Boca Raton London New York Washington, D.C.

Library of Congress Cataloging-in-Publication Data

Hippenstiel, Ralph Dieter, 1941-
Detection theory : applications and digital signal processing / Ralph D. Hippenstiel.
p. cm.
Includes bibliographical references and index.
ISBN 0-8493-0434-2 (alk. paper)
1. Signal processing—Digital techniques. 2. Signal detection—Mathematics. 3. Electronic surveillance. 4. Detectors. 5. Signal theory (Telecommunication)—Mathematics. I. Title.

TK5102.9 .H58 2001
621.382′2—dc21 2001043859

Direct all inquiries to CRC Press LLC, 2000 N.W. Corporate Blvd., Boca Raton, Florida 33431.

International Standard Book Number 0-8493-0434-2
Library of Congress Card Number 2001043859
Printed in the United States of America 2 3 4 5 6 7 8 9 0
Printed on acid-free paper

Preface

This book has its origin in lecture notes created at the Naval Postgraduate School (NPS), whose student and faculty support made this work possible. The book is written for senior undergraduates and first-year graduate students. Practicing engineers should especially find the book useful in obtaining the basic knowledge of detection and estimation theory. The main part of the text deals with detection aspects, while the estimation part is presented as needed.

Many modern signal processing ideas are embedded in the approach to detect events. Many aspects of the fast Fourier transform (FFT) and the wavelet transform (WT) are examined and discussed in the context of detection and parameter estimation. Digital signal processing has had a tremendous influence on radar, sonar, digital communications, and the related detection and estimation area. To this end, where possible, the signal processing ideas are incorporated into the text.

Where possible, the use of complex envelopes is avoided, even though acquaintance with the discrete Fourier transform (DFT) makes the concept simple. In Appendix B, the complex signal notation is used and related to the DFT. Also, Chapter 9, the applications chapter, uses complex valued expressions.

Chapter 1 introduces the concept of detection and estimation. Chapter 2 reviews deterministic and random concepts, which are also addressed in Appendices A, B, and D. Basic signal processing, focusing on the FFT and the WT are discussed in Chapter 3. Filtering, FIR filters, and the periodogram are examined in detail. Appendix E contains more information related to the wavelet transform. Chapter 4 explores hypothesis testing, using many examples. Non-parametric and sequential detection are discussed in Chapter 5. Detection of dynamic signals embedded in white Gaussian noise is presented in Chapter 6. Also, the notion of coherent and incoherent averaging is discussed. Chapter 7 deals with the detection of signals embedded in colored Gaussian noise. Approaches using series type expansion are presented. Estimation is presented in a very rudimentary form in Chapter 8. Chapter 9 is dedicated to a variety of detection, parameter estimation, and classification problems. Basic topics that are introduced are: Periodogram, Spectrogram, Correlation, Instantaneous Correlation Function, Wigner-Ville Distribution, Spectral Correlation, Ambiguity Function, Cyclo-Stationary Processing, Higher Order Moments and Poly-Spectra, Coherence Processing, Wavelet Processing, and Adaptive Techniques.

A typical class reviews aspects of Appendices A–C and Chapter 2. Chapters 1, 3–6, and 8, and Sections 7.1-7.7 and 9.2 are completely covered in the eleven weeks of a teaching quarter with four hours of lecture per week. Some MATLAB-based projects are assigned to solidify the understanding of the lecture material. Chapter 3 uniquely tailors the operation of the Fourier and wavelet transforms (FFT and WT) to the intended audience.

I want to thank Professors Herschel H. Loomis, Jr., Roberto Cristi, and Monique P. Fargues of the Naval Postgraduate School (NPS) for their helpful review and thoughtful comments on some of the chapters. I also want to thank numerous students at NPS that provided feedback for early versions of Chapters 1, 2, 4–7, and 8. In particular, I want to acknowledge the helpful comments by students: Brian T. Alexander, Christos Athanasiou, Jaime C. L. Briggs, Robert D. Broadston, Daniel B. Copeland, Athanasios Konsolakis, Dimitrios Koupatsiaris, Kyle E. Kowalske, Mitchell Shipley, Ah Tuan Tan, and Craig A. Wilgenbusch. Special thanks go to Jim Allen for the line drawings and the typesetting of this work.

The author welcomes feedback and suggestions. No doubt, with probability one, some errors did occur. The author can be reached at: rdhippen@nps.navy.mil or hipp@montereybay.com.

Finally, I want to thank my immediate family, my wife Sylvia, my daughters Claudia, Patricia, and Linda as well as my mom Sofie, to whom the book is dedicated, for their encouragement and understanding. The dedication of this book is made realizing that words alone cannot make up for the lost time in their lives.

Ralph D. Hippenstiel
Monterey, California

Author

Ralph D. Hippenstiel is currently an Associate Professor of Electrical and Computer Engineering at the Naval Postgraduate School in Monterey, California. A graduate of the University of Texas at El Paso, the author holds an MS degree in Electrical Engineering from the University of Texas at El Paso, an MSEE degree in Information Science from the University of California at San Diego and a Ph.D. in Electrical Engineering from the New Mexico State University at Las Cruces. Professor Hippenstiel has taught at the University of Texas at El Paso, Hartnell College in Salinas, CA, and the Naval Postgraduate School in Monterey, CA. He has worked for eight years as an Electrical Engineer at the Naval Ocean Systems Center (NOSC) in San Diego, CA and was a visiting professor at the Universitat der Bundeswehr in Neubieberg, Germany.

Contents

Chapter 1

Introduction

1.1 GENERAL PHILOSOPHY

In the age of modern warfare the theory of detection and estimation has become a very important topic. The development of radar, sonar, digital communications, and digital signal processing has immensely stimulated these areas. With the advent of the flying machine, which can deliver ordinance to any point at any time, early detection has become essential. The roots of modern detection theory can be found in the desire to protect valuable resources by detecting and destroying enemy airplanes and missiles under all types of environmental conditions. A large body of detection-related articles and books can be found in the statistics area. Hypothesis testing relates to a topic that historically deals with statistics. This statistics topic focuses on the detection and/or estimation of certain phenomena and usually includes the computation of the errors associated with these procedures.

World War II has a rich warfare history that demonstrates the importance of detecting enemy attacks reasonably early to allow for countermeasures or at least for some evasive action. Detection must be sufficiently early to minimize the loss of life. Of course, to obtain a timely warning, early detection is essential and hence, detection must usually be accomplished at very low signal-to-noise ratios (SNRs). In recent history, the development of nuclear, chemical, and biological armament and the many covert ways of delivering this type of ordinance have made detection even more important. For example, it is relatively easy to hide a submarine, a potential missile deployment system, in the ocean. Passive and active sonar systems allow the detection, identification, and localization of submerged platforms. Another example is the ballistic missile system that can be hidden in movable containers. Ballistic missiles, for example the scud missile in the conflict with Iraq, can be detected by radar, optical or imaging systems.

In each one of these scenarios, automated detection allows scanning of large data sets for the purpose of detecting, localizing, and identifying particular targets of interest. We realize, if there is no distortion (i.e., no noise or interference) when observing data then one can readily deduce the presence (or absence), and when appropriate, the type of the signal emitter.

Typically, throughout this text, we assume that the noise is of the additive type, that is, the observed (received) component consists of a signal or signals embedded in additive noise. With the exception of Chapter 4, which discusses hypothesis testing, most of the noise will be of the Gaussian type. During the first part of the text, we will deal with delta correlated noise (i.e., white noise), while later sections will address colored noise.

Since the 1960s modern digital signal processing has influenced all areas of signal conditioning and signal processing. Some detector implementations will draw on this topic. To this extent, some introductory digital signal processing is included, especially as it relates to the use and interpretation of the fast Fourier transform (FFT) and the wavelet transform (WT), both of which have revolutionized many signal processing applications.

Many books have been written that cover detection related topics [1–11]. The reference list is, by no means, exhaustive, but it is meant to provide additional helpful material. At this time, references [1–5,7–11] remain in print. These books are excellent references but are sometimes a little difficult to read.

Our text attempts to question why things are set up in a particular way and how the results are derived. Our students, primarily military officers, are interested in the topics from a practical point of view because good understanding will potentially affect the operation and success of some of their future missions.

The mathematical level is kept at a minimum, with most of the necessary ideas developed in the text or given in the appendices. If that is not feasible, at least ample references are provided. Mathematics is kept at a level where the typical engineer can follow and understand the material. The mathematical development is appropriate for engineers and physicists and in no way is meant to be rigorous.

The presentation is oriented toward Hilbert spaces and projections (inner products) in these spaces. This allows a general description and solutions for different scenarios. With this approach, the extensions to other basis functions such as Karhunen-Loève is a natural one, allowing a straightforward interpretation.

1.2 DETECTION AND ESTIMATION PHILOSOPHY

The emphasis of this text is on the detection aspect, that is, the declaring of the occurrence of a particular event with some measure of confidence. Estimation can be interpreted as an extension of the detection part. It answers the question as to how much of a particular item of interest there is (i.e., a fine localization in the parameter space) and provides a measure of the accuracy or confidence.

1.2.1 Detection

Some typical detection examples are

(a) Radar: passive, active, or bi-static (i.e., it uses an antenna and tests for target presence)

(b) Sonar: bi-static, active, or passive (i.e., it uses hydrophones and tests for target presence)

(c) Digital communication: coded binary words (symbols), (i.e., one wants to detect transmitted, possible encoded, message bits consisting of binary zeros and ones)

(d) Acoustic detection (i.e., it uses microphones): for intrusion alarm, emitter detection (i.e., gunnery, helicopter sound detection/localization, etc.)

(e) Seismic detection (i.e., it uses geophones): to detect an earthquake, tunnel digging, nuclear testing, etc.

In general, we obtain data (also called measurements or observations) which is thought to consist of one or several signal components embedded in additive noise. The noise may be natural (i.e., the environment, medium, channel, electronics, etc.) or man-made (i.e., jammer, power lines, shipping noise, oil exploration activities in an ocean environment, etc.). Based on the received data, we try to decide whether or not a particular event has occurred. Note, there is also a subclass of problems. For instance, see Example 4.17, where the prevailing probability density or a statistical moment indicates the occurrence of a particular event. If convenient, we shall use as the start time $t = 0$, realizing that it can easily be changed to $t = t_0$, if the need arises.

We separate the detection problem into three types of classes:

- **Class I:** Known signals in additive noise.
- **Class II:** Signals with unknown parameters in additive noise.

Signal shape, type, etc., is known except for some signal-related parameter or parameters. Typical examples are unknown carrier frequency, amplitude, or phase.

- **Class III:** Random signals in additive noise.

  Signals are described statistically. That is, the unknowns are described statistically (i.e., via probability density functions, or moments, or correlation functions).

(a) **Class I** (known signals in additive noise): One will decide, based on the received data $r(t)$, whether or not the message (or signal) s_0 or s_1 was transmitted.

Example 1.1

$$\begin{aligned} s_0(t) &= \cos(\omega_0 t) \;\; \text{(i.e., a signal with carrier frequency } \omega_0\text{)} \\ s_1(t) &= \cos(\omega_1 t) \;\; \text{(i.e., a signal with carrier frequency } \omega_1\text{)} \\ r(t) &= s_i(t) + n(t) \;\; \text{for } 0 \le t \le T \quad \text{and } i = 0, 1 \end{aligned}$$

where $r(t)$ is the received, noise corrupted signal and $n(t)$ is an additive noise process.

Example 1.2

$$\begin{aligned} s_0(t) &= A_0 \;\; \text{(i.e., a DC level of value } A_0\text{)} \\ s_1(t) &= A_1 \;\; \text{(i.e., a DC level of value } A_1\text{)} \\ r(t) &= s_i(t) + n(t) \;;\; \text{for } 0 \le t \le T \;;\; i = 0, 1 \end{aligned}$$

where $r(t)$ is the received, noise corrupted signal and $n(t)$ is an additive noise process.

Example 1.3

$$\begin{aligned} s_0(t) &= \cos(\omega_c t) \;\; \text{(i.e., a signal with carrier frequency } \omega_c\text{)} \\ s_1(t) &= \sin(\omega_c t) \;\; \text{(i.e., a phase shifted version of } s_0(t)\text{)} \\ r(t) &= s_i(t) + n(t) \;;\; \text{for } 0 \le t \le T \text{ and } i = 0, 1 \end{aligned}$$

where $r(t)$ is the received, noise corrupted signal and $n(t)$ is an additive noise process.

(b) **Class II** (signals with unknown parameters in additive noise): One will decide, based on the received data $r(t)$, whether or not message (or signal) s_0 or s_1 was transmitted.

Example 1.4

$$r(t) = cos(\omega_i t + \theta_i) + n(t) \quad \text{for } i = 0, 1 \text{ and for } 0 \leq t \leq T$$

where $r(t)$ is the received, noise corrupted signal, θ_0 and θ_1 are unknown deterministic phase values, and ω_0 and ω_1 are known.

Example 1.5

$$\begin{aligned} s(t) &= \cos[\omega_c(t)] \quad \text{for } 0 \leq t \leq T \quad \text{(radar, sonar problem)} \\ r(t) &= \begin{cases} a\cos[(\omega_c + \omega_D)(t-\tau) + \theta] + n(t) \\ & \text{for } \tau \leq t \leq \tau + T \\ n(t) \end{cases} \end{aligned}$$

where a represents signal attenuation; the coefficient θ is an unknown phase shift, $n(t)$ is the additive noise component, ω_D represents the unknown Doppler shift, and τ is the unknown time delay (proportional to the round trip distance to the target).

(c) **Class III** (random signals in additive noise): One will decide, based on the received data $r(t)$, whether or not the message (or signal) $s_w(t)$ is present or not.

Example 1.6

$$r(t) = \begin{cases} s_w(t) + n(t) & ; \text{ if the target is present} \\ n(t) & ; \text{ if the target is not present for } 0 \leq t \leq T \end{cases}$$

where $s_w(t)$ is a realization of a random process, and a probabilistic description of $s_w(t)$ is available.

1.2.2 Estimation

Some typical examples are

(a) Frequency estimation (i.e., determine the actual Doppler shift)

(b) Differential time delay estimation (i.e., determine the distance or differential distance to a target)

(c) Amplitude estimation (i.e., target strength, size, distance)

(d) Phase and/or bandwidth estimation (i.e., target identification)

(e) Spectral estimation (i.e., target identification)

(f) Analog wave form estimation (i.e., original noise-free recovery of the original wave form using optimal filtering)

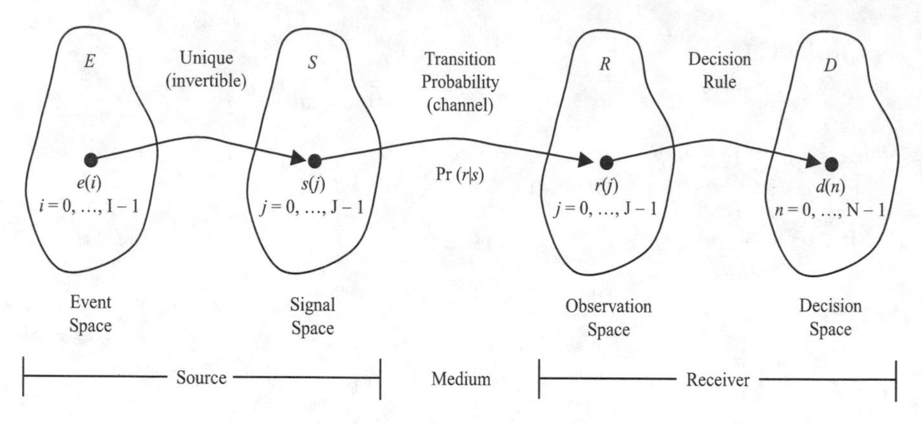

Figure 1.1: Decision model.

1.3 DESCRIPTION OF SPACES INVOLVED IN THE DECISION

The general signal/data flow for detection/estimation is shown in Figure 1.1. The general signal flow shows how the information moves from the event space, via the signal and observation spaces to the decision space, with the spaces denoted by E, S, R, and D, respectively. The events are encoded as signals (i.e., quantities that propagate through the channel), where one signal typically corresponds to one particular event. In some cases, the number of signals (J) can be different from the number of events (I). An example of this type is given by the following scenario. A single event "target present" may be represented by several different signal components (i.e., the emission of a distorted sinusoidal signal manifests itself as a basic sinusoid and additional harmonically related components). Hence, one event could be translated into or represented by several signals.

The signals are coupled via the channel, which introduces noise, to the receiver which also is called the detector. The receiver takes samples of the data and with some (optimal) processing makes a decision regarding the original event. If we test for the occurrence of a particular event, then we are addressing a detection problem. Conversely, if we evaluate the size of or the quantity relevant to the event, then we are addressing an estimation problem.

E: Event Space

In the event space E, one of the I possible events can happen. We may or may not know *a priori* probabilities of the event e_i (i.e., Pr $\{e_i\}$ for $i = 0, \cdots, I-1$). Typical events may be

(a) Messages: alphabetical, numeric, station keeping, Morse code

(b) Parameters: frequency, phase angle, object moving or not, hence, object speed, etc.

(c) Targets: target present/target absent, friend/foe, etc.

Note: The message may be in a discrete or continuous form, may be a random variable, random vector, or a realization of a random process.

S: Signal Space

The events are converted into representative signals. Of course, one mapping might use the original data (events) as the signal directly.

Example 1.7 *Phase Modulation*

Events : Signals :

$$\begin{array}{lllllllll} e_0 & = & 0^\circ & = & \theta_0 & \rightarrow & s_0(t) & = & A\cos(\omega_0 t + \theta_0) \\ e_1 & = & 180^\circ & = & \theta_1 & \rightarrow & s_1(t) & = & A\cos(\omega_0 t + \theta_1)\ ; \quad \text{for } 0 \leq t \leq T \end{array}$$

Example 1.8 *Event Encoding*

Events : Signals :

$$\begin{array}{lllllll} A & = & e_0 & \rightarrow & s_0(t) & = & \text{signal representing event } A \\ B & = & e_1 & \rightarrow & s_1(t) & = & \text{signal representing event } B \\ C & = & e_2 & \rightarrow & s_2(t) & = & \text{signal representing event } C \\ & \vdots & & & & \vdots & \qquad ; \text{ for } 0 \leq t \leq T \end{array}$$

Example 1.9 *Radar/Sonar Problem*

Figure 1.2 provides a typical target detection scenario. The signal $s(t)$ of duration T is transmitted. A target is R units in range away. The observation $r(t)$ is received by the receiver. The delay time is $t_R = 2R/C$; where C is the propagation velocity. No noise or Doppler is considered.

Events : Signals :

$$\begin{array}{llll} e_0(\text{no target}) & \rightarrow & \tau(t) = s_0(t) = 0 & ; \quad \text{for } 0 \leq t \\ e_1(\text{target present}) & \rightarrow & \tau(t) = s_1(t) = a\, s(t - t_R) & ; \quad \text{for } t_r \leq t_R \leq t_R + T \end{array}$$

where a is the signal loss coefficient.

Example 1.10 *Passive Detection*

Narrow band detection (passive detection addresses the question: is a particular spectral component at a certain frequency present or not ?).

Events : Signals :

$$\begin{array}{lll} e_0(\text{no target}) & \rightarrow & s_0(t) = 0 \\ e_1(\text{target present}) & \rightarrow & s_1(t) = A\cos(\omega_c t + \theta_c)\ ; \ \text{for } t_0 \leq t \leq t_0 + T \end{array}$$

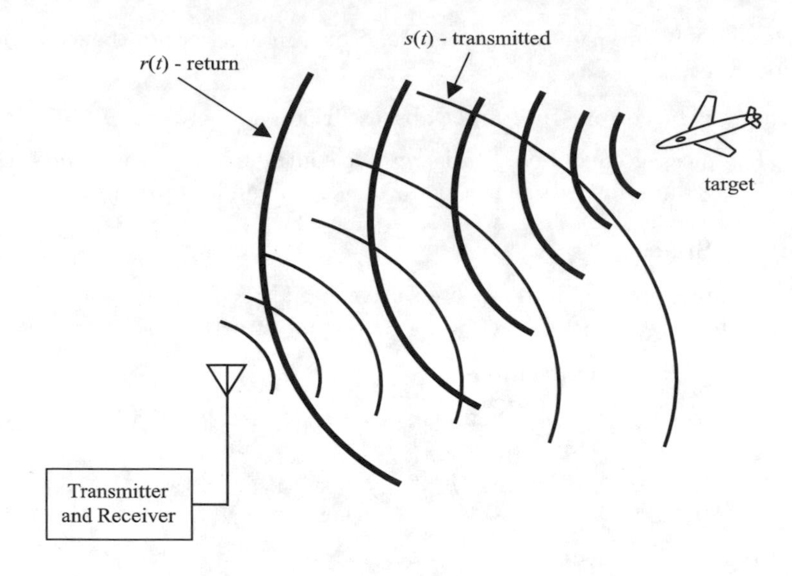

Figure 1.2: Typical target detection problem.

Example 1.11 *Communication System*

Binary communication problem (i.e., frequency shift keying (FSK))

Events :		Signals :	
e_0	$\rightarrow$	$s_0(t) = A\sin(\omega_0 t)$	; for $0 \leq t \leq T$
e_1	$\rightarrow$	$s_1(t) = A\sin(\omega_1 t)$	; for $0 \leq t \leq T$

Example 1.12 *Intrusion Alarm (i.e., motion detector for $0 \leq t$)*

Events :		Signals :
e_0(no intruder)	$\rightarrow$	$s(t) = A\cos(\omega_0 t + \theta)$
e_1(intruder present)	$\rightarrow$	$s(t) = A\cos(\omega_0 t + \omega_D(t)) + \theta)$

where $\omega_D(t)$ is a time dependent frequency shift.

R: Observation Space

The signals are observed (i.e., obtained) after having passed through some channel (i.e., the medium). Usually, this accounts for the additive noise component $n(t)$, which may be due to the medium, the electronics, and may also potentially be due to a jammer (i.e., man-made noise). In general, the received data is the transmitted signal plus an additive noise component: $r(t) = s(t) + n(t)$.

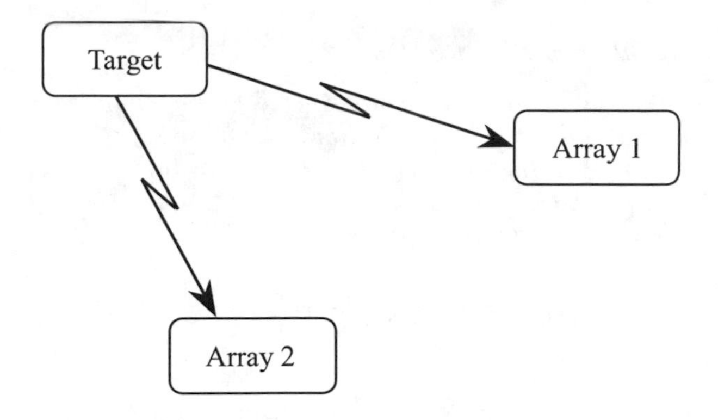

Figure 1.3: Two array processing.

Example 1.13 *Target Detection*

Events :		Received Data :
e_0(no target)	$\rightarrow$	$r_0(t) = n(t)$; for $0 \leq t \leq T$
e_1(target present)	$\rightarrow$	$r_1(t) = a\,s(t - t_R) + n(t)$; for $t_R \leq t \leq T + t_R$

where a is the signal loss coefficient (i.e., no Doppler shift assumed in this particular example) and t_R is the propagation delay, which accounts for the round trip to and from the target.

Example 1.14 *Two Array Processing*

Two arrays are receiving data from a particular geographical area. The target may or may not be present and may be or not be observed at either or both locations. Details can be seen in Figure 1.3.

Denote by $r_{ij}(t)$ the data received at array i given H_j, the jth hypothesis, is true.

Events :		Received Data :
e_0(no target)	$\rightarrow$	$r_{10}(t) = n_1(t)$; for $0 \leq t \leq T$
		$r_{20}(t) = n_2(t)$; for $0 \leq t \leq T$
e_1(target present)	$\rightarrow$	$r_{11}(t) = a_1 s(t - t_1) + n_1(t)$; for $0 \leq t \leq T$
		$r_{21}(t) = a_2 s(t - t_2) + n_2(t)$; for $0 \leq t \leq T$

where a_i is the loss coefficient for the signal component at the ith array, assuming a zero Doppler shift, and t_i represents the delay to reach the ith array, for $i = 1, 2$, and T is the duration of the snapshot used in the decision.

Example 1.15 *Binary FSK Problem*

This example represents an FSK signal over a fading channel (multiplicative fading) in additive noise. The multiplier $m(t)$ represents the multiplicative fading term and $n(t)$ is the additive observation noise.

$$
\begin{array}{lll}
\text{Events}: & & \text{Received Data}: \\
e_0 & \rightarrow & r_0(t) = \sum_n m_n(t) \, \cos\left[\omega_0 \left(t - \tau_n(t)\right)\right] + n(t) \\
e_1 & \rightarrow & r_1(t) = \sum_n m_n(t) \, \cos\left[\omega_1 \left(t - \tau_n(t)\right)\right] + n(t)
\end{array}
$$

where $m_n(t)$ is the time-varying attenuation factor associated with the nth propagation path and $\tau_n(t)$ is the corresponding propagation delay. In this simple model, it is assumed that no inter-symbol interference (ISI) is present.

D: Decision Space

The dimension of D is usually the same as the dimension of E, but not always. If $\dim(D) = \dim(E)$, then we try in a best, that is in an optimal, fashion to estimate the original message. An example for $\dim(D)$ to be less than $\dim(E)$ is the estimation of target parameters leading to the decision yes or no, based on multiple pieces of information, such as multiple spectral lines in the passive SONAR detection problem.

Our text will focus on "how to obtain a mapping from R to D." This rule (mapping) must be such that:

(a) An unambiguous decision is made.

(b) Every outcome must lead to some decision.

Example 1.16 *Multiple Decision (see Figure 1.4)*

d_i: *choose H_i, the hypothesis that the event e_i occurred which means that the observation must lie in the space (region) R_i, for $i = 0, 1, 2, 3$.*

1.4 SUMMARY

In Section 1.1, a general introduction to the detection topic is given and related reference material is pointed out. At the same time a historical connection to modern signal processing is made. Section 1.2 provides insight into detection and estimation problems using examples. The signal detection problem is classified into three types of classes: known parameters, unknown

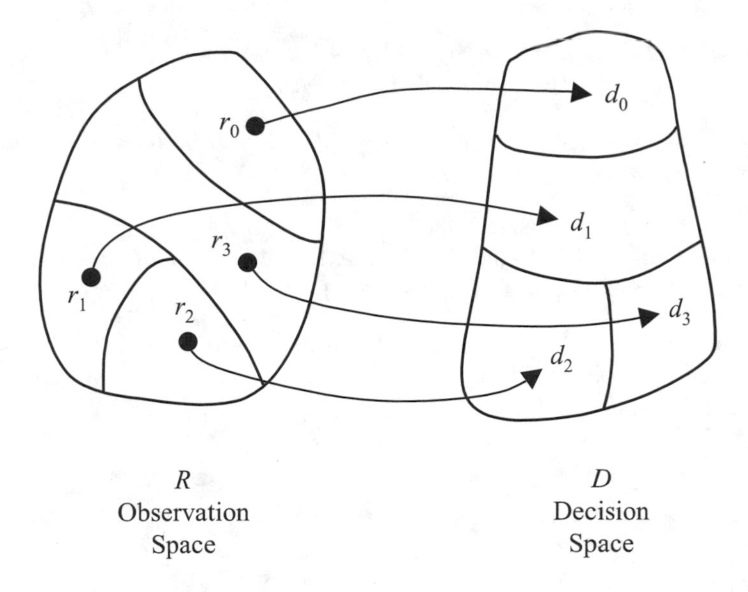

Figure 1.4: Decision regions for Example 1.16.

parameters, and random parameters. The last section provides a generic description of information flow and the spaces involved in the generation, transmission, reception, and processing of the data.

References

[1] Whalen, A., *Detection of Signals in Noise*, Orlando, FL: Academic Press, 1971.

[2] Van Trees, H.L., *Detection, Estimation, and Modulation Theory*, Part 1, New York: John Wiley & Sons, 1968.

[3] Kazakos, D., and Papantoni-Kazakos, P., *Detection and Estimation*, New York: Computer Science Press, 1990.

[4] Poor, V.H., *An Introduction to Signal Detection and Estimation*, New York: Springer-Verlag, 1988.

[5] Helstrom, C.W., *Elements of Signal Detection and Estimation*, Englewood Cliffs, NJ: Prentice-Hall, 1995.

[6] Barkat, M., *Signal Detection and Estimation*, Norwood, MA: Artech House, 1991.

[7] Weber, C.L., *Elements of Detection and Signal Design*, New York: McGraw-Hill Book Co., 1968 and Springer-Verlag, 1987.

[8] Melsa, T.L., and Cohn, D.L., *Decision and Estimation Theory*, New York: McGraw-Hill, 1978.

[9] Srinath, M.D., and Rajasekaran, P.K., *An Introduction to Statistical Signal Processing with Applications*, New York: John Wiley & Sons, 1979.

[10] McDonough, R.N., and Whalen, A.D., *Detection of Signals in Noise*, 2nd Ed., San Diego, CA: Academic Press, 1995.

[11] Srinath, M.D., Rajasekaran, P.K., and Viswanathan, R., *Introduction to Statistical Signal Processing with Applications*, Englewood Cliffs, NJ: Prentice-Hall, 1996.

Chapter 2

Review of Deterministic and Random System and Signal Concepts

2.1 SOME MATHEMATICAL AND STATISTICAL BACKGROUND

This chapter reviews some signal characteristics and system properties. The review helps to visualize the implementation of the basic ideas and the corresponding relationships among the building blocks. It also serves to interpret a given signal and/or system scenario. Most problems addressed will be of the additive noise type, that is the signal, if it is present, has an additive noise component. Conversely, if there is no signal component, then only noise is present. As a matter of interest, in typical situations we want to detect and track targets (i.e., signals) when the target is still far away. This always sets up a noisy scenario. If no noise is present, then the detection and estimation of parameters (i.e., target characteristics) are usually very simple. The data typically consists of:

$x(t) = s(t) + n(t)$; where (t) denotes the continuous time dependency

$x(n) = s(n) + n(n)$; where (n) denotes discrete time at intervals nT and T is the sampling interval. The T dependency is usually suppressed.

Here, $s(\cdot)$ is the signal component which may also have a parametric dependency, where the parameter may be deterministic or random, and $n(\cdot)$ is the noise component which may have a parametric description in terms of some model parameters.

Some problems are given and solved in an analog fashion (i.e., time-continuous problems) while some are given and approached in a digital fashion (i.e., discrete time problems). This requires some background in terms of system theory and random processes for both representations.

The minimum amount of structure that is needed is that of a normed inner product space (Hilbert space). Here we can compute the projection of signals and noise onto the prevailing coordinate axes. For additional background, the book by Franks [1] is recommended.

In general, we can think of the data space of being decomposable into two orthogonal subspaces, consisting of a signal and a noise only space. Depending on how we want to represent the data in this space, we may choose a particular set of basis functions. Theoretically, an infinite number of choices is available, but in practice only a few sets are used.

A well-known expansion uses as basis set, the orthogonal direction components of the signal. Here we choose as the primary basis function one of the signals. After the normalization of this chosen basis function, we select as a second basis function the orthogonal component of a second signal. After normalizing the second basis function, we repeat the process as often as necessary to span the total signal space. Hence, sequentially we design every basis function in an ortho-normal fashion relative to all previously selected basis functions (i.e., the Gram-Schmidt decomposition, see Chapter 7, Section 7.4).

We could use any orthogonal decomposition of the signal as a basis function. Actually, the most commonly used one is the Fourier type decomposition. But, we realize that a host of other decompositions (i.e., Hermite, Legendre, Chebbycheff, Laguerre, etc.) is available.

In the discrete type problems, the eigenvalue decomposition and the singular value decomposition (SVD) have found many applications [2,3,19]. These allow decomposition of the data space based on eigenvectors where the selection is done by using the magnitude of the eigenvalue. Here the argument is used that the sum of the squares of the eigenvalues corresponds to the energy of the system. Hence, in typical scenarios, the signal components correspond to the dominant eigenvalues (eigenvectors), while the noise only components tend to be those having the smallest eigenvalue.

There is one particularly useful decomposition where the basis functions are determined by the noise covariance matrix (i.e., the Karhunen-Loève expansion). In this expansion, we assume that we have correct knowledge about the noise covariance function. Each one of the approaches discussed so far has its own characteristics, merits, and disadvantages, which we try to point out as we go along.

Since we work with projection onto certain basis functions, we need to establish the characteristics of the general vector space (Hilbert space) that we are using. The vector space must have all the properties assigned to a linear vector space. We will indicate vector or matrix quantities by resorting to bold faced letters.

Definition 2.1: Vector space over the field of real or complex numbers. We need a set K of objects, called vectors, together with a vector addition on K and a scalar multiplication of vectors by numbers with properties.

Vector Addition:

(1) Vector addition is commutative and associative.

(2) There is a unique vector 0 (zero) in K such that $0 + \mathbf{x} = \mathbf{x}$ for all $\mathbf{x}$ in K.

(3) There is a unique vector $-\mathbf{x}$ in K such that $\mathbf{x} + (-\mathbf{x}) = 0$.

Scalar Multiplication:

(4) Must satisfy:

  (a) $(ab)\mathbf{x} = a(b)\mathbf{x}$;

  (b) $1\mathbf{x} = \mathbf{x}$

  and be related to vector addition by the two distributive laws:

(5) (a) $c(\mathbf{x} + \mathbf{y}) = c\mathbf{x} + c\mathbf{y}$, for all $\mathbf{x}$ and $\mathbf{y}$ in K.

  (b) $(a + b)\mathbf{x} = a\mathbf{x} + b\mathbf{x}$, for all $\mathbf{x}$ and $\mathbf{y}$ in K.

In addition, the vector space must have a distance (norm) and an angle (inner product) measure. So if we take a complete inner product space (i.e., a vector space), all of these properties hold. Since the majority of problems and applications deal with discrete-time problems, more attention is brought to these types of problems. Where appropriate, the continuous time type problems are addressed. The hierarchy for the vector space is as follows:

group $\rightarrow$ ring $\rightarrow$ field $\rightarrow$ vector space $\rightarrow$ algebra $\rightarrow$ complete normed space (i.e., Banach space) $\rightarrow$ complete inner product space (real or complex vector space with inner product, i.e., Hilbert space).

That is, as one imposes more and more structure, one moves from the specification of a group, to those of a ring, to those of a field, and so on. Some additional information can be found in Appendix C and in references [1,3,4].

A summary of many of the probability properties is presented in Appendix A. Most of our work deals with statistical moments. The first moment is the DC component, while the second moment describes the total power of the random variables in question. The variance of the random variable X is given by the expression variance $= E(X - m_X)^2 = E(X^2) - m_X^2$. It describes the spread of the density function of the random variable about its mean. These concepts are extended to provide auto-correlation and cross-correlation functions of random processes (or random sequences). Many of the properties of the correlation functions and their Fourier transforms (i.e., the power spectral density (PSD)) are given in Appendix A. Chapter 9, Section 9.2 examines power spectral density estimation will provide some more insight.

2.2 SYSTEMS AND SIGNALS (DETERMINISTIC AND RANDOM)

This section reviews linear system theory and system behavior due to deterministic and to random inputs as well as some related topics. In general, we usually assume that the continuous and discrete time systems are linear and time invariant (LTI), or linear and shift invariant (LSI), as the case may be. The random processes and random sequences are assumed to be wide sense stationary (w.s.s.). If we deal with non-linear or non-stationary entities, we shall point it out where appropriate.

Linear Systems:

It is well known that the response of a linear time invariant system can be expressed in the time domain (i.e., convolution) and in one of the transform domains (i.e., product of the respective transforms). In the time domain, we have in general

$$y(t) = \int_{-\infty}^{\infty} h(t-\tau)\, x(\tau) d\tau \tag{2.1}$$

while for a causal system and a causal input, we have

$$y(t) = \int_{0}^{t} h(t-\tau)\, x(\tau) d\tau \tag{2.2}$$

here $h(t)$ denotes the system impulse response, $x(t)$ is the input, and $y(t)$ is the output.

Equivalently, for discrete-time systems, we have in general

$$y(n) = \sum_{m=-\infty}^{\infty} h(n-m)\, x(m) \tag{2.3}$$

while for a causal system and causal input, we have

$$y(n) = \sum_{m=0}^{n} h(n-m)\, x(m) \tag{2.4}$$

where $h(n)$, $x(n)$, and $y(n)$ is the sampled impulse response, input, and output (sampled in the Nyquist sense), respectively. In all discrete-time scenarios, we assume proper analog-to-digital (A/D) conversion has been obtained.

In the transform domain (say the Laplace or Fourier transform domain for the continuous time case and the Z-transform or discrete time Fourier transform domain for the sampled data case), we have

$$\begin{aligned} Y(\mathrm{s}) &= H(\mathrm{s})\ X(\mathrm{s}) \\ Y(\omega) &= H(\omega)\ X(\omega) \end{aligned}$$

and

$$\begin{aligned} Y(\mathrm{z}) &= H(\mathrm{z})\ X(\mathrm{z}) \\ Y\left(\exp[j\omega]\right) &= H\left(\exp[j\omega]\right)\ X\left(\exp[j\omega]\right) \end{aligned} \tag{2.5}$$

It is very convenient to use a related letter for the transformed variables, i.e., the Laplace transform of $h(t)$ is denoted by $H(\mathrm{s})$. When convenient, we shall use the notation $x(n)$ and x_n interchangeably.

There are other linear transformations that could be used, but the ones mentioned above are the most common ones. We note that for causal quantities, only one-sided transforms need to be used (i.e., the regions of convergence are one-sided). The output of a linear time invariant system is also of interest when a random input is present. Then, statistical descriptions, i.e., moments, are used to describe the responses. We deal primarily with first and second order moments which are all based on the quantity y (time) or Y (frequency) as described earlier. One can spend much more time on systems concepts but we assume that the reader is aware of the concepts or is willing to pursue one of the references listed at the end of the chapter [1,5,7,13–18,20].

For random inputs, we examine the output in terms of first and second order moments. The first order moment, in general, is given by

$$Ey(t) = \int_{-\infty}^{\infty} h(t-\tau)\ Ex(\tau)d\tau \tag{2.6}$$

and

$$Ey(n) = \sum_{m=-\infty}^{\infty} h(n-m)\ Ex(m) \tag{2.7}$$

The second order moments are obtained by forming the appropriate products and then taking an expectation. For example, for the continuous time system, assuming the input is wide sense stationary, we have

$$\begin{aligned} Ey(t)y(u) &= \int_{-\infty}^{\infty}\int_{-\infty}^{\infty} h(t-\tau)h(u-\sigma)\ E(x(\tau)x(\sigma))d\tau d\sigma \\ &= \int_{-\infty}^{\infty}\int_{-\infty}^{\infty} h(t-\tau)h(u-\sigma)\ R_X(\tau-\sigma)d\tau d\sigma \end{aligned} \tag{2.8}$$

That is, $R_{YY}(t-u)$ is the convolution of the input correlation function with the impulse response and a convolution with the time reversal of the impulse response. The power spectral density is the Fourier transform of the correlation function and can be expressed as

$$S_{YY}(\omega) = S_{XX}(\omega)\ |H(j\omega)|^2 \tag{2.9}$$

where $H(j\omega)$ can also be written as $H(\omega)$. A similar expression holds in the sampled, w.s.s. data case.

$$\begin{aligned} R_{YY}(n-m) &= E(y(n)y(m)) \\ &= \sum_{k=-\infty}^{\infty}\sum_{\ell=-\infty}^{\infty} h(n-k)h(m-\ell)\ E(x(k)x(\ell)) \\ &= \sum_{k=-\infty}^{\infty}\sum_{\ell=-\infty}^{\infty} h(n-k)h(m-\ell)\ R_X(k-\ell) \end{aligned} \tag{2.10}$$

More background can be found in [4,5,7,13,20]. The power spectral density is the Fourier transform of the correlation function and can be expressed as

$$S_{YY}(e^{j\omega}) = S_{XX}(e^{j\omega})\ \left|H(e^{j\omega})\right|^2 \tag{2.11}$$

2.3 TRANSFORMATION OF RANDOM VARIABLES

Some probability density functions (PDFs) occur frequently in communication/detection related work. We will examine some of the standard probability density functions. The probability density plots contained in Chapter 2 are created using MATLAB® Version 5.3 [12]. Most of these densities and related topics can be found in very enjoyable form in Whalen [6] and in Papoulis [7]. Most densities are derived using the Gaussian PDF as generating densities, where the different random variables involved are assumed to be jointly Gaussian and independently identically distributed (i.i.d.).

For many typical physical phenomena, the Gaussian assumption is a reasonable approximation of actual behavior, that is, the central limit theorem [7] and extensions of the central limit theorem [22], which do not require i.i.d. random variables, tend to justify the use of the Gaussian distribution.

In many instances, we are interested in computing the area of a particular density function to the right of a given threshold T_0. For example, in the radar/sonar problem, the area of the density function has the following interpretation: When the target is present (when we say the so-called H_1 hypothesis is true), then the area of the density function to the right of a given threshold represents the probability of detection P_D. The statistical literature refers to P_D as the power of the test. When no target is present (that is the noise only condition is true, we say that the so-called H_0 hypothesis is true), then the area to the right of a particular threshold of the density function represents the probability of false alarm (P_{FA}, also called the size of the test). Of course one can also compute the area to the left of a given threshold T_0, which is just the ones-complement of the density area computed for the area to the right of the threshold T_0. This probability is also of interest when describing the detector performance. In the appropriate section, we will address these quantities. We will also briefly discuss the quantities: the power and size of the test, for some of the commonly occurring densities.

2.3.1 Gaussian Density

Gaussian random variables are closed under scalar multiplication and typical linear transformations. These properties are easily verified using the transformation of random variables or by using the characteristic function. This implies that filtering operations, convolution, integration, differentiation, as well as the common transforms (i.e., Hilbert, Fourier, Laplace, and Z-transform, to mention a few) retain the character of Gaussian random variables, that is, the transformed random variables stay Gaussian. If data is filtered and the effective integration time of the filter is relatively long compared to the correlation time of the noise, the PDF of the output compared to the input variables tends to be closer to a Gaussian PDF. This is a consequence of the central limit theorem (or one of its more general versions) in that the filtering operation constitutes a summation of many random variables when interpreting the integration using Riemann sums.

The density function for a single Gaussian (normal) random variable $x \sim N(m_X, \sigma^2)$ is given by:

For real x [7]

$$f_X(x) = \frac{1}{\sqrt{2\pi\sigma^2}} \exp - \frac{(x - m_X)^2}{2\sigma^2} \tag{2.12}$$

For complex x [13]

$$f_X(x) = \frac{1}{\pi\sigma} \exp - \frac{|x - m_X|^2}{\sigma^2} \tag{2.13}$$

The area to the right of a threshold T_0 for a zero mean, real valued, normalized Gaussian random variable is

$$Q(T_0) = \frac{1}{\sqrt{2\pi}} \int_{T_0}^{\infty} \exp - \left(\frac{x^2}{2}\right) dx \tag{2.14}$$

Some textbooks, for example see [6], refer to the integral expressed in (2.14) as the complementary error function (i.e., $\text{erfc}(T_0)$). Figure 2.1 shows a typical Gaussian PDF and the area referred to in (2.14).

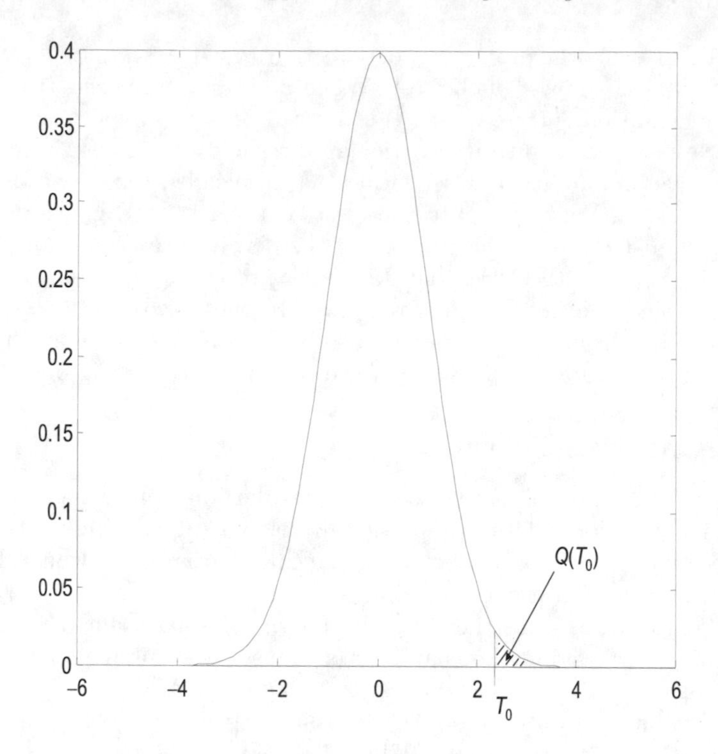

Figure 2.1: Gaussian density function.

The fourth order moment of the zero mean real valued Gaussian, which is very useful in computing the variability of the second order moment of Gaussian based variables, can easily be computed and is of the form:

$$Ex_1x_2x_3x_4 = Ex_1x_2 \; Ex_3x_4 + Ex_1x_3 \; Ex_2x_4 + Ex_1x_4 \; Ex_2x_3 \tag{2.15}$$

Even if $x_1 = x_2 = x_3 = x_4 = x$, then $Ex^4 = Ex^2 \; Ex^2 + Ex^2 \; Ex^2 + Ex^2 \; Ex^2 = 3{\sigma_X}^4$. We note, if linear operations are performed on Gaussian random variables, the resultant stays Gaussian. This result is especially useful when dealing with discrete time systems, such as those used when working with digital processing implementations. A typical example is the averaging procedure. Suppose we take a boxcar averager, that is a device or algorithm that sums up N contiguous samples of x as shown in Figure 2.2, and uniformly weights the data as described by

$$y = \sum_{i=0}^{N-1} x_i$$

This expression is sometimes normalized by the number of terms involved and is then called the sample mean. We know that if the input samples are

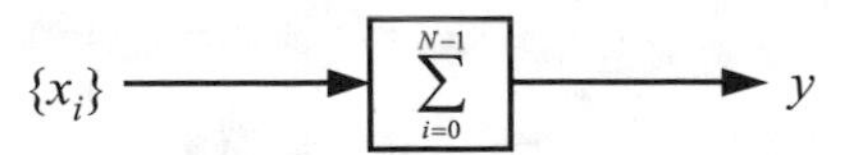

Figure 2.2: Boxcar averager.

Gaussian, the output y will be Gaussian too. Hence, all that needs to be established is the new mean of y and its variance. Suppose the input x_i are i.i.d. Gaussian random variables with mean m_X and variance σ_X^2 then the output y is Gaussian with mean

$$\begin{aligned} Ey &= E\sum_{i=0}^{N-1} x_i = \sum_{i=0}^{N-1} Ex_i \\ &= \sum_{i=0}^{N-1} m_X = Nm_X \end{aligned}$$

with second moment

$$\begin{aligned} Ey^2 &= E\sum_{i=0}^{N-1}\sum_{j=0}^{N-1} x_i x_j = \sum_{i=0}^{N-1}\sum_{j=0}^{N-1} Ex_i x_j \\ &= (N-1)Nm_X^2 + \sum_{i=0}^{N-1} Ex_i x_i \\ &= (N-1)Nm_X^2 + N(\sigma_X^2 + m_X^2) \\ &= N^2 m_X^2 + N\sigma_X^2 \end{aligned}$$

and variance

$$\sigma_Y^2 = N\sigma_X^2$$

Hence, $y \sim N(Nm_X, N\sigma_X^2)$.

2.3.2 Rayleigh Density

This density occurs naturally in processing that deals with envelopes, magnitudes, or distances defined in two dimensions. Given two independent identically distributed Gaussian random variables, say x and y with $x \sim N(0, \sigma^2)$ and $y \sim N(0, \sigma^2)$. The envelope is formed by

$$z = \sqrt{x^2 + y^2}$$

Given that

$$f_{XY}(x, y) = \frac{1}{(2\pi\sigma^2)} \exp -\frac{(x^2 + y^2)}{2\sigma^2}$$

then

$$f_Z(z) = \frac{z}{\sigma^2} \exp \frac{-z^2}{2\sigma^2} U(z) \tag{2.16}$$

where $U(z)$ denotes the unit step function.

The area to the right of a given threshold T_0 is given by

$$\int_{T_0}^{\infty} \frac{z}{\sigma^2} \exp -\frac{z^2}{2\sigma^2} dz \tag{2.17}$$

The variable z is the output of a typical envelope detector (under the noise only conditions). We note that the variable z can also denote the magnitude of a complex variable, where x and y are the real and imaginary components, or the Euclidean distance in a two-dimensional space.

Some time ago, a convenient (analog) electronic circuit was the envelope detector, typically implemented with a one way rectifier diode and an RC filter (see Figure 2.3(a)). Equivalently, we can take the data (after digitization) and apply an envelope forming operation (Figure 2.3(b)). Many times, it turns out that a video output which is actually an envelope squared type output is created by the processing technique. A typical example is the quadrature demodulator, followed by a squaring and summing operation (Figure 2.3(c)). Or as in the case of the fast Fourier transform (FFT) implementation, the square root of the sum of the real and imaginary output squared (Figure 2.3(d)) makes up the envelope. Here the signal $r(t)$ is given

$$\begin{aligned} r(t) &= z(t)\cos(\omega_c t + \phi(t)) \\ &= x(t)\cos\omega_c t - y(t) sin \omega_c t \end{aligned}$$

where $r(t)$ has a Rayleigh PDF, $\phi(t)$ has a uniform PDF, and $x(t)$ and $y(t)$ are i.i.d. Gaussian random variables for each fixed point in time. One can easily show, i.e., via the marginal density, that the Rayleigh random variable is independent of the uniform random variable. The PDFs of two different Rayleigh variates are shown in Figure 2.4, with the area to the right of an arbitrary threshold indicated.

2.3.3 Cauchy Density

For completeness sake, we will discuss the transformation of random variables that leads to the Cauchy density. This is particularly useful as a stepping stone to derive the probability density function of the phase of a Gaussian bandlimited noise process. Let t be the ratio of the two independent identically distributed Gaussian random variates, then for either $t = x/y$ or $t = y/x$ we obtain the following density function

$$f_T(t) = \frac{1}{\pi(1+t^2)}\ ; \quad \text{for all } t \tag{2.18}$$

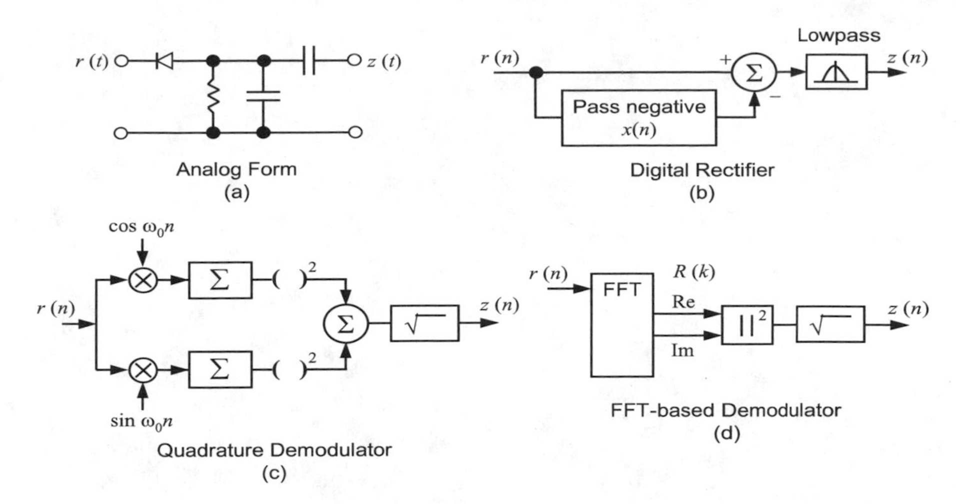

Figure 2.3: Envelope detectors.

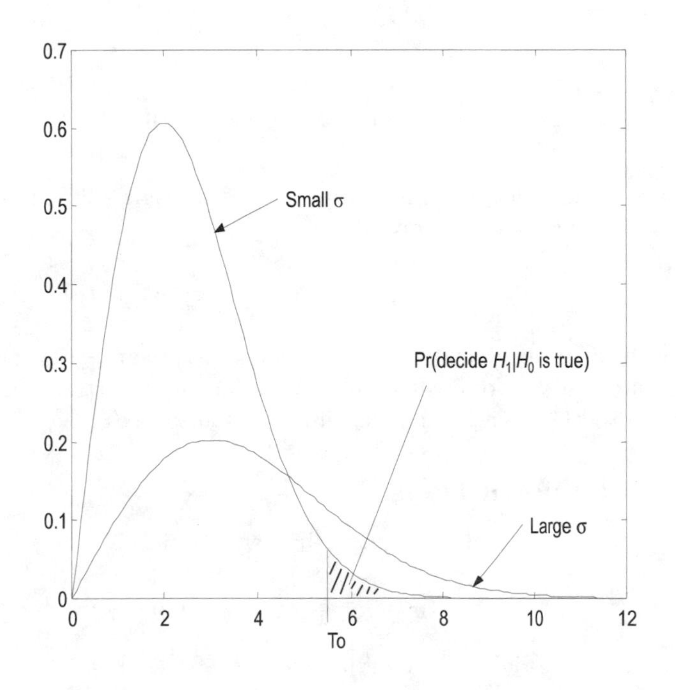

Figure 2.4: Rayleigh PDF.

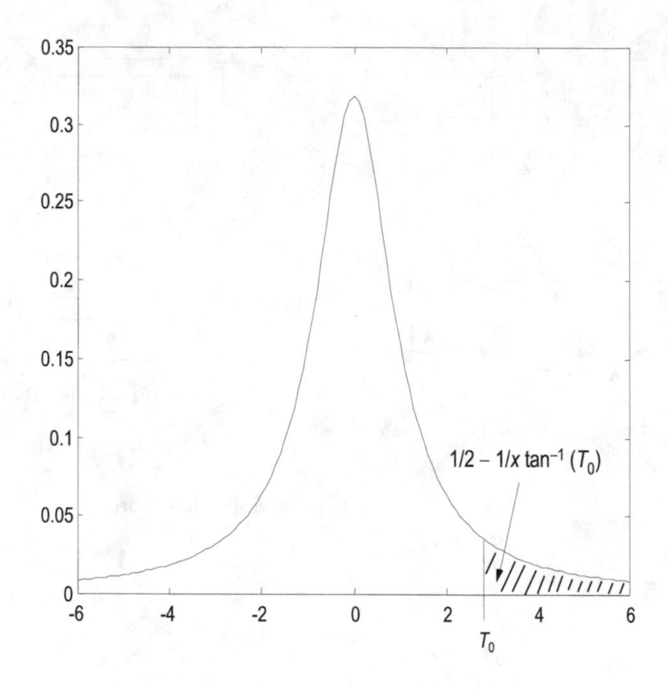

Figure 2.5: Cauchy PDF.

The probability density function is plotted in Figure 2.5 and the area to the right of a threshold T_0 is given by:

$$\int_{T_0}^{\infty} \frac{1}{\pi(1+t^2)} dt = \frac{1}{2} - \frac{1}{\pi} \tan^{-1}(T_0) \tag{2.19}$$

From the Cauchy density, the uniform density can easily be derived. The uniform density is frequently used to model the uncertainty in the phase of a bandlimited Gaussian or a sinusoidal random process.

2.3.4 Uniform Density

Let $w = \arctan(t) = \arctan(y/x)$, then we obtain

$$f_W(w) = \frac{1}{(2\pi)} \; ; \qquad |w| < \pi \tag{2.20}$$

The area to the right of a threshold T_0 is

$$\int_{T_0}^{\pi} \frac{1}{(2\pi)} dw = \frac{1}{2\pi}(\pi - T_0) \tag{2.21}$$

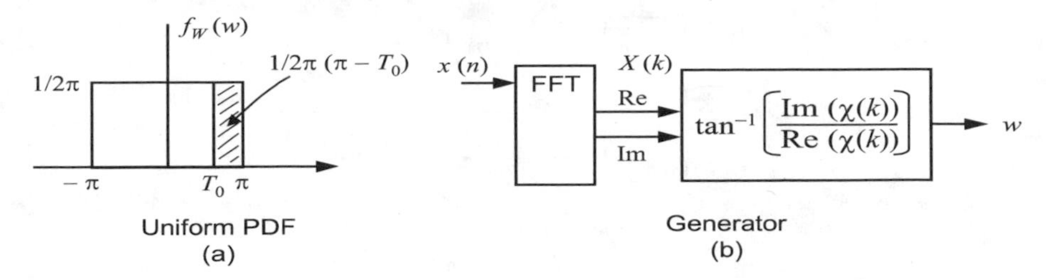

Figure 2.6: (a) Uniform PDF; (b) generator.

Figure 2.6(a) shows the uniform PDF while Figure 2.6(b) shows a typical phase estimation scheme using the kth bin of the FFT which when dealing with Gaussian noise, leads to the uniform PDF. We note that a narrowband Gaussian noise process can equivalently be expressed as a random process with a Rayleigh distributed envelope and a uniform random phase for each point in time or as a process consisting of quadrature components centered at a carrier frequency.

2.3.5 Chi-Squared Density

The PDF of the sum of N-independent squared identically distributed Gaussian random variables is chi-squared with N-degrees of freedom, where N is the number of variates involved in the squaring operation and $x_i \sim N(0, \sigma^2)$ for $i = 1, 2, \cdots, N$ [7].

Let the variable C be a chi-squared random variable with 2 degrees of freedom. Hence,

$$c = x_1^2 + x_2^2 \tag{2.22}$$

$$f_C(c) = \frac{1}{2\sigma^2} \exp \frac{-c}{2\sigma^2} U(c) \tag{2.23}$$

where $U(c)$ is the unit step function. Note this is also called the exponential probability density function. The area to the right of a threshold T_0 is

$$\int_{T_0}^{\infty} \frac{1}{2\sigma^2} \exp \frac{-c}{2\sigma^2} dc = \exp - \left(\frac{T_0}{2\sigma^2} \right) \tag{2.24}$$

This density occurs typically at the output of a spectrum analyzer under the (Gaussian) noise only condition. This is the equivalent to Figure 2.3(d) without the use of the square root operation. In general, if the input noise density is Gaussian or we can claim the summation of many independent random variables form two independent identically distributed Gaussian variables, then the output density tends to be exponential (i.e., chi-squared with two degrees of freedom). A typical processing scheme is the arrangement given

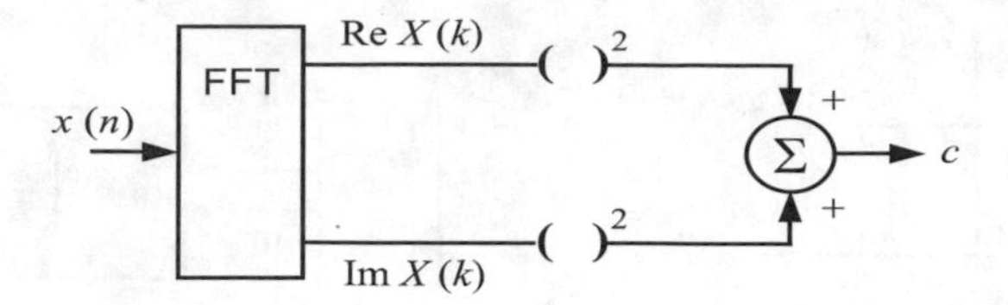

Figure 2.7: Spectrum analyzer.

in Figure 2.7. This version of spectral estimation is called the periodogram, which due to its robustness, ease of interpretation, and efficiency, is used most of the time when dealing with the detection of spectral components or the estimation of the power spectral density. To obtain a properly scaled power spectral density, the variable c should be normalized by the number of terms in the FFT.

In modern processors, this is accomplished by using the FFT simply because it is so fast, reliable, and accurate. Once the data has been properly sampled (i.e., the A/D conversion obeys the Nyquist criterion), it is simply processed as given by

$$FFT\{x_n\} = \sum_{n=m}^{m+N-1} x_n \, e^{-j2\pi k(n/N)}$$

where N is the transform length (also called the coherent integration time), k corresponds to the frequency (i.e., $f = k\, f_{\mathsf{sampling}}/N$), n is the time index ($t = \Delta t\, n$), Δt is the sample spacing in seconds, m is the start time in samples, and $f_{\mathsf{sampling}} = 1/\Delta t$. If we denote the output of the FFT by $X(k)$, then the spectral estimate $P_X(k)$ is denoted by

$$P_X(k) = \frac{1}{N}\,|X(k)|^2 \;=\; \frac{1}{N}\left|\sum_{n=m}^{m+N-1} x_n e^{-j2\pi n(k/N)}\right|^2$$

This estimate of the spectral density is called the periodogram [14,15], as earlier mentioned. If no special data weighting is used, we can easily show that the mean of the estimate equals to one standard deviation of the estimate when only white Gaussian noise is present (i.e., the H_0 hypothesis is true). This is the Achilles heel of the periodogram, namely the variance is relatively large and does not decrease as the integration time (N) increases. A typical exponential density, using a σ of 1, is shown in Figure 2.8. In general, if c is the sum of N i.i.d., Gaussian squared random variables

$$c = \sum_{i=1}^{N} x_i^2 \; ; \qquad x_i \text{ are i.i.d.} \tag{2.25}$$

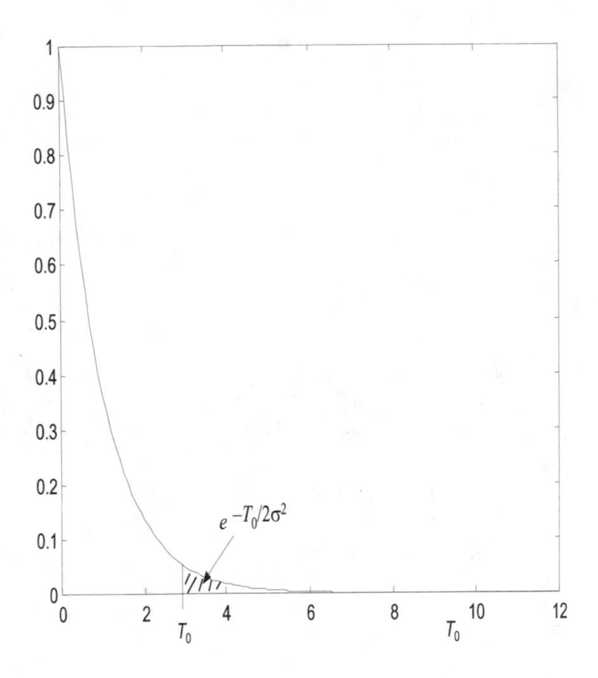

Figure 2.8: Exponential PDF.

where the $x_i \sim N(0, \sigma^2)$. The PDF then becomes

$$f_C(c) = \frac{1}{2^{N/2}\sigma^N \Gamma(N/2)} c^{(N/2)-1} \exp - \left(\frac{c}{2\sigma^2}\right) U(c) \tag{2.26}$$

Some typical members of this PDF, using a σ of 1, are shown in Figure 2.9.

We note that this corresponds to incoherent, also called power averaging of periodograms. Typically, one seeks to improve the variance of the basic periodogram by averaging sequential spectral outputs. These outputs are sequential in time and, for the purpose of our discussion, assumed to be independent. Asymptotically, for large numbers of averages, one expects about a 1.5 dB (decibel) improvement per doubling of the number of terms used in the averaging procedure. The basic filter bandwidth of the spectrum analyzer (FFT) remains constant (roughly f_{sampling}/N) providing some robustness. On the other hand, using coherent averaging would have resulted in a 3 dB gain per doubling of the number of terms in the FFT operation, but then the filter bandwidth shrinks (f_{sampling}/N) with the increasing N. Depending on what the signal of interest does, it may be lost by being over-resolved, that is the signal spills into many adjacent spectral bins. One needs to carefully evaluate the trade-offs between variance reduction, bandwidth,

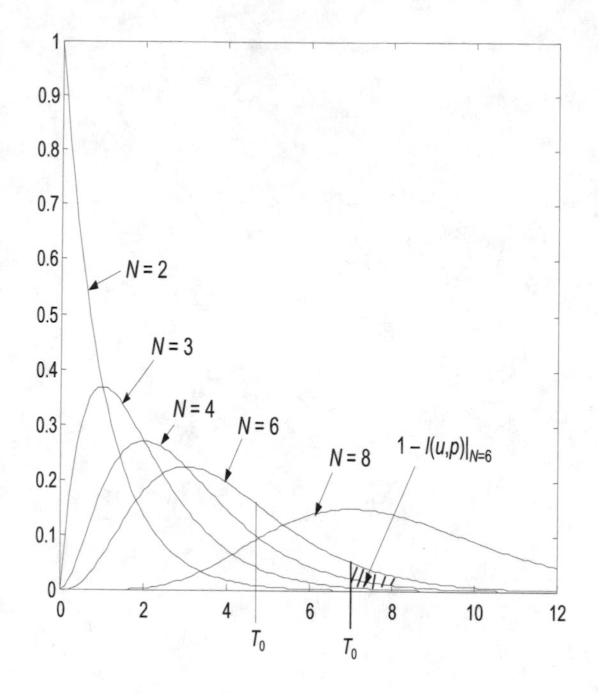

Figure 2.9: Chi-squared PDF with N-degrees of freedom.

processing gain, and over-resolving to extract spectral information.

The area to the right of a threshold T_0 is given by

$$\int_{T_0}^{\infty} \frac{1}{2^{N/2}\sigma^N \Gamma(N/2)} c^{(N/2)-1} \exp - \left(\frac{c}{2\sigma^2}\right) dc = 1 - I(u, p) \tag{2.27}$$

where $I(u, p)$ is Pearson's incomplete γ function [23],

$$\frac{1}{\Gamma(N/2)} \int_0^{T/2} e^{-w}\, w^{(N/2-1)} dw = I\left(T\,(2N)^{-1/2}, N/2 - 1\right)$$

and

$$\begin{aligned} p &= \frac{N}{2} - 1 \\ u &= \frac{T_0}{\sigma^2 2 (p+1)^{1/2}} \end{aligned}$$

2.3.6 Rician Density (Non-Central Rayleigh)

The Rician density is a modified Rayleigh density which is obtained as the envelope of two independent Gaussian random variables that have one or two signal related components in one or both variables, respectively. This corresponds to the density of the output of an envelope detector which is fed by the signal component(s) embedded in Gaussian bandpass noise. S.O. Rice, after which the density is named, is credited with the original work [9]. The detector form is the same as in Figure 2.3. The density is expressed as

$$f_U(u) = \frac{1}{\sigma^2}\, e^{-(u^2+m^2)/2\sigma^2}\, u\, I_0\left(\frac{um}{\sigma^2}\right) \tag{2.28}$$

for positive u where

$$u = \sqrt{(x^2+y^2)}$$

and

$$f_{XY}(x,y) = \frac{1}{2\pi\sigma^2} \exp - \frac{(x-m)^2+y^2}{2\sigma^2}$$

Some typical density function examples are given in Figure 2.10.

The area to the right of a given threshold T_0 is

$$P_D = \int_{T_0}^{\infty} \frac{1}{\sigma^2}\, e^{-(u^2+m^2)/2\sigma^2}\, u\, I_0\left(\frac{um}{\sigma^2}\right) du = Q\left(\frac{m}{\sigma}, \frac{T_0}{\sigma}\right) \tag{2.29}$$

where $Q(\ ,\ )$ is the Marcum Q-function, which is tabulated in [24].

2.3.7 Non-Central Chi-Squared Density

The mechanism to create this density is the same as for the regular chi-squared density. The difference comes from the embedded signal components that shift the central density to a non-central one.

For a block diagram description, consult Figure 2.3(c) or 2.3(d) (removing the square root operation). The density is created by the following operation, where the x_i are i.i.d. $\sim N(0, \sigma^2)$.

$$\begin{aligned} v &= \sum_{i=1}^{N} (A_0 + x_i)^2 \\ f_V(v) &= \frac{1}{2\sigma^2}\left(\frac{v}{\lambda}\right)^{(N-2)/4} \exp - \left(\frac{\lambda+v}{2\sigma^2}\right) I_{((N/2)-1)}\left[\frac{(v\lambda)^{1/2}}{\sigma^2}\right] \end{aligned} \tag{2.30}$$

for $v \geq 0$; with $\lambda = A_0^2 N$. A typical set of PDFs is shown in Figure 2.11.

The area to the right of a given threshold T_0 is given by

$$P_D = \int_{T_0}^{\infty} \frac{1}{2\sigma^2}\left(\frac{v}{\lambda}\right)^{(N-2)/4} \exp - \left(\frac{\lambda+v}{2\sigma^2}\right) I_{((N/2)-1)}\left[\frac{(v\lambda)^{1/2}}{\sigma^2}\right] dv$$

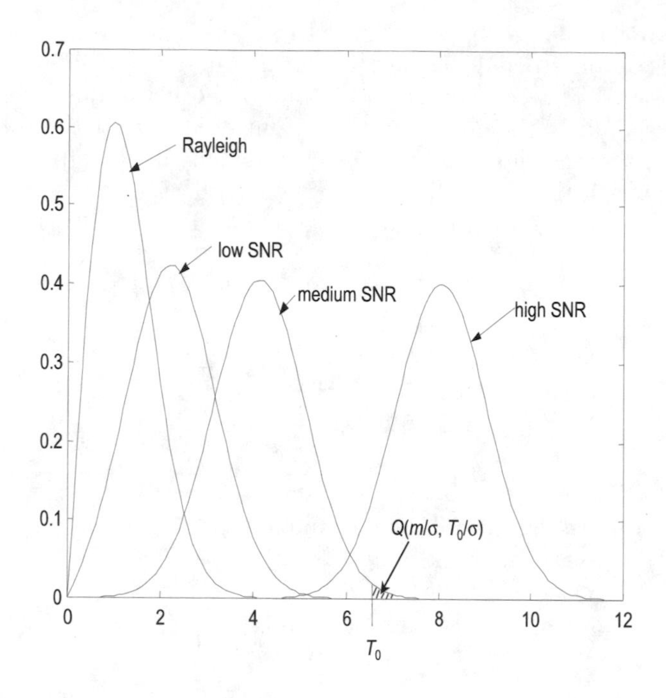

Figure 2.10: Rician PDF.

$$= Q_M\left(\sqrt{NA_0{}^2}, \sqrt{T_0/\sigma^2}\right) \tag{2.31}$$

where $Q_M(\ ,\ )$ is the generalized Marcum Q-function [6,8,25] with $M = N/2$. This type of density is typical for the output of a spectrum analyzer based on the periodogram, that is, the magnitude squared of the Fourier transform of signal embedded in Gaussian noise. In this particular case N is 2, therefore M is 1. Higher order densities are achieved when averaging incoherently (i.e., power averaging) the spectral bins of the periodogram.

The densities described so far are the ones that most frequently occur in communication (detection) related problems. In many problems, we deal with sums or weighted sums of Gaussian variates. This, of course, retains the Gaussian PDF, with possible changes in the mean, variance, and correlation properties. For many of our detection problems, the underlying premise is that of an additive Gaussian type perturbation (i.e., Gaussian noise). A few results are available for some problems where the Gaussian assumption cannot be used. The results are not of a general nature.

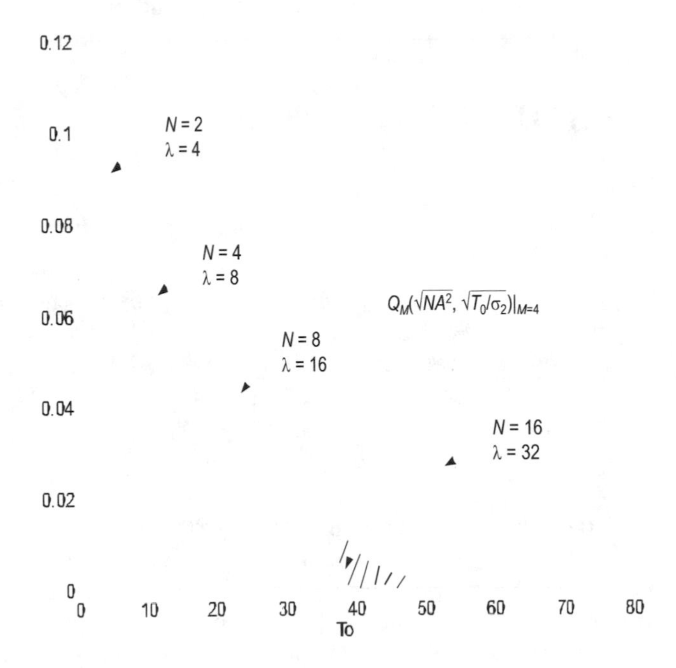

Figure 2.11: Non-central chi-squared PDF.

Some results are available using higher order moments or cumulants. Cumulants of order three and higher are zero for true Gaussian variates, allowing the estimation and identification of the degree of Gaussianity (i.e., how much of the noise is Gaussian) [21]. Recently, several papers appearing in the *IEEE Signal Processing Transactions* deal with these types of problems [10,11].

We note also that the types of densities described earlier in this chapter are typical for those observed at the output of FFT-type processors which are used to obtain spectral detection (i.e., power spectrum) or spatial detection (i.e., intensity or power out of a beam former as a function of look direction).

2.4 SUMMARY

This chapter provides some system, signal, and statistical background. The first section introduces the notion of vector spaces. Section 2.2 discusses some aspects of system theory in terms of correlation functions and power spectral densities. Section 2.3 introduces the transformation of Gaussian random variables. In particular, the random variables of interest in communications, detection theory, and signal processing are discussed. An effort is made to show how these PDFs might be generated. Some of the appendices provide more details and summaries of the basic probability properties.

2.5 PROBLEMS

1. Computer Exercise

   (a) Create uniform random variables $u(n)$ for $n = 1, \cdots, 12,000$ (or more) $u \sim U(-0.5, 0.5)$.

   (b) Create Gaussian random variables by using the uniform random variables, generated in i) above, $g(n)$, for $n = 1, \cdots, 1000$. Make sure that none of the random variables $u(n)$ is used more than once.

$$g(n) = \sum_{i=1}^{12} u(i)$$

   (c) Create Rayleigh random variables $r(n)$, for $n = 1, \cdots, 500$ by using i.i.d. Gaussian random variables (r.v.s) having a zero mean and a variance of 4. One way to do this is to scale the random variables obtained in step (b), or to use the Gaussian noise generator (randn in MATLAB) directly. Make sure that none of the random variables $g(n)$ is used more than once.

$$r(n) = \sqrt{g(i)^2 + g(j)^2} \qquad i \neq j$$

   (d) Create Rician random variables by using two jointly Gaussian independent r.v. one with mean zero, one with mean 2, variance = 4. For $w(n)$, for $n = 1, \cdots, 500$. Make sure that none of the random variables $g(n)$ is used more than once.

$$w(n) = \sqrt{g(i)^2 + g(j)^2} \qquad i \neq j$$

   (e) Create chi-squared random variables (with 2 degrees of freedom) using the same parameters for the mother densities as in part (c). Note, you are allowed to square the variables generated by problem (c), but say why this is proper. Obtain 500 random variables.

   (f) Create non-central chi-squared random variables (with 2 degrees of freedom) using the same parameters for the mother densities as in problem (d), i.e., Gaussian, mean 2, and variance 4. Obtain 500 random variables.

   (g) For each one of the six different random variables, do the following:

      (i) Plot the histogram.

(ii) Plot the theoretical density function.

(iii) Compute the first sample moment and the theoretical mean.

(iv) Compute the sample variance and theoretical variance.

(v) Comment on your results.

The following notation is used: given two independent identically distributed Gaussian random variables, say x and y, with $x \sim N(0, \sigma^2)$ and $y \sim N(0, \sigma^2)$

$$x \sim N(m_X, \sigma^2) \implies f_X(x) = \frac{1}{\sqrt{2\pi\sigma^2}} \exp - \frac{(x - m_X)^2}{2\sigma^2}$$

$$y \sim N(m_Y, \sigma^2) \implies f_Y(y) = \frac{1}{\sqrt{2\pi\sigma^2}} \exp - \frac{(x - m_Y)^2}{2\sigma^2}$$

2. Find $f(z)$, where z is defined as

$$z = \sqrt{x^2 + y^2}$$

$$f_{XY}(x, y) = \frac{1}{(2\pi\sigma^2)} \exp - \frac{(x^2 + y^2)}{(2\sigma^2)}$$

where x and y are jointly Gaussian, with $x \sim N(0, \sigma^2)$ and $y \sim N(0, \sigma^2)$.

3. Let t be the ratio of the two i.i.d. Gaussian random variables (as defined in problem 2 above), so $t = x/y$. Find $f(t)$.

4. Let $w = \arctan(t) = \arctan(x/y)$. Find $f(w)$, where x and y are jointly Gaussian random variables (as defined in problem 2 above).

5. Let e be the sum of 2 i.i.d. squared jointly Gaussian random variables as defined in problem 2, for $i = 1, 2$, $x_i \sim N(0, \sigma^2)$. Find $f(e)$, where

$$e = x_1^2 + x_2^2$$

6. Let v be $\sqrt{x^2 + y^2}$; where x and y are jointly Gaussian and $x \sim N(m, \sigma^2)$ and $y \sim N(0, \sigma^2)$. Find $f(v)$.

References

[1] Franks, L.E., *Signal Theory*, Englewood Cliffs, NJ: Prentice-Hall, 1969.

[2] Van Trees, H.L., *Detection, Estimation, and Modulation Theory*, Part 1, New York: John Wiley & Sons, 1968.

[3] Noble, B., and Daniel, J.W., *Applied Linear Algebra*, Englewood Cliffs, NJ: Prentice-Hall, 1988.

[4] Scharf, L.L., *Statistical Signal Processing: Detection, Estimation, and Time Series Analysis*, New York: Addison-Wesley, 1991.

[5] Fante, R.L., *Signal Analysis and Estimation: an Introduction*, New York: John Wiley & Sons, 1988.

[6] Whalen, A.D., *Detection of Signals in Noise*, New York: Academic Press, 1971, Chapter 4.

[7] Papoulis, A., *Probability, Random Variables, and Stochastic Processes*, New York: McGraw-Hill, 1965.

[8] McDonough, R.N., and Whalen, A.D., *Detection of Signals in Noise*, 2nd Ed., San Diego, CA: Academic Press, 1995.

[9] Rice, S.O., "Mathematical analysis of random noise," *Bell Syst. Tech. J.*, Vol. 23, pp. 282–332, 1994.

[10] Mendel, J.M., "Tutorial on higher-order statistics (spectra) in signal processing and system theory: Theoretical results and some applications," *Proc. IEEE*, Vol. 79, No. 3, pp. 278–305, March 1991.

[11] Nikias, C.L., and Raghuveer, M.R., "Bispectrum estimation: A digital signal processing framework," *Proc. IEEE*, Vol. 75, No. 7, pp. 869–891, July 1987.

[12] The Mathworks, Inc., Natick, MA.

[13] Therrien, C.W., *Discrete Random Signals and Statistical Signal Processing*, Englewood Cliffs, NJ: Prentice-Hall, 1992.

[14] Kay, S.M., *Modern Spectral Estimation, Theory and Application*, Englewood Cliffs, NJ: Prentice-Hall, 1988.

[15] Marple, S.L., *Digital Spectral Analysis*, Englewood Cliffs, NJ: Prentice-Hall, 1987.

[16] Chen, C.-T., *Linear Systems Theory and Design*, New York: Holt, Rinehart, and Winston, 1984.

[17] Mortensen, R.E., *Random Signals and Systems*, New York: John Wiley & Sons, 1987.

[18] Kailath, T., *Linear Systems*, Englewood Cliffs, NJ: Prentice-Hall, 1980.

[19] Golub, G.H., and VanLoan, C.F., *Matrix Computations*, 2nd Ed., Baltimore and London: The Johns Hopkins University Press, 1989.

[20] Shanmugan, K.S., and Breipohl, A.M., *Random Signals: Detection, Estimation and Data Analysis*, New York: John Wiley & Sons, 1988.

[21] Nikias, C.L., and Petropulu, A.P., *Higher-Order Spectra Analysis: A Nonlinear Signal Processing Framework*, Englewood Cliffs, NJ: Prentice-Hall, 1993.

[22] Larson, H.J., and Shubert, B.O., *Probabilistic Models in Engineering Sciences: Volume 1, Random Variables and Stochastic Processes*, New York: John Wiley & Sons, 1979.

[23] Pearson, K., *Tables of the Incomplete* Γ*-Function*, London and New York: Cambridge University Press, 1965.

[24] Marcum, J.I., "Table of Q-Functions," Rand Corporation, Report RM-339, January 1950.

[25] Proakis, J.G., *Digital Communications*, 3rd Ed., New York: McGraw-Hill, 1995.

Chapter 3

Introduction to Signal Processing

3.1 INTRODUCTION

This chapter provides an introduction to some of the basic aspects of digital signal processing (DSP). It can also be used for a quick review of some of the essential ideas. The basic mathematical background required is an exposure to summations, Z-transforms, and discrete-time Fourier transforms (DTFTs). After finishing the first part of this chapter, the reader will have been exposed to the concepts of convolution, filtering, correlation, and the power spectral density, and will have gained some appreciation for the DTFT, the discrete Fourier transform (DFT), and its fast cousin the fast Fourier transform (FFT). The FFT is the main tool in many signal processing applications and will be examined from several points of view. The second part of the chapter addresses the wavelet transform (WT) at a very elementary level, allowing the reader to become familiar with some of the important attributes of the WT, at least as far as their application is concerned.

3.2 DATA STRUCTURE AND SAMPLING

We realize that once analog data (continuous time information) has been sampled, a sequence of data points is obtained. Usually, the points are evenly spaced in time and one assumes that a sufficient number of bits is available so that the amplitude of the data samples accurately models the amplitude of the analog signal at the sampling times. We assume that the analog-to-digital (A/D) conversion is properly accomplished, obeying the Nyquist sampling

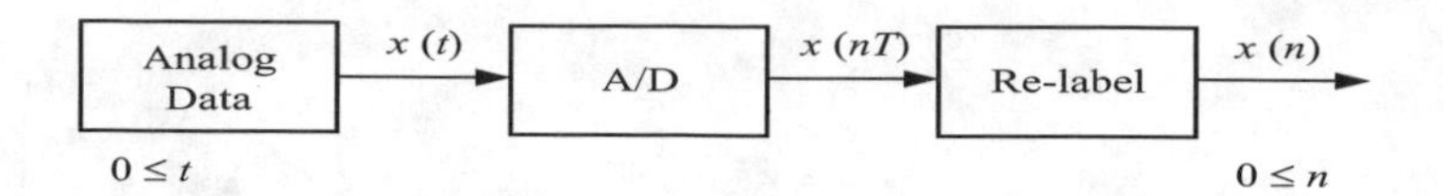

Figure 3.1: Sequence generation.

theorem [1]. The data sequences can also be represented in vector form. A typical data flow scenario is shown in Figure 3.1. For more information, the reader is encouraged to consult [1,2,3].

Suppose that we have the sequence

$$\{x_0, x_1, x_2, \cdots, x_{N-1}\} = \{x_i\}_{i=0}^{N-1} \tag{3.1}$$

then we can also represent the sequence using the vector notation

$$\mathbf{x}^T = \{x_0, x_1, \cdots, x_{N-1}\} \tag{3.2}$$

If the intent is clear, we can write the sequence as

$$\{x_0, x_1, \cdots, x_{N-1}\} = \{x(n) \ : \ n = 0, 1, \cdots, N-1\}$$

or in short hand notation as $x(n)$ or x_n by dropping the curly brackets. Given two equal length sequences x_n and u_n, we can operate on them in the following ways:

Product: (point by point) $\quad z_i = x_i \, u_i, \quad \text{for } i = 0, 1, \cdots, N-1$

Sum: (point by point) $\quad w_i = x_i + u_i, \quad \text{for } i = 0, 1, \cdots, N-1$

Convolution:
$$y_n = \sum_i x_i \, u_{n-i} = x(n) \underset{\text{conv}}{*} u(n) \tag{3.3}$$

Correlation:
$$c_n = \sum_i x_i \, u_{i+n} = x(n) \underset{\text{corr}}{*} u(n)$$

Inner product: (or Scalar product) $\quad c = \mathbf{x}^T\mathbf{u}, \quad$ size (1×1), where $\mathbf{x}$ and $\mathbf{y}$ are of dimension $(N \times 1)$

Outer product: $\quad \mathbf{Q} = \mathbf{x}\mathbf{u}^T, \quad$ size $(N \times N)$, where $\mathbf{x}$ and $\mathbf{y}$ are of dimension $(N \times 1)$

3.3 DISCRETE-TIME TRANSFORMATIONS

The bilateral Z-transform of the sequence $x(n)$ is defined by

$$Z\{x(n)\} = X(z) = \sum_{n=-\infty}^{\infty} x(n)\, z^{-n} \tag{3.4}$$

where $z = re^{j\omega}$, $\omega \in [-\pi, \pi]$, and r is a positive (real) number. We note that in circuit and system analysis, when going to the transform domain, convolution and correlation operations are reduced to product operations. Of course, one has to perform both the forward and inverse transformations. The inverse is given by

$$x(n) = Z^{-1}\{X(z)\} = \frac{1}{2\pi j} \oint_c X(z)\, z^{n-1}\, dz \tag{3.5}$$

Most engineers perform the inverse, preferably in terms of standard expressions, via partial fraction expansion. These expressions are easily inverted using tables as can be found in [1,3]. Note that in the bi-linear transformation the time variable n ranges over all time indexes. This admits the existence of a non-causal sequence. In circuit analysis, the variable n will take on only non-negative values (the exceptions are the possible non-zero initial conditions). For bilateral quantities, such as correlation functions, probability density functions, power spectral densities, and of course spatial variables (i.e., position in x, y, z) one needs to resort to the two-sided transform as defined in (3.4).

The DTFT pair is defined by the forward transform

$$X(e^{j\omega}) = \sum_{n=-\infty}^{\infty} x(n)\, e^{-j\omega n} \tag{3.6}$$

which is just the Z-transform evaluated on the unit circle (i.e., $z = r\exp^{-j\omega}$, with $r = 1$), while the inverse transform is given by

$$x(n) = \frac{1}{2\pi} \int_{-\pi}^{\pi} X(e^{j\omega})\, e^{j\omega n}\, d\omega$$

If the data sets are finite in duration $\{x(n) : \ n = 0, \cdots, N-1\}$ then the DTFT pair becomes the discrete Fourier transform (DFT) pair given by

$$X(k) = \sum_{n=0}^{N-1} x(n)\, e^{-j(2\pi/N)kn} \tag{3.7}$$

and

$$x(n) = \frac{1}{N} \sum_{k=0}^{N-1} X(k)\, e^{+j(2\pi/N)kn}$$

where $(2\pi/N)k$ corresponds to the kth frequency component also expressed as ω_k. The DFT is the discrete time analog of the continuous time Fourier series expansion, that is $X(k)$ is the projection of $x(n)$ onto a complex sinusoid which has exactly k periods over N sequential points in time. To simplify the mathematical expressions we sometimes replace $\exp(-j2\pi k/N)$ with the symbol W_N^k. The DFT and its inverse can then be written as

$$X(k) = \sum_{n=0}^{N-1} x(n)\, W_N^{kn}$$

and

$$x(n) = \frac{1}{N} \sum_{n=0}^{N-1} X(k)\, W_N^{-kn}$$

The normalization by $1/N$ can be applied in either domain or can be split (i.e., $1/\sqrt{N}$) and be applied in both domains.

There is a brute force way to evaluate the DFT (3.7) and a fast way. Traditionally, if N is a power of 2 (i.e., $N = 2^\nu$, for ν = positive integer) then (3.7) is computed using the FFT. Relative to the brute force way, which requires N^2 multiplications when manipulating N input data points, the FFT is faster (i.e., the typical cost is $N/2 \log_2 N$ multiplications), while preserving exactly the same accuracy. There are many other ways to implement the FFT operation (i.e., the Winograd transform, number theoretic transform, etc.) [4].

3.4 FILTERING

One of the main signal conditioning and signal processing operations performed is the filtering operation. Roughly speaking, given a data set with some particular characteristics, we want to isolate (block or pass) certain types of characteristics. Usually we deal with a spectral discussion, in which we talk about a spectral region (frequency region(s)) that we want to shape, that is, annihilate or pass. A simple example relates to the radio spectrum, say the FM portion, in where we want to pass, that is to listen to, one given FM radio station. Typically, we tune a filter (bandpass) to the spectral region (say for example, 100.7 MHz) and pass signals that exist in the neighborhood of the carrier frequency (100.7 MHz). For commercial FM stations, the neighborhood is ±75 KHz (i.e., bandwidth) and the carrier spacing (i.e., potentially, the minimum spectral distance to another FM station) is 200 KHz.

Typical filter functions are of a low pass, high pass, bandpass, or bandstop nature. As the name implies, low frequency, high frequency, and intermediate frequency components are passed or intermediate frequency components are

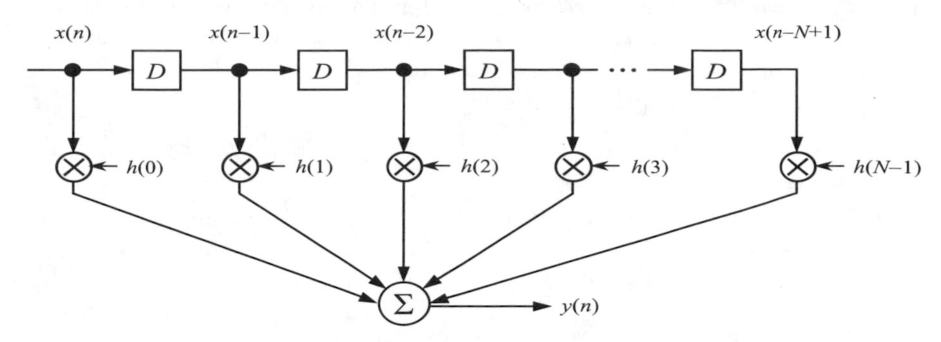

Figure 3.2: N-sized FIR filter.

removed. In some applications, the filter frequency response in the pass band region is functionally related to the signal and noise power spectral densities (i.e., the matched filter).

3.5 FINITE IMPULSE RESPONSE FILTER

Finite impulse response (FIR) filters have an impulse response that will be zero after some finite time, that is, $h(n) = 0$, for $n \geq N$. A typical N-sized FIR filter is shown in Figure 3.2.

An inspection of the diagram reveals that at time n, the output $y(n)$ is given by the convolutional sum

$$\begin{aligned} y(n) &= x(n)h(0) + x(n-1)h(1) + \cdots + x(n-N+1)h(N-1) \\ &= \sum_{i=0}^{N-1} x(n-i)\, h(i) \end{aligned} \tag{3.8}$$

If we inject a Kronecker-delta function, say $x(n) = \delta(n)$, into the relaxed network, (i.e., the initial conditions are zero), then the output will be

$$\{h(0), h(1), h(2)\cdots, h(N-1),\ 0,\ 0,\ 0, \cdots\} = \{h(0), h(1), h(2)\cdots, h(N-1)\}$$

This demonstrates why the name FIR filter is attached to the structure of Figure 3.2 and also illustrates the convolutional properties (i.e., (3.8) is the discrete time convolution) of the input and the impulse response. We note there is no feedback, hence the filter will always be stable. An equivalent way of noting this is to observe that the transfer function of the filter has no poles (other than at the origin), but has zeros only or that a bounded input results in a bounded output (bounded input/bounded output (BIBO) stability).

Classical analog filter theory has memory (feedback) giving rise to poles in the transfer function. There are also digital filters of this form, i.e., IIR filters. IIR (infinite impulse response) filters have an impulse response that goes on forever (i.e., $h(n) = \alpha^n U(n)$, $0 < \alpha < 1$). For more background on IIR filters, the reader is encouraged to consult [1–3].

We can take the Fourier transform of (3.8)

$$Y(e^{j\omega}) = X(e^{j\omega})\, H(e^{j\omega})$$

and obtain the output in the time domain via the inverse transform given by

$$y(n) = F^{-1}\left\{Y(e^{j\omega})\right\} \tag{3.9}$$

If this is done using FFTs, we call the operation of (3.9)

$$\begin{aligned} y(n) &= FFT^{-1}\left(Y(k)\right) \\ &= FFT^{-1}\left\{FFT\{x(n)\}\; FFT\{h(n)\}\right\} \end{aligned} \tag{3.10}$$

the fast convolution. For long filters (i.e., N is large), it may be much more economical to take the required forward and inverse transforms to obtain the time domain (i.e., $y(n)$) data. Some considerations must be given to avoid the wrap-around (modulus N, also called mod(N)) problem, which is caused by the circular convolutions [1,3].

Example 3.1 *Boxcar Averager*

This building block is also called an integrate and dump filter. Figure 3.3 shows the averager in (a) block diagram and (b) FIR filter form, while (c) depicts the frequency response.

By inspecting Figure 3.3(c), the impulse response is the sequence

$$\{\overbrace{1, 1, \cdots, 1}^{N\text{-times}}\}$$

Its frequency response is

$$\begin{aligned} H(e^{j\omega}) &= \sum_{n=0}^{N-1} 1\, e^{-j\omega n} = \frac{1 - e^{-j\omega N}}{1 - e^{-j\omega}} \\ &= e^{-j\omega(N/2-1/2)}\, \frac{\sin(\omega N/2)}{\sin(\omega/2)} \end{aligned} \tag{3.11}$$

The ratio of the two sinusoidal functions is also known as the digital sinc function. The magnitude squared frequency response becomes

$$\left|H(e^{j\omega})\right|^2 = \frac{\sin^2(\omega N/2)}{\sin^2(\omega/2)}$$

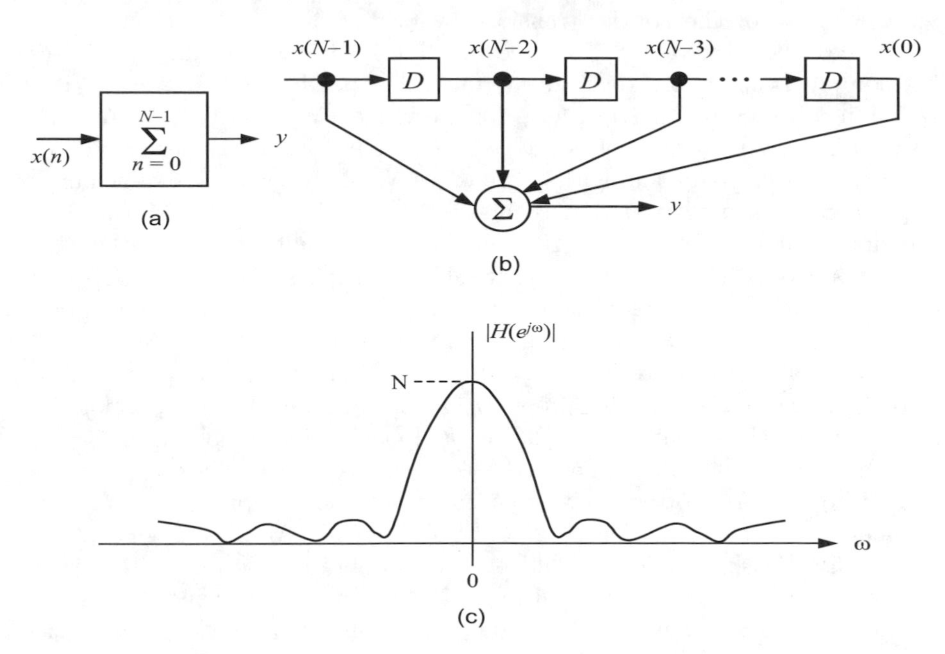

Figure 3.3: Boxcar averager (integrate and dump filter): (a) block diagram, (b) FIR filter form, and (c) magnitude of the frequency response.

which is plotted in Figure 3.3(c). It shows that the boxcar averager is a low pass filter, since frequency components around $\omega = 0$ are passed with power gain equal to N^2, and frequency components away from $\omega = 0$ (i.e., DC) are attenuated. The attenuation tends to increase with frequency (i.e., the further away from $\omega = 0$, the larger the attenuation). In terms of passband definition, one could use the 3 dB points, the null-to-null distance, or the concept of noise equivalent bandwidth. Of all possible low pass filter implementations, the boxcar averager tends to be the worst one in terms of side lobe leakage and roll off in the transition band.

3.6 THE FAST FOURIER TRANSFORM

The fast Fourier transform (FFT), which is just a fast implementation of the DFT, as introduced in (3.7) has many desirable attributes and many interesting and useful interpretations. If we visualize two black boxes, one containing the DFT and the other containing the FFT, and we evaluate the output of each box at spectral locations $2\pi k/N$ (rad) one cannot determine which box (i.e., which algorithm) is responsible for the output under observation. The responses are indistinguishable. One can associate the speed (or conversely, the slowness) of the Fourier transform algorithm (3.7) with the number of multiples (complex valued). We see that the DFT implementation requires N complex valued ($4N$ real valued) multiplies per frequency location (i.e., per bin). To obtain the transform for all frequencies ($k = 0, 1, \cdots, N-1$), N^2 complex valued multiplies are required. The FFT implementation (see Ludeman [3] or Strum and Kirk [1]) requires $N/2 \log_2 N$ complex valued multiplies. Suppose we have a data sequence of length $N = 1{,}024$, then the DFT version would require $1024^2 \approx 10^6$ complex valued multiplies, while the FFT requires only $512 \cdot 10 = 5{,}120$ complex valued multiplies. An increase in speed by more than two orders without a loss in accuracy is obtained. The larger the transform size, the larger the speed up will be relative to the time a DFT implementation would require. Speedups in the order of 2 to 3 (i.e., 100 to 1,000) are not unusual. This property has made the FFT the most popular tool in spectral estimation, correlation, and many filtering operations.

One could have used different basis functions in decomposing the data sequence [4]. Any orthogonal expansion would do. Typical examples are Legendre, Hermite, and the Laguerre polynomials. Also rectangular basis functions (Walsh functions) could be used to obtain a decomposition with just the use of additions (i.e., multiplication by plus or minus one amounts to projecting the data onto the rectangular basis functions). The FFT, because of its sinusoidal basis functions, is a natural for representation (decomposition, synthesis, analysis) of data sequences. Many physical responses of interest can be interpreted via a differential, or difference equation (DE). Solutions to a second order DE typically lead to a sinusoidal solution. Sinusoids are also eigenfunctions of linear time invariant systems [1]. That is, once a complex sinusoid (i.e., $\exp(-j\omega t)$) is injected into a linear time invariant (LTI) (or linear shift invariant (LSI)) system, the output will be the complex sinusoid, possibly changed in magnitude and most likely shifted in-phase. This also applies to linear combinations of complex sinusoids (i.e., sine and cosine) as well as to linear combinations of sinusoids (i.e., weighted sums of sines and cosines). The statistical analysis of DFT/FFT type algorithms tends to be very simple for white Gaussian noise. We, the human observers, usually prefer to deal with sinusoidal signal interpretations since we can relate to sinusoidal based representations more readily than to non-sinusoidal decompositions. In the next seven subsections, we will examine different interpretations of the DFT/FFT. Some of the interpretations are described in [7].

3.6.1 FIR Filter with Complex Valued Weights Interpretation

Figure 3.4 depicts the FIR filter interpretation of the DFT/FFT where

$$W_N^k \hat{=} e^{-j(2\pi/N)k}$$

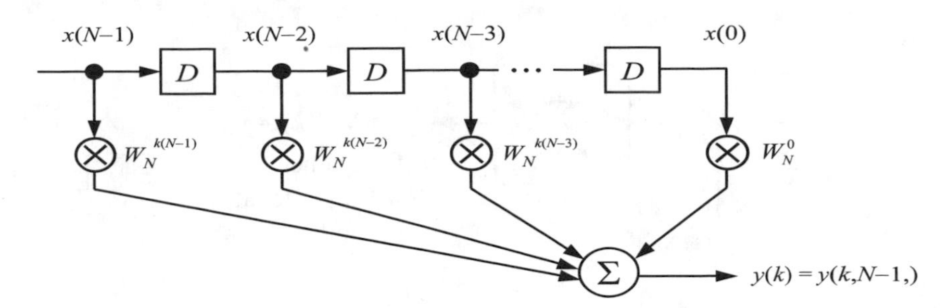

Figure 3.4: FIR filter interpretation.

We fixed the time index at n, but it should be obvious we could ask for the output at time $n+i$, in which case all data samples in the delay line move "i" places to the right (i.e., the youngest data sample would be $x(n+i)$, while the oldest data sample would be $x(i)$). Usually, data is processed in segments (blocks) of size N, where N can be the length of a data segment or a shorter piece of it. One could also zero pad a shorter segment, say we use M data points, where $M < N$, and append the data vector to size N using $N - M$ zeros. For more details, see Chapter 2, Section 2.3.5, and Chapter 9, Section 9.2. The actual data length may be much longer than the DFT/FFT size, i.e., real time processing, so one uses segments of the data in a sequential manner. If the segments are contiguous (that is non-overlapped with no data point being omitted), then the FIR filter output rate (at each spectral bin k) is $1/N$ times the input sample rate. In this sense, the FFT (DFT) corresponds to a multi-rate filter (i.e., filter and decimate in time, see Section 3.9, Chapter 9, Section 9.8, and Appendix E [22].

3.6.2 Complex Demodulator Interpretation

If we look at (3.7), we can see that it is the product of two sequences followed by a summation device. Figure 3.5 presents this type of interpretation. We note that $e^{-j(2\pi/N)kn}$ is periodic in N (i.e., modulus N), hence we can look at it as the output of an oscillator (i.e., $e^{-j(2\pi/N)kn} = \cos(2\pi/N)kn - j\sin(2\pi/N)kn)$ where every k corresponds to a complex valued sinusoidal signal that has exactly k periods over N data points. If we interpret the N data points as one period of a larger duration data segment, then we see that this interpretation is that of the discrete time Fourier series decomposition. The setup in Figure 3.5 is the same as in any conventional heterodyne receiver, the difference is in the nature of the oscillator signal (a complex valued sinusoid rather than a real valued sinusoid). The multiplication creates the sum and difference frequency terms, while the summer (a low pass filter, see Example 3.1) removes the high (i.e., sum) frequency terms.

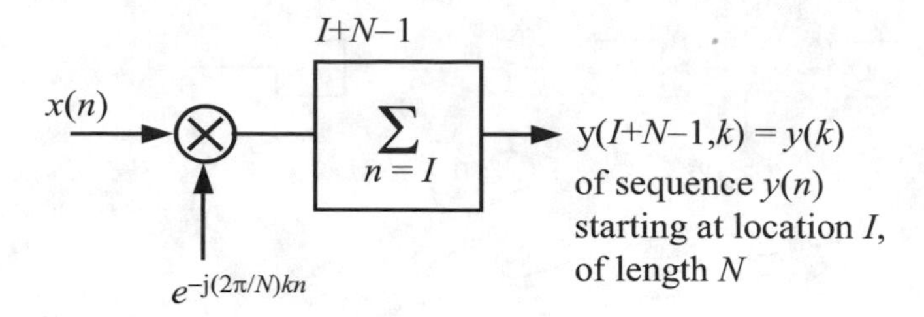

Figure 3.5: Complex demodulator interpretation.

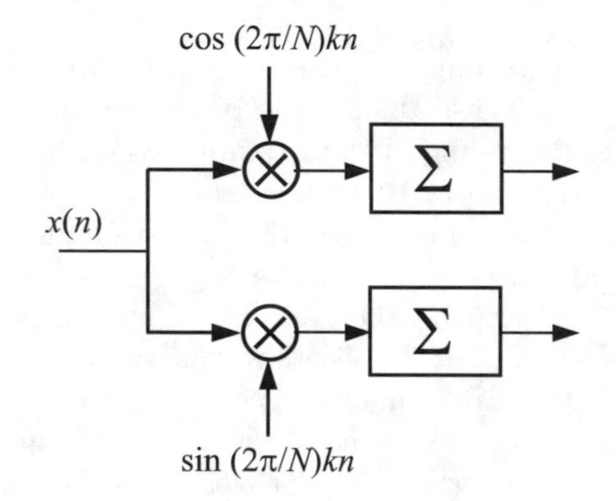

Figure 3.6: I-Q demodulator.

3.6.3 I-Q Demodulator (In-Phase Quadrature Demodulator) Interpretation

If one plots the real and imaginary component sections of Figure 3.5 separately, one obtains the form as shown in Figure 3.6. The top leg corresponds to the in-phase term (I-term), while the bottom leg corresponds to the quadrature term (Q-term). So we see the real and imaginary parts of the FFT (DFT) correspond to the conventional (discrete time) I-Q demodulator. The demodulators are followed by decimators [22] since N data samples result in one output sample per processing branch.

3.6.4 Correlator Interpretation

The expression given in (3.7) can also be sketched as shown in Figure 3.7. Figure 3.7 is the classic correlator (cross-correlator) that evaluates the cross-correlation between the data and a (known or assumed reference) sinusoidal function. The correlator does not compute correlation values other than for

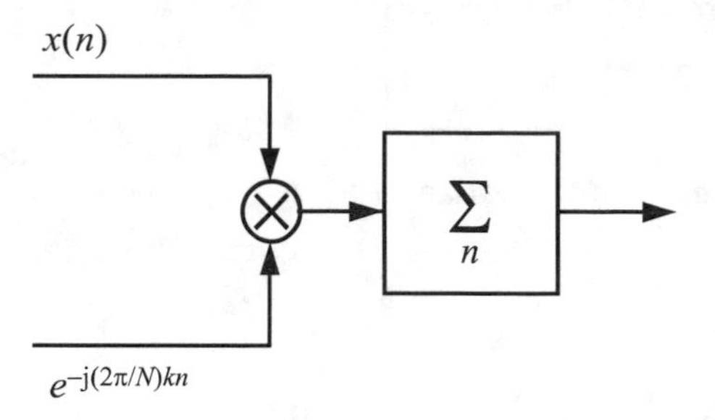

Figure 3.7: Correlator.

the zero lag, but it could easily be modified to do so. We also note that the cross-correlation is just the inner product, which is the projection of the data onto a set of orthogonal sinusoidal functions. This operation is also called a scalar product.

3.6.5 Convolver Interpretation

In general, the output of a linear time invariant filter can be written as

$$y(n) = \sum_{i=0}^{n} x(i)\, h(n-i) \tag{3.12}$$

Looking at (3.7), we see if we let $X(k)$ be the output of the filter at time $N-1$, denoted by

$$\begin{aligned} X(k) &= X(k, N-1) = y(N-1) \\ &= \sum_{n=0}^{N-1} x(n)\, e^{-j(2\pi/N)kn} \end{aligned} \tag{3.13}$$

We can also write (3.12) as

$$y(n) = \sum_{i=0}^{n} h(i)\, x(n-i) \tag{3.14}$$

If we let $h(i) = e^{-j(2\pi/N)k(N-1-i)}$ and $n = N-1$, then (3.14) becomes

$$y(N-1) = \sum_{i=0}^{N-1} e^{-j(2\pi/N)k(N-1-i)}\, x(N-1-i) \tag{3.15}$$

which equals (3.7) by using a change of variables and reversing the upper and lower limits of the summation. This interpretation is given in Figure 3.8.

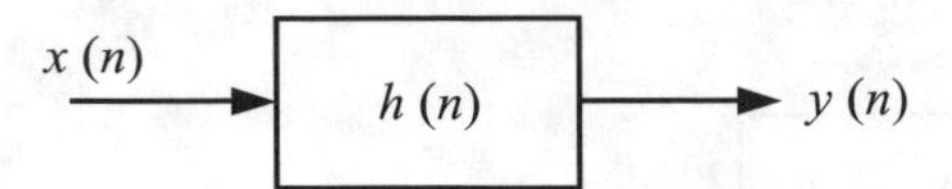

Figure 3.8: Convolver.

3.6.6 Matched Filter Interpretation

As will be shown in Chapter 6, the optimum detector for a known signal embedded in white Gaussian noise is the matched filter, where the matched filter response is the time reversed (i.e., mirrored) signal. We see that the $h(n)$ defined in Section 3.6.5 denotes a (complex valued) sinusoid, hence the FFT (DFT) is the optimal receiver (detector) for sinusoids embedded in white Gaussian noise. As such, the FFT/DFT constitutes a bank of matched filters with frequencies (bins) located at spectral locations $f_s k/N$, where k is the bin index ($0 \leq k \leq N-1$), N is the transform size, and f_s is the sampling frequency.

3.6.7 Coordinate Transformation Interpretation

We can also use linear algebra (vectors and matrices) to discuss the FFT operation. Let $\mathbf{x}^T = [x_0, x_1, \cdots, x_{N-1}]$

$$\mathbf{W} = \begin{bmatrix} 1 & 1 & 1 & \cdots & 1 \\ 1 & e^{-j(2\pi/N)} & e^{-j(2\pi/N)2} & \cdots & e^{-j(2\pi/N)(N-1)} \\ 1 & e^{-j(2\pi/N)2} & e^{-j(2\pi/N)4} & \cdots & e^{-j(2\pi/N)2(N-1)} \\ \vdots & \vdots & \vdots & & \vdots \\ 1 & e^{-j(2\pi/N)(N-1)} & e^{-j(2\pi/N)(N-1)2} & \cdots & e^{-j(2\pi/N)(N-1)(N-1)} \end{bmatrix}$$

with

$$\mathbf{X}^T = [X(0), X(1), \cdots, X(N-1)]$$

then

$$\mathbf{X} = \mathbf{W}\mathbf{x} = \text{FFT}\{x(n)\} \tag{3.16}$$

and

$$\mathbf{x} = \frac{1}{N}\mathbf{W}^{-1}\mathbf{X} = \text{FFT}^{-1}\{X(k)\} \tag{3.17}$$

Equation (3.16) is in the form of any standard matrix equation $\mathbf{y} = \mathbf{A}\mathbf{x}$, where $\mathbf{y}$ is a linear transformation of $\mathbf{x}$ [6,8].

3.7 FAST CORRELATION

Correlating two data sequences provides a measure of likeness between the two sequences. If one normalizes the correlation outputs, the values take on a value between minus and plus one. The normalization is a division of the square root by the product of the energy of the two sequences involved. A minus one and plus one correspond to -100 percent and $+100$ percent, respectively. When the normalized cross-correlation value is $+1$, the two sequences are identical. Conversely, a value of -1 indicates that the two sequences are identical in a magnitude sense, but differ in-phase by 180°. A correlation of a value of zero indicates that the two sequences are uncorrelated. Assuming that the sequences are of equal length, a typical correlation expression, disregarding the normalization, is given by

$$R_{XY}(\ell) = \sum_{n} x(n)\, y(n+\ell) \tag{3.18}$$

The data sequences, as used in (3.18), do not have to be of identical length. To get meaningful results when using a numerical evaluation routine, we would zero pad each sequence to be of length equal to or larger than the sum of length (x) plus length $(y) - 1$. Some software (i.e., XCORR in MATLAB) automatically takes care of this requirement. This operation also corresponds to the projection of one of the vectors onto the second vector. We can also obtain this result by

$$R_{XY}(\ell) = \mathrm{FFT}^{-1}\left\{X(k)\, Y^{*}(k)\right\} \tag{3.19}$$

where

$$\begin{aligned} X(k) &= \mathrm{FFT}\left\{x(n)\right\} \\ Y(k) &= \mathrm{FFT}\left\{y(n)\right\} \end{aligned}$$

and we assume that the lengths of the data sequences are a power of two, or that they have been zero padded to meet this criterion. The symbol * denotes conjugation. The procedure given in (3.19) is called the fast correlation. Other than the suggested normalization (i.e., none), normalization can be used to obtain other estimates of the cross-correlation function such as biased or un-biased estimates. When using the FFT-based approach, one also needs to properly zero pad the $x(n)$ and $y(n)$ sequences to ensure that no fold over takes place or will cause difficulties. For example, correlating two sequences of length N results in a correlation function of length $2N - 1$. Hence, both sequences should be zero extended at least to a length of $2N - 1$. This is in addition to the zero padding required by the FFT operations.

3.8 PERIODOGRAM (POWER SPECTRAL DENSITY ESTIMATE)

A very common way to obtain an estimate of the power spectral density of the sequence $x(n)$ is via the following expression

$$P_X(k) \hat{=} \frac{1}{N} \left| \sum_{n=0}^{N-1} x(n)\ w(n)\ e^{-j(2\pi/N)kn} \right|^2 \tag{3.20}$$

where N is the data length (power of two) and $w(n)$ is a data window (i.e., Hamming, Gaussian, etc. [9]). This estimate of the power spectral density is called the periodogram. If one uses no window, then one actually uses a rectangular window of size N. If a data segment is not equal in length to a power of two (or is chosen not to be), the segment is zero padded to a power of two (possibly the next power of two, but maybe even to a larger power of two). Zero padding produces an interpolated spectrum. The explicit use of a window will broaden the main lobe response (i.e., the width of the spectral peak of a corresponding sinusoid). The rectangular window (i.e., no explicit window) has the most narrow spectral main lobe response, but also has the poorest side lobe (i.e., leakage) response.

For a cross-spectral density expression, disregarding windows, (3.20) is modified to

$$P_{XY}(k) = \frac{1}{N} \left(\sum_{n=0}^{N-1} x(n)\ e^{-j(2\pi/N)kn} \right) \left(\sum_{n=0}^{N-1} y(n)\ e^{-j(2\pi/N)kn} \right)^* \tag{3.21}$$

More application-oriented material can be found in Chapter 9. There are many other ways to obtain a spectral estimate. Some of these are auto-regresive (AR), moving average (MA), and auto-regressive moving average (ARMA) modeling, subspace methods, and Prony's and Capon's method [10–14]. Extensions that deal with time varying scenarios, such as the spectrogram and the Wigner-Ville distribution, etc., are introduced in Chapter 9.

3.9 WAVELETS

3.9.1 Introduction

Wavelets have caught the attention of the signal processing and communications community on a large scale. Ingrid Daubechies [15,16] advocated the use of basis functions having a short support (duration). The effective length of the support is different at different frequencies (scales). Her research followed earlier work by Morlet, Grossmann, Meyer, Mallat, and others (for more detail, the reader is encouraged to consult [17]). If the reader is interested in a

DSP related discrete time wavelet reference, the books by Strang and Nguyen [18] and Vetterli and Kovačević [19] are recommended. We cannot do justice to all the books and papers that relate to wavelets since we must limit ourselves to a few references. One of the problems with the wavelet literature is that most authors claim to have a description that can be followed easily by the typical engineer. Despite some very enticing book titles, this is just not the case. We make an attempt to relate wavelet concepts to ideas engineers are already exposed to and leave it up to the reader to go on to the references (some of them are listed in here) to obtain a deeper understanding.

Wavelet processing is also known as multi-rate filtering, octave band processing, constant Q-filtering, wavelet series expansion, multi-resolution signal processing, and bandwidth proportional processing. One could follow the references under each one of these headings to get more information about these and related topics. For the newcomer, there are many references (mainly reprints) available by doing an Internet search on the topics mentioned. The wavelet community also maintains an informative e-mail based newsletter. Details can be found at `http://www.wavelet.org`. Earlier copies of the newsletter are archived, and easily made available when using an Internet browser (i.e., Netscape Navigator, Internet Explorer, or similar).

We are interested in wavelets in the context of detection and also parameter estimation. There are other groups and individuals that are interested in wavelet based concepts to solve partial differential equations, compress data (i.e., pictures, video, music, speech), or denoise data (i.e., remove unwanted disturbances), just to mention a few. Each one of these areas has a large body of reference material. We will develop the basic ideas of wavelet based processing, first in a brute force way and then in terms of a fast version (similar to the notion of classical DFT and FFT processing).

As far as the wavelet topic is concerned, we use FIR filters (linear and time invariant) that are typically of short length and are followed by a decimation (integer resampling) procedure. More often than not, the multipliers of the FIR filter are real valued.

3.9.2 Revisiting FIR Filters and the DFT/FFT

Let us look at a data sequence $x(n)$, $n \geq 0$ and a network (filter) with impulse response $h(n) = \sin \omega_0 n$, for $0 \leq n \leq N-1$, where $\omega_0 = (2\pi\ell)/N$ and ℓ is some fixed integer.

The function $h(n)$, if plotted, will go exactly through ℓ complete periods over N data points. The frequency response $H(e^{j\omega})$ is the Fourier transform of $h(n)$ denoted by $F\{h(n)\}$. This is the same scenario as in Figure 3.3(c), except that the filter has its center frequency at locations plus and minus ℓ and, of course, the spectral height (i.e., energy density) will be scaled. Let us idealize the response by assuming it rolls off very fast at the edges of the bandpass and that the side lobes are negligible (i.e., close to zero).

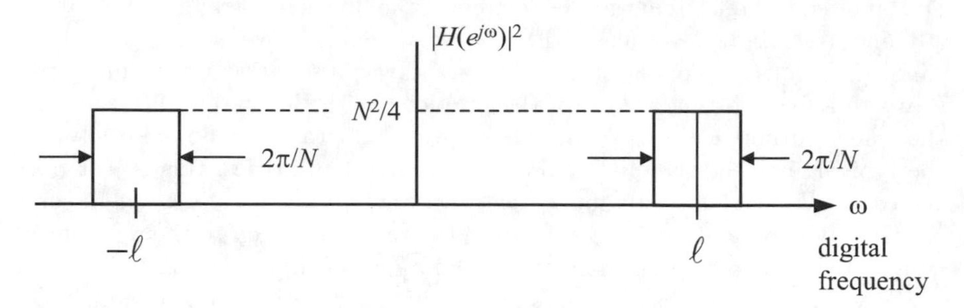

Figure 3.9: Bandpass (ideal) filter.

Then the spectral response is approximated as shown in Figure 3.9. We note that the filter response is given by $\left|H(e^{j\omega})\right|^2$, where the center frequency is $\omega = \pm(2\pi\ell)/N$. If we evaluated the output (of this filter) at its center frequency, we can refer to $\left|H(e^{j(2\pi\ell)/N})\right|^2$ or $|H(\ell)|^2$, with the understanding that the integer ℓ corresponds to the spectral location $(2\pi\ell)/N$ or ω_ℓ.

The FFT (see Section 3.6.1) is a collection of FIR bandpass filters, with complex valued impulse responses, whose outputs are decimated in time. Each filter has, when we idealize the performance, a response just as the filter discussed in Sections 3.6.5 and 3.6.6. The result is a filter bank, with the filter's center frequencies positioned at locations ℓ ($\ell = 0, 1, \cdots, N-1$). The zero location has no complementary spectral term (i.e., $\pm$ zero is zero) and has the same spectral width as any of the one-sided bandpass filters. The magnitude squared response is the same at all spectral locations except at locations $\ell = 0$ and $\ell = N/2$. Since all locations (time and frequency) are cyclic (i.e., modulus (N)), spectral locations $N - k$ correspond to negative frequency locations $(-k)$. For real valued input data, we expect Hermitian symmetry in the amplitude spectrum, that is $X(k) = X^*(-k) = X^*(N-k)$. For complex valued input data there is no symmetry at all in the amplitude spectrum.

If we segment the data, then every N data point segment provides a single output point (one from every filter). Sliding the segment by one data point along the data sequence again produces one data point from each filter location making the output rate equal the input rate. Of course, in a given spectral bin the output data would be heavily correlated. If the segments are chosen to be non-overlapping and contiguous, then the output data rate is $1/N$ times the data sampling rate. Different overlap factors between successive transforms lead to different output data rates. For example, overlap factors of 50 percent (2:1 overlap) and 75 percent (4:1 overlap) lead to output rates of $(f_s2)/N$ and $(f_s4)/N$, respectively. In this case, the DFT/FFT is a bank (combination) of bandpass filters followed by a decimation scheme (see Figure 3.10). Figure 3.10(b) attempts to show that the DFT/FFT is a bandpass FIR filter bank followed by a decimator that throws away $N-1$ output data points and keeps the Nth output data point. If more segments are to be processed, then from each filter one obtains one output data point per segment. From Figure 3.10(a) and Section 3.6, it is clear that each filter has the same bandwidth.

3.9.3 FIR Filters and Wavelet Transforms

Suppose that we let the bandwidth of each filter be proportional to its center frequency and we resample the output at a rate governed by the bandwidth. The filters (filters in a bank) are arranged such that the spectral region is contiguously covered, having no holes and no redundant coverage. The majority of wavelet filters are FIR filters having a real valued impulse response. We can construct wavelets that do have a complex valued impulse response, but we will only use it to describe a poor man's version of the wavelet transform. To illustrate the discussion, we shall choose a small data size, say $N = 64$. Figure 3.11(a) and (b) shows the desired arrangements for size $N = 64$.

If one were to execute one FFT of length 64, 64 spectral bin (filter) outputs would be obtained, each one consisting of one data point. The number of data points is preserved; we started with 64 time samples and we ended up with 64 (filtered) output points. A similar conclusion can be drawn from Figure 3.11. If we sum up the output data points $(32+16+8+4+2+1+1=64)$ we again preserve the number of data points. The scheme in Figure 3.11 has been around for a long time and is known as proportional bandpass filtering, octave filtering, and constant Q-filtering. Q is the ratio of center frequency to bandwidth. It is also called the quality factor in the electronic circuit literature. The circle symbol with the down-pointing arrow and integer I represents the decimation operation (i.e., keep every I^{th} sample). Note, while our example uses a data size of $N = 64$ it can be any arbitrary power of two. The number of bandpass filters will always be $\log_2 N$, i.e., $\log_2 64 = 6$, while there is only one low pass filter.

We have established a wavelet transformer consisting of $\log_2 N$ (i.e., six) bandpass filters and one low pass filter. Let us use the example described in Figure 3.11 to make some observations. The first one is that the number of data points (i.e., a coordinate transformation) is preserved. Needless to say, if we properly invert the seven available filter outputs, we will get back the original data sequence. The filter band width becomes progressively smaller, starting with the half-band filter (top leg in Figure 3.11(b)). It is cut in half in each successive stage (i.e., progressing down the filter bank). Since the bandpass filters reduce the bandwidth of their respective outputs, we can resample (i.e., change the sampling rate as indicated). The first output (i.e., the top of Figure 3.11(b)), denoted by SC1, has an output sampling rate one half of its input sampling rate. The second output, denoted by SC2, has an

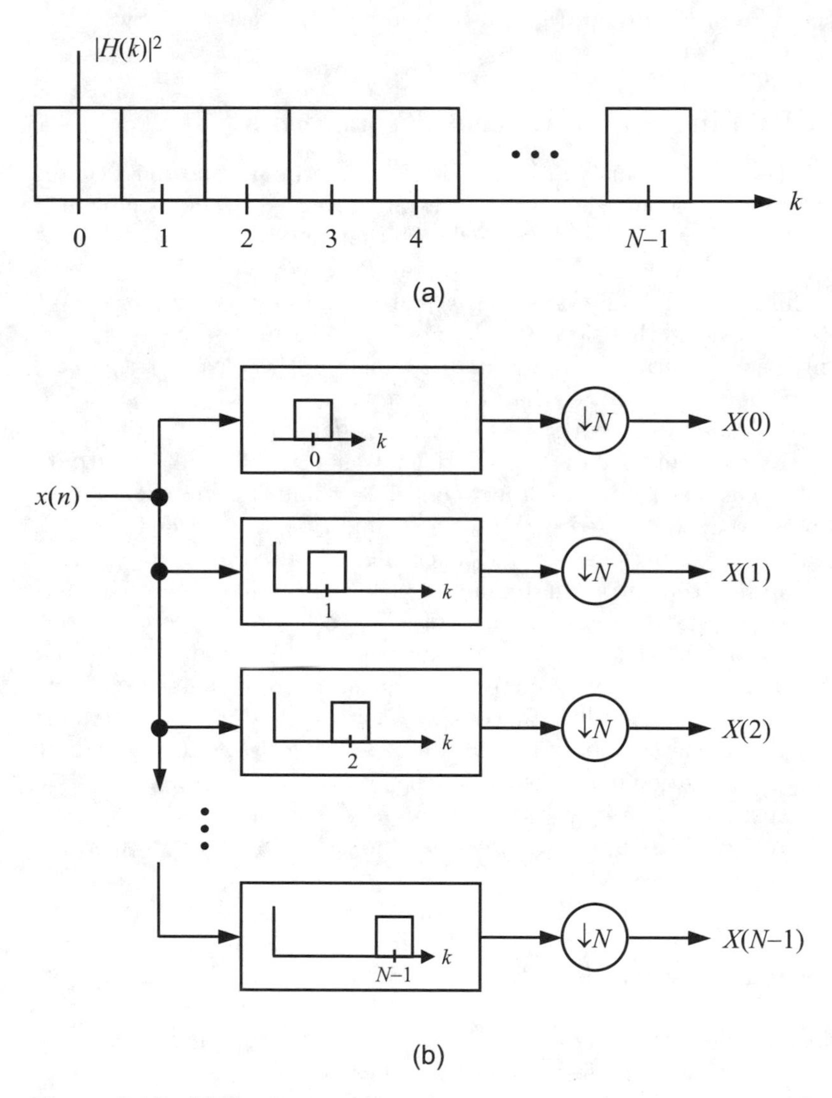

Figure 3.10: (a) Spectral regions of a bank of bandpass filters and (b) bandpass filters followed by decimation.

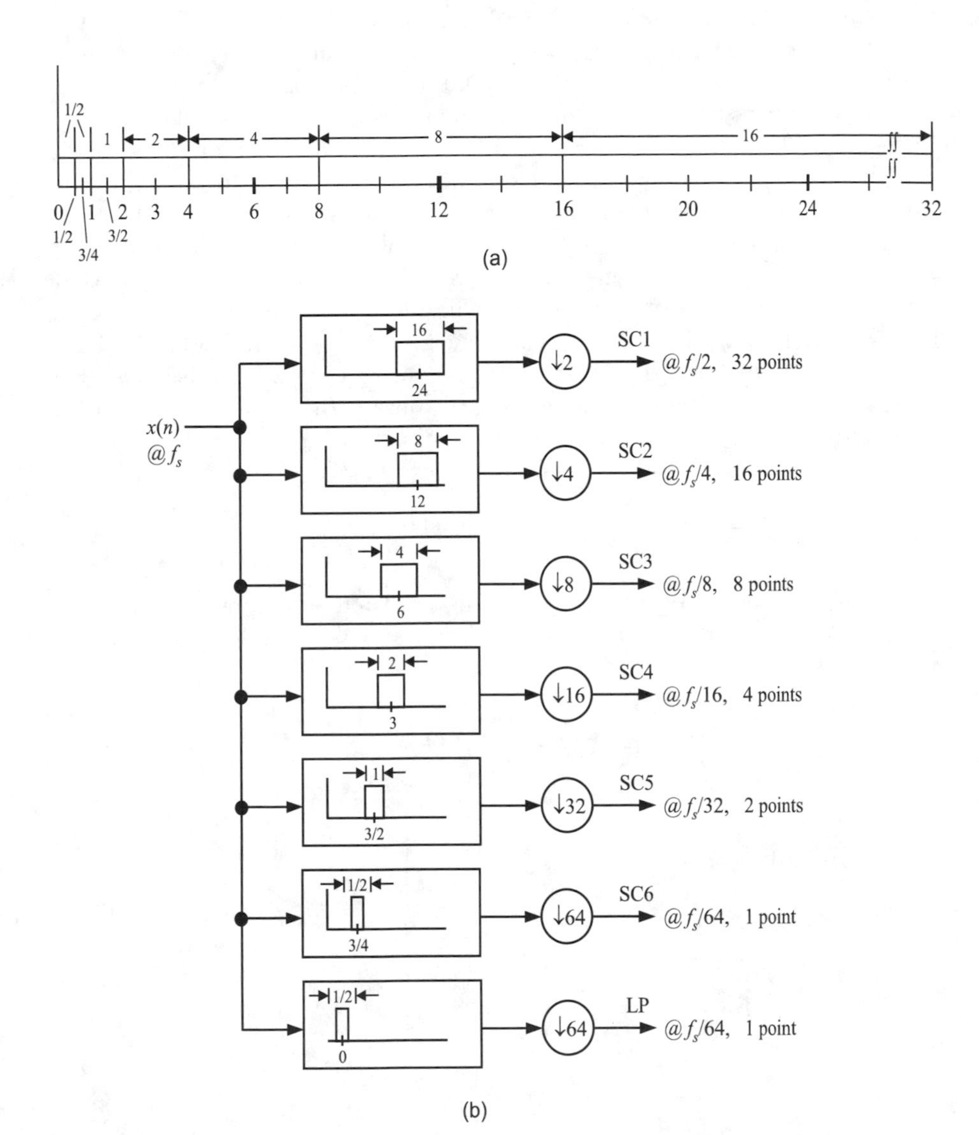

Figure 3.11: (a) Spectral regions of a band of proportional bandpass filters ($N = 64$) and (b) six bandpass filters and one low pass filter followed by decimators.

output sampling rate one fourth of the input sampling rate. The notation SC refers to what is commonly called scales, with the widest spectral bandwidth associated with scale 1. Rather than using a bin (bin numbers) as in the DFT, we label the filter as scales, usually starting with the number one for the highest frequency and increasing in count as the frequency location becomes smaller. We note that some authors label the scales in the same direction as the frequency (i.e., the filter output on top of Figure 3.11(b) would be labeled number 6). This enhances the presentation when using a vector space based approach (i.e., a larger number corresponds to a larger vector space). The scale outputs are also referred to as detail functions, while the lowpass output is called the approximation. The brute force implementation of Figure 3.11(b) is computationally very expensive. However, there is a cascaded version that will speed up the operations which we will address in the next section (i.e., the fast wavelet transform). If we implement the scheme in Figure 3.11(b), we could do it with a with a Fourier transform kernel, that is, use the DFT as a filter to obtain the desired results. To obtain the lowpass output (one data point) we sum over the whole data set as given by

$$LP(0) = \sum_{n=0}^{63} x(n) \qquad \text{(one output point)}$$

To obtain the scale outputs, we sum up weighted data points where the start, stop, and summation times depend on the scale. The scale outputs are obtained as given by

$$SC6(0) = \sum_{n=0}^{63} x(n)\ e^{-j(2\pi(3/4)/64)n} \qquad \text{(1 output point)}$$

$$\left.\begin{aligned} SC5(0) &= \sum_{n=0}^{31} x(n)\ e^{-j(2\pi(3/2)/64)n} \\ SC5(1) &= \sum_{n=32}^{63} x(n)\ e^{-j(2\pi(3/2)/64)n} \end{aligned}\right\} \qquad \text{(2 output points)}$$

$$\left.\begin{aligned}
SC4(0) &= \sum_{n=0}^{15} x(n)\ e^{-j(2\pi 3/64)n} \\
SC4(1) &= \sum_{n=16}^{31} x(n)\ e^{-j(2\pi 3/64)n} \\
SC4(2) &= \sum_{n=32}^{47} x(n)\ e^{-j(2\pi 3/64)n} \\
SC4(3) &= \sum_{n=48}^{63} x(n)\ e^{-j(2\pi 3/64)n}
\end{aligned}\right\} \quad \text{(4 output data points)}$$

If we use a compact notation, we can write the filter outputs as

$$SC3(i) = \sum_{n=i\cdot 8}^{(i+1)8-1} x(n)\ e^{-j(2\pi 6/64)n} \qquad i = 0, 1, \cdots, 7 \ \text{(8 output points)}$$

$$SC2(i) = \sum_{n=i\cdot 4}^{(i+1)4-1} x(n)\ e^{-j2\pi(12/64)n} \qquad i = 0, 1, \cdots, 15 \ \text{(16 output points)}$$

and finally

$$SC1(i) = \sum_{n=i\cdot 2}^{(i+1)2-1} x(n)\ e^{-j2\pi(24/64)n} \qquad i = 0, 1, \cdots, 31 \ \text{(32 output points)}$$

where i corresponds to the segment number (also called tile number). That is, scale 6 has only one segment, that is one output data point. Scale 5 has two segments, that is, two output data points. Scale 4 has four segments, that is, four output data points. Scale 3 has eight segments, that is, eight output data points. Scale 2 and scale 1 have 16 and 32 segments (i.e., data output points), respectively.

This particular example was chosen, since we believe most readers can relate to the Fourier transform in a natural way. We could have used any type of FIR filter, as long as it can be construed to be the bandpass filters with the desired parameters. Actually, most wavelet transforms use real valued weights in their FIR filters. This reduces the computational burden relative to the filters we have chosen to make the illustration. At his point, we realize that the wavelet transform is just a collection of proportional bandwidth, linear time invariant bandpass filters with a cleverly chosen bandwidth and an optimal decimation of the output data rate. The spectral response of the bandpass filters is typically not very good. This is a direct consequence of making these FIR filters relatively short in duration (i.e., few filter weights therefore few processing multiplies). In Sections 3.9.3 and 3.9.4 we assumed to have a unit gain in the passband region. This is usually not true, rather than that, the gain increases as the frequency bands step towards the DC region. A white input noise process would tend to have the same power (energy) in each detail sequence.

3.9.4 Mallat's Algorithm

Just as in the case of the DFT, there is a clever way to speed up the computation of the wavelet transform. Again, we use Figure 3.11 to help with the description. We note that extensions to any size N is straightforward (i.e., just add more bandpass filters, hence scales). We start with a bandpass and low pass filter section that splits the spectral region into equal segments. Figure 3.12 shows the details of the algorithm, which is also known as Mallat's algorithm [20,21]. All high pass (HP) filters are identical and all low pass (LP) filters are identical. The number of filter weights is an even number. The HP filter is just the LP filter spectrally relocated to the fold over frequency. That is, the low pass filter transfer function is modulated to the fold over frequency and the weights are time reversed (i.e., mirrored)

$$h_{HP}(n) = (-1)^n \; h_{LP}(M - 1 - n)$$

for $n = 0, 1, \cdots, M - 1$, given that the FIR filters have M weights. Typically, the filters have very few weights which, for a given wavelet transform, do not change in number nor in value. The number of weights is at least two and maybe as high as 20 to 40. The larger the number of weights, the better the filter's spectral characteristics but the poorer its ability to localize a position in time. There are constraints on the filter weights if reconstruction (i.e., the inverse wavelet transform) is desired [15–18]. For the purposes of this book, good characteristics in the wavelet domain are desirable, while we are less interested in using the inverse wavelet transform (IWT). In most cases, we use a real valued impulse response for the filters, but we note that there are several wavelet candidates allowing complex valued weights. We want to also check out the computational burden of the WT. From the previous discussion it should be clear that one needs $\log_2 N$ stages, where each stage consists of a bandpass and a low pass filter. Each filter at the start of the chain (tree) processes N samples, using M real valued coefficients in both the HP and LP filters. The computational cost in the first stage is $2 \cdot M \cdot N$. As we proceed to the next stage, we notice that the filter sizes stay the same but the number of samples has been reduced to about $N/2$. Hence, the cost of multiplication of the second stage is $2 \cdot M \cdot N/2$. Adding up all the costs (multiplications) for the whole tree, we get, in general

$$\text{Cost}_{\text{ WT}} = 2MN \left(1 + \frac{1}{2} + \frac{1}{4} + \frac{1}{8} + \cdots + \frac{1}{N}\right) < 4MN$$

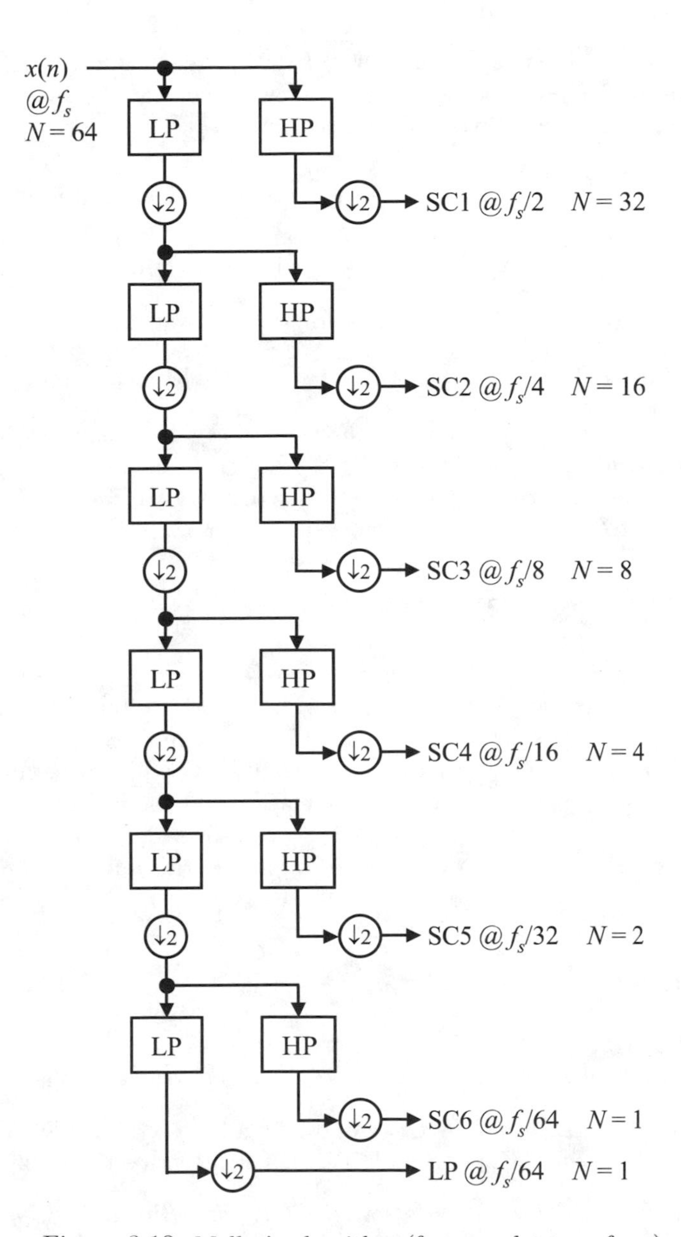

Figure 3.12: Mallat's algorithm (fast wavelet transform).

For our case we have

$$\text{Cost}_{\text{WT}} \leq 4M64 = 256M$$

The general result is valid for any size of data set, assuming it is inputed in segments of size as a power of two. If we compare the multiplication processing cost of the FFT versus the fast WT, we see that for the FFT

$$\begin{aligned} \text{Cost}_{\text{FFT}} &= \left(\frac{\log_2 N}{2}\right) N \qquad \text{(FFT, complex valued multiplications)} \\ &= (2\log_2 N)N \qquad \text{(FFT, real valued multiplications)} \end{aligned}$$

where N = FFT size = data size, while for the fast wavelet transform (WT)

$$\text{Cost} = (4M)N \qquad \text{(WT, real valued multiplications)}$$

where M is filter size, N = data size. For small M (the smallest M possible is two, corresponding to the Haar wavelet, which is also called Daubechies of order 2) and large N, the WT can easily outperform the FFT. We will get more exposure to the WT in Chapter 9 and in Appendix E.

3.10 SUMMARY

This chapter introduces discrete time signals and typical processing tools. In particular, linear data processing is examined. FIR filtering is addressed in Section 3.5. The Fourier transform (DFT/FFT) are discussed in detail in Section 3.6, providing different useful interpretations of this processing operation. Sections 3.7 and 3.8 address fast correlation and the fast power spectral density estimation techniques. The final section provides a layman's introduction to wavelet processing and contrast the Fourier and WT processing methods.

References

[1] Strum, R.D., and Kirk, D.E., *First Principles of Discrete Systems and Digital Signal Processing*, New York: Addison-Wesley Publishers, 1988.

[2] McClellan, J.H., Schafer, R.W., and Yoder, M.A., *DSP First: A Multimedia Approach*, Upper Saddle River, NJ: Prentice-Hall, 1998.

[3] Ludeman, L.C., *Fundamentals of Digital Signal Processing*, New York: Harper & Row, 1986.

[4] Elliott, D.F., and Ramamohan-Rao, K., *Fast Transforms: Algorithms, Analysis, Applications*, Orlando, FL: Academic Press, 1982.

[5] Burrus, C.S., Gopinath, R.A., and Guo, H., *Introduction to Wavelets and Wavelet Transforms: A Primer*, Upper Saddle River, NJ: Prentice-Hall, 1998.

[6] Strang, G., *Linear Algebra and Its Applications*, 2nd ed., New York: Academic Press, 1980.

[7] Harris, F.G., "The discrete Fourier transform applied to time domain signal processing," *IEEE Communications Magazine*, May 1982, pp. 13–22.

[8] Noble, B., and Daniel, J.W., *Applied Linear Algebra*, 3rd ed., Englewood Cliffs, NJ: Prentice-Hall, 1988.

[9] Harris, F.G., "On the use of windows for harmonic analysis with the discrete Fourier transform," *Proc. IEEE*, Vol. 66, No. 1, pp. 51–83, Jan. 1978.

[10] Kay, S.M., *Modern Spectral Estimation, Theory, and Applications*, Englewood Cliffs, NJ: Prentice-Hall, 1988.

[11] Marple, S.L., *Digital Spectral Analysis*, Englewood Cliffs, NJ: Prentice-Hall, 1987.

[12] Stoica, P., and Moses, R., *Introduction to Spectral Analysis*, Upper Saddle River, NJ: Prentice-Hall, 1997.

[13] Naidu, P.S., *Modern Spectrum Analysis of Time Series*, Boca Raton, FL: CRC Press, 1996.

[14] Therrien, C.W., *Discrete Random Signals and Statistical Signal Processing*, Englewood Cliffs, NJ: Prentice-Hall, 1992.

[15] Daubechies, I., "Ortho-normal bases of compactly supported wavelets," *Communications on Pure and Applied Mathematics*, Vol. 41, pp. 909–997, Nov. 1988.

[16] Daubechies, I., "Ten lectures on wavelets," *SIAM*, Philadelphia, PA, 1992.

[17] Burrus, C.S., Gopinath, R.A., and Guo, H., *Introduction to Wavelets and Wavelet Transforms, A Primer*, Upper Saddle River, NJ: Prentice-Hall, 1998.

[18] Strang, G., and Nguyen, T., *Wavelets and Filter Banks*, Wellesley, MA: Wellesley-Cambridge Press, 1996.

[19] Vetterli, M., and Kovačević, J., *Wavelets and Subband Coding*, Englewood Cliffs, NJ: Prentice-Hall, 1995.

[20] Mallat, S.G., “A theory for multi-resolution signal decomposition: The wavelet representation,” *IEEE Trans. on Pattern Recognition and Machine Intelligence*, Vol. 7, No. 11, pp. 674–693, July 1989.

[21] Mallat, S.G., “Multi-resolution approximation and wavelet ortho-normal bases of L^2,” *Trans. of Amer. Math. Soc.*, Vol. 315, pp. 69–87, 1989.

[22] Vaidyanathan, P.P., *Multirate Systems and Filter Banks*, Englewood Cliffs, NJ: Prentice-Hall, 1992.

Chapter 4

Hypothesis Testing

4.1 INTRODUCTION

Hypothesis testing, as presented in this chapter, deals with a finite number of samples. The samples are also called data or observations. The data is used to decide which hypothesis, that is, which signal or symbol is true (i.e., which one is present) and hence was transmitted. Throughout Sections 4.1–4.8 we assume that all parameters are known. That is, we know where and when to look for the signals. In the first seven sections, the binary case is addressed, that is, only one of two signal conditions can be true. This is typical in radar/sonar detection and binary digital communication problems. Section 4.8 addresses multiple hypothesis testing, while Section 4.9 deals with composite hypothesis testing. Composite hypothesis testing allows us to deal with situations when some uncertainty of the parameters is involved. Section 4.10 addresses receiver operating characteristics (ROC) curves and performance descriptions.

The problems in this chapter can be separated into two distinct types. The first type of problem addresses testing for the presence of one of two distinct signals which are obscured by additive noise. The solution provides an answer to the question of which signal is immersed in the additive noise.

The second type of problem addresses the testing for the presence of one of two distinct signals, where each signal has a unique statistical description (i.e., its own PDF). The solution provides an answer to the question of which probability density function governs the observed data or equivalently which signal is present.

Problems in this chapter are not limited to Gaussian type probability density functions. In both types of problems, the statistical part (the additive noise or the original population density) is allowed to be governed by any probability law. Any type of probability density function is allowed. In most problems, we want to detect the presence or absence of a signal with a certain measure of reliability. Of course, usually detection and estimation are performed when additive noise is present. If it was not for the noise disturbance, then it would be no problem to decipher the message symbol (yes-no, zero-one, etc.). One would simply need to decide which one of the two hypotheses, H_1 or H_0 is true (i.e., which signal is present) by just reading it off the data.

In all cases, a detection scheme in terms of the simplest function or functional of the observation (i.e., the received data) is attempted. This leads in many cases to what is called a sufficient statistic, which contains all the information as far as the problem of interest is concerned but is numerically and analytically easier to deal with than the original variable or variables. Reduction to a simple form of the observation also ensures that the computational effort to obtain the detection variable is minimal.

Initially, the Bayes' detector is derived and then its structure is used as the basic building block to motivate other detection schemes. These other detection schemes are based on the maximum *a priori* (MAP), the maximum likelihood (ML), the minimum probability of error, the Min-Max, or the Neyman-Pearson criterion. Sometimes the phrases: criterion, strategy, philosophy, technique, method, algorithm, detector, or receiver are used interchangeably. These detectors are also appropriate if each hypothesis (i.e., signal) is governed by a different probability density function. In these cases, one establishes (decides) which generating PDF (event, hypothesis) produces the observation (data).

Before the basic detector structure is derived, the concept of the "*a posteriori* probability" is examined. Suppose two different events are possible, such as a binary "one" or "zero" (i.e., target present or not). The letter i ($i = 0, 1$) is used to indicate which hypothesis (event) is true. The probability of the event "i" (i.e., hypothesis H_i), given the observation $\mathbf{y}$ is described by the "*a posteriori* probability" $\Pr\{H_i|\mathbf{y}\}$. The "*a priori* probability" of the event "i" (i.e., probability of hypothesis H_i) is given by $\Pr\{H_i\}$. The situation is best illustrated using the general signal flow diagram of Figure 4.1.

Figure 4.1 depicts the source (generating two possible symbols), two error-free and two error transmission paths, and the receiver (i.e., the decision element). The path accounts for the channel or medium used to convey the message from the transmitter to the receiver. A binary "one" (1) or a binary "zero" (0) is transmitted with probability P_1 or P_0, respectively. These probabilities are "*a priori* probabilities," known prior to transmitting either one of the symbols. The term "transition probability" is used to denote $\Pr\{\text{observed data(symbol)}|\text{actual symbol transmitted}\}$. The quantity of interest is the probability $\Pr\{\text{symbol transmitted}|\text{observed data}\}$.

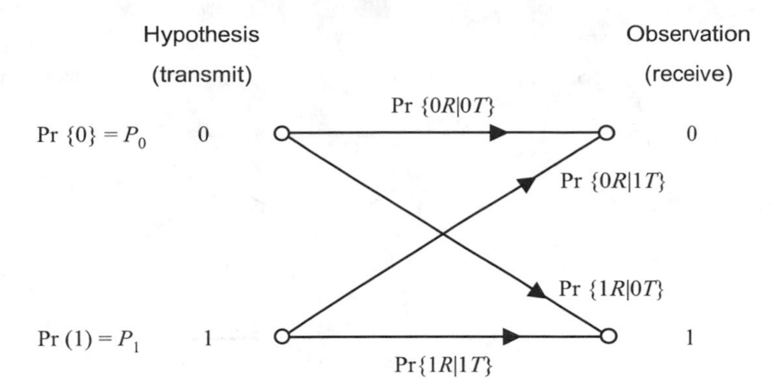

Figure 4.1: Binary transmission channel.

Using Bayes' rule, this is easily obtained as

$$\Pr[H_i|\mathbf{y}] = \frac{\Pr[\mathbf{y}|H_i]}{\Pr[\mathbf{y}]}\Pr[H_i] \tag{4.1}$$

where $\mathbf{y}$ *is given by* $\mathbf{y} = (y_1, y_2, \cdots, y_N)^T$, the observation vector.

Example 4.1 *Binary transmission. The probability of transmitting a "zero" (0) is 0.7. This makes* Pr{one transmitted} *equal to 0.3. The probability of receiving a zero, given that a zero was transmitted, is 0.8 and the probability of receiving a one, given that a one was transmitted, is also 0.8. The error probabilities are therefore 0.2 for each one of the two possible errors. That is the probability of receiving a one, given that a zero was transmitted, is the same as the probability of receiving a zero, given that a one was transmitted. This type of transmission channel is also called a symmetric channel. The signal flow diagram is given in Figure 4.2.*

$$\begin{aligned}
\Pr\{0R\} &= \Pr\{0R|1T\}\Pr\{1T\} + \Pr\{0R|0T\}\Pr\{0T\} \\
&= 0.2 \times 0.3 + 0.8 \times 0.7 = 0.62 \\
\Pr\{1R\} &= \Pr\{1R|0T\}\Pr\{0T\} + \Pr\{1R|1T\}\Pr\{1T\} \\
&= 0.2 \times 0.7 + 0.8 \times 0.3 = 0.14 + 0.24 = 0.38 \\
\Pr\{1T|1R\} &= \Pr\{1R|1T\}\Pr\{1T\}/\Pr\{1R\} \\
&= 0.8 \times 0.3/0.38 = 0.632 \\
\Pr\{0T|0R\} &= \Pr\{0R|0T\}\Pr\{0T\}/\Pr\{0R\} \\
&= 0.8 \times 0.7/0.62 = 0.903
\end{aligned}$$

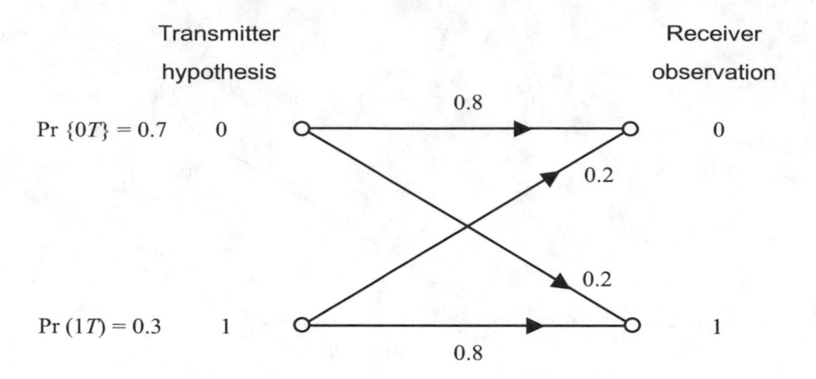

Figure 4.2: Signal flow diagram for Example 4.1.

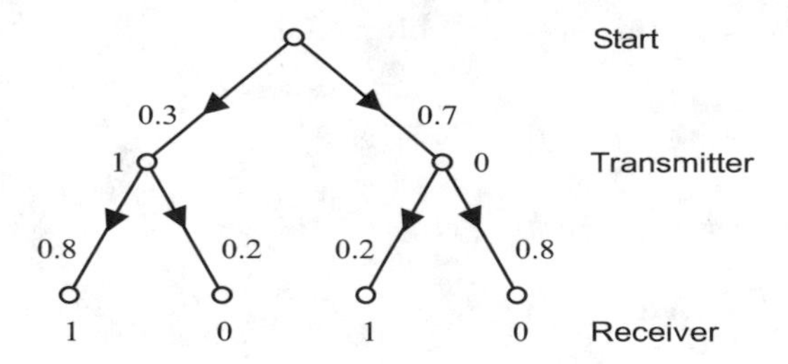

Figure 4.3: Trellis representation.

Note:

$$\Pr\{1T, 1R\} = \Pr\{1T|1R\}\Pr\{1R\} = 0.632 \times 0.38 = 0.24$$

$$\Pr\{0T, 0R\} = \Pr\{0T|0R\}\Pr\{0R\} = 0.903 \times 0.62 = 0.56$$

Note: Another way of obtaining the results is by using a trellis representation as shown in Figure 4.3. For example, the probability of receiving a one is $\Pr\{1R\} = 0.2 \times 0.7 + 0.8 \times 0.3$ *and of course as before, we can compute* $\Pr\{1T|1R\} = \Pr\{1R|1T\}\Pr\{1T\}/\Pr\{1R\} = 0.8 \cdot 0.3/0.38 = 0.632$.

Usually, one needs to evaluate how well a detection scheme (detector) works. In the binary detection situation, four probabilities are of interest. There are two proper and two improper detections. The proper detection of the "one" hypothesis is denoted by P_D. The two possible errors are denoted by P_{FA} and P_M. P_{FA} corresponds to the detection of the "one" event, given that the "zero" event is true and P_M corresponds to detecting the "zero" event, given that the "one" event is true. The sum of P_D and P_M must

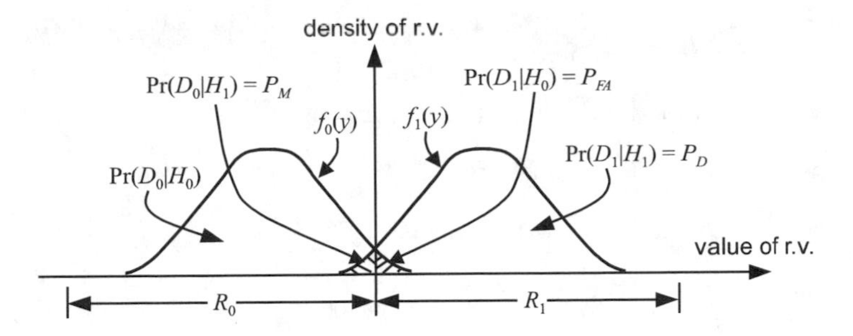

Figure 4.4: Probabilities of binary detection.

equal 1. In the statistical literature, P_{FA}, P_M, and P_D are also denoted by the error of the first kind, error of the second kind and power of the test, respectively. To illustrate these definitions, we assume that we have a single observation y which has likelihood functions $f_0(y)$ and $f_1(y)$ when events 0 and 1, respectively are true (see Figure 4.4). Suppose the observation y is directly used as a detection statistic and a threshold T_0 is established at the crossover point of the two likelihood functions. The area to the right of T_0 is the decision region R_1 (i.e., decide on D_1), while the area to the left of T_0 is the decision region R_0 (i.e., decide on D_0). P_{FA} is the area under $f_0(y)$ in the region R_1. P_M is the area under $f_0(y)$ in the region R_0 and P_D (power of the test) is the area under $f_1(y)$ in region R_1. The area under $f_0(y)$ in region R_0 denoted by P_{FA} is also known as the size of the test. The derivation of the detectors is not unique to this book; one can find similar expressions and additional examples in references [6–8].

4.2 BAYES' DETECTION

Given the observation (data) $y(t)$,we assume that one binary signal or symbol is transmitted during the interval $(0 \leq t \leq T)$. To obtain a vector of length N, the data segment $y(t)$ has been properly converted from analog to discrete time (A/D). So for a noisy reception we have $y_n = s_n + n_n$, for $n = 1, \cdots, N$. The labeling, i.e., the first value of the count variable "n" can be arbitrarily set to zero or one. A vector type representation usually prefers one as the first index, while FFT type processing prefers zero as the first index. The sampling (i.e., A/D conversion), of course, is done in a way that will not introduce errors (i.e., the Nyquist criterion is not violated and the amplitude discretization does not create errors). We denote the sequence by the vector $\mathbf{y} = (y_1, y_2, \cdots, y_N)^T$, the prior probability $\Pr\{H_i\}$ by P_i, and the likelihood function by $f_{\mathbf{Y}|\mathbf{H_i}}(\mathbf{y}|H_i) = f_i(\mathbf{y})$ = density of $\mathbf{y}$ under the "i^{th}" hypothesis, for $i = 0, 1$. The sum of the P_i must equal one.

The possible outcomes, accounting for the true symbol that is transmitted, in a binary detection experiment are

(a) Choose H_0; H_0 is true (correct decision)

(b) Choose H_1; H_0 is true (mistake)

(c) Choose H_1; H_1 is true (correct decision)

(d) Choose H_0; H_1 is true (mistake)

Only (a) and (c) are correct decisions, while (b) and (d) are errors. Let C_{ij} be the cost associated with choosing hypothesis "i" when actually hypothesis "j" is true. So the costs for our problem become

(a) C_{00} (choose 0, 0 is true)

(b) C_{10} (choose 1, 0 is true)

(c) C_{11} (choose 1, 1 is true)

(d) C_{01} (choose 0, 1 is true)

The average (i.e., expected) value of the cost function (also called the risk or the penalty function) will be computed. The average cost is a weighted sum of cost terms C_{ij}. The weights are the probabilities corresponding to the events, designated by the indices of the cost terms. This average cost C must be minimized. Hence, the average cost is defined as

$$\begin{aligned} C &= C_{00} \Pr(\text{choose } 0, 0 \text{ is true}) \\ &\quad + C_{10} \Pr(\text{choose } 1, 0 \text{ is true}) \\ &\quad + C_{11} \Pr(\text{choose } 1, 1 \text{ is true}) \\ &\quad + C_{01} \Pr(\text{choose } 0, 1 \text{ is true}) \end{aligned}$$

Using Bayes' rule for probabilities, $\Pr(A, B) = \Pr\{A|B\} \Pr\{B\}$, the average cost can be written as

$$\begin{aligned} C &= C_{00}\, P_0 \Pr(\text{choose } 0|0 \text{ is true}) \\ &\quad + C_{10}\, P_0 \Pr(\text{choose } 1|0 \text{ is true}) \\ &\quad + C_{11}\, P_1 \Pr(\text{choose } 1|1 \text{ is true}) \\ &\quad + C_{01}\, P_1 \Pr(\text{choose } 0|1 \text{ is true}) \end{aligned}$$

In the radar/sonar problem, the second probability term represents the probability of false alarm (P_{FA}), the third probability term represents the probability of detection (P_D), while the fourth probability term represents the probability of a miss (P_M). The first probabilistic term (i.e., $1 - P_{FA}$), is sometimes called the probability of proper dismissal.

But these conditional probabilities are conditional densities integrated over the appropriate decision region. These decision regions are denoted by R_0 and R_1. These are the regions in which one decides that H_0 or H_1 is true. We note that some authors like to use the symbols D_0 and D_1 to indicate the decision corresponding to these two regions (see Figure 4.4). The average cost can be expressed as

$$\begin{aligned} C &= C_{00}\, P_0 \int_{R_0} f_0\,(\mathbf{y})\, d\mathbf{y} \\ &+ C_{10}\, P_0 \int_{R_1} f_0\,(\mathbf{y})\, d\mathbf{y} \\ &+ C_{11}\, P_1 \int_{R_1} f_1\,(\mathbf{y})\, d\mathbf{y} \\ &+ C_{01}\, P_1 \int_{R_0} f_1\,(\mathbf{y})\, d\mathbf{y} \end{aligned} \tag{4.2}$$

Note, for an N-dimensional observation space, the integrals are of dimension N (i.e., N-fold). Equation (4.2) can be expressed in terms of one single region of integration, say R_0. This eventually allows work with one unknown, that is a single region. One makes the observation that the cost of a bad decision is always larger than the cost of a good decision and that all costs are by definition non-negative quantities. The observation space R is the direct sum, $R = R_1 \oplus R_0$, where $R_1 \cap R_0 = \phi$ (the empty set). This implies that $R_0 = R - R_1$ and allows the average cost to be expressed as

$$\begin{aligned} C &= C_{00}\, P_0 \int_{R_0} f_0\,(\mathbf{y})\, d\mathbf{y} \\ &+ C_{10}\, P_0 \int_{R-R_0} f_0\,(\mathbf{y})\, d\mathbf{y} \\ &+ C_{11}\, P_1 \int_{R-R_0} f_1\,(\mathbf{y})\, d\mathbf{y} \\ &+ C_{01}\, P_1 \int_{R_0} f_1\,(\mathbf{y})\, d\mathbf{y} \end{aligned} \tag{4.3}$$

where the integrals are of the same dimension as the dimension of the vector $\mathbf{y}$. The conditional densities can be integrated over the total probability region R resulting in

$$\int_R f_i\,(\mathbf{y})\, d\mathbf{y} = 1\ ; \quad \text{for } i = 0, 1$$

where the integrals are of the same dimension as the dimension of the vector $\mathbf{y}$. The equation for the average cost reduces to

$$\begin{aligned} C &= C_{00}\,P_0 \int_{R_0} f_0(\mathbf{y})\,d\mathbf{y} \\ &\quad + C_{10}\,P_0 \left(1 - \int_{R_0} f_0(\mathbf{y})\,d\mathbf{y}\right) \\ &\quad + C_{11}\,P_1 \left(1 - \int_{R_0} f_1(\mathbf{y})\,d\mathbf{y}\right) \\ &\quad + C_{01}\,P_1 \int_{R_0} f_1(\mathbf{y})\,d\mathbf{y} \end{aligned} \tag{4.4}$$

Equation (4.4) can be rewritten as

$$C = \overbrace{C_{10}P_0 + C_{11}P_1}^{\text{fixed cost}} + \underbrace{\int_{R_0} [P_1(C_{01} - C_{11})f_1(\mathbf{y}) - P_0(C_{10} - C_{00})f_0(\mathbf{y})]\,d\mathbf{y}}_{\text{variable cost to be optimized}} \tag{4.5}$$

Every individual term in (4.5) is non-negative with the first two terms constituting a given fixed positive cost. The optimal strategy to minimize the overall (average) cost is to select R_0 to make the contribution from the integral minimal, or if possible, negative. This is accomplished by assigning $\mathbf{y}$ to R_0 (hence, say hypothesis H_0 is true) when the negative term in the integral is dominant, reducing the fixed cost. Conversely, if the first term in the integrand is dominant, $\mathbf{y}$ is assigned to belong to region R_1 (say hypothesis H_1 is true); hence, no additional cost is added to the fixed cost. The decision criterion becomes

$$P_1\;[C_{01} - C_{11}]\;f_1(\mathbf{y}) \underset{H_0}{\overset{H_1}{\gtrless}} P_0\;[C_{10} - C_{00}]\;f_0(\mathbf{y})$$

Or in more conventional form, the Bayes' formulation becomes

$$\frac{f_1(\mathbf{y})}{f_0(\mathbf{y})} \underset{H_0}{\overset{H_1}{\gtrless}} \frac{P_0\;[C_{10} - C_{00}]}{P_1\;[C_{01} - C_{11}]} \qquad \text{(Bayes' detector)} \tag{4.6}$$

Note all prior probabilities and all terms within the square brackets are positive, hence the rearrangement across the inequalities is simple. The quantity on the left-hand side is called the likelihood ratio (LR) and is denoted by

$\Lambda(\mathbf{y})$, while the right-hand quantity is denoted by λ, a positive scalar. So we can write

$$\Lambda(\mathbf{y}) \underset{H_0}{\overset{H_1}{\gtrless}} \lambda \tag{4.7}$$

where

$$\lambda = \frac{P_0 (C_{10} - C_{00})}{P_1 (C_{01} - C_{11})}$$

Equation (4.7) is interpreted as follows: If $\Lambda(\mathbf{y}) > \lambda$, decide on H_1, conversely if $\Lambda(\mathbf{y}) < \lambda$, decide on H_0. In the unlikely event that there is an equality (i.e., the left-hand side of (4.7) equals the right-hand side) then one arbitrarily can assign any hypothesis or let nature decide by flipping an unbiased coin and assign the decision according to the coin toss. Note both sides of the inequality are non-negative; hence, if one takes the logarithm of both sides (i.e., monotonic function), the inequality is preserved. The natural logarithm of the likelihood ratio (LR) is denoted by LLR.

$$\ln \Lambda(\mathbf{y}) \underset{H_0}{\overset{H_1}{\gtrless}} \ln \lambda \tag{4.8}$$

Example 4.2 *Under H_1, a constant positive voltage, say "m" volts, is transmitted over a wire. Under H_0, nothing (i.e., 0 volt) is transmitted. N samples are used in making a decision. Additive Gaussian noise, $n_n \sim N(0, \sigma^2)$, i.i.d., is introduced by the channel (see Figure 4.5 for a typical likelihood function plot when $N = 1$)*

$$H_0 : y_n = n_n \ ; \qquad \text{for } n = 1, \cdots, N$$

$$H_1 : y_n = m + n_n \ ; \qquad \text{for } n = 1, \cdots, N$$

The joint PDF of the noise is given by

$$f_{\mathbf{N}}(\mathbf{n}) = \frac{1}{(2\pi\sigma^2)^{N/2}} \exp - \sum_{n=1}^{N} \frac{n_n^2}{2\sigma^2}$$

Looking at y_n, we realize that it is the sum of the random variable n_n and the deterministic quantity m, hence y_n has a Gaussian distribution. All that we need to find is the mean and variance of y_n to have a complete statistical description for this random variable. The mean of y_n is given by

$$E\ y_n = E(m + n_n) = m$$

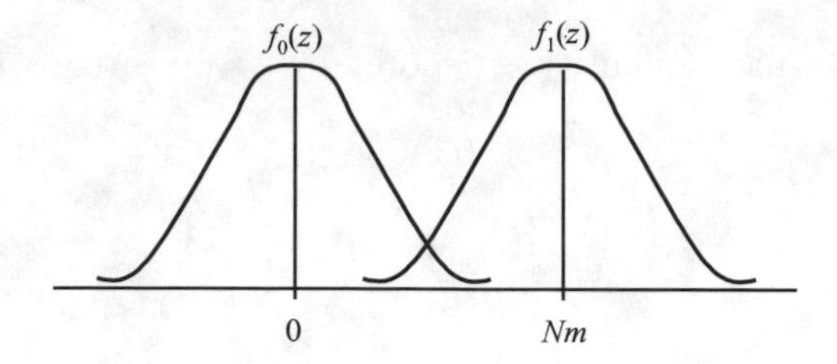

Figure 4.5: Likelihood functions of Example 4.2.

The variance is given by

$$\sigma_{y_n}^2 = E(y_n - Ey_n)^2 = E(m + n_n - m)^2 = E{n_n}^2 = \sigma^2$$

Hence, the individual PDFs for the nth *sample given* H_1 *or* H_0 *are true, are given by*

$$\begin{aligned} f_1(y_n) &= \frac{1}{(2\pi\sigma^2)^{1/2}} \exp - \frac{(y_n - m)^2}{2\sigma^2} \\ f_0(y_n) &= \frac{1}{(2\pi\sigma^2)^{1/2}} \exp - \frac{y_n^2}{2\sigma^2} \end{aligned}$$

Accounting for all the observations, this becomes

$$\begin{aligned} f_1(\mathbf{y}) &= \prod_{n=1}^{N} \frac{1}{(2\pi\sigma^2)^{1/2}} \exp - \frac{(y_n - m)^2}{2\sigma^2} \\ f_0(\mathbf{y}) &= \prod_{n=1}^{N} \frac{1}{(2\pi\sigma^2)^{1/2}} \exp - \frac{(y_n)^2}{2\sigma^2} \end{aligned}$$

The likelihood ratio test (LRT), becomes

$$\begin{aligned} \Lambda(\mathbf{y}) &= \frac{\prod_{n=1}^{N} \frac{1}{\sqrt{2\pi\sigma^2}} \exp - \frac{1}{2\sigma^2}(y_n - m)^2}{\prod_{n=1}^{N} \frac{1}{\sqrt{2\pi\sigma^2}} \exp - \frac{1}{2\sigma^2} y_n^2} \\ &= \frac{\exp \frac{-1}{2\sigma^2} \sum_{n=1}^{N} (y_n - m)^2}{\exp - \frac{1}{2\sigma^2} \sum_{n=1}^{N} y_n^2} \end{aligned}$$

$$= \exp -\frac{1}{2\sigma^2}\sum_{n=1}^{N}(-2my_n + m^2) \underset{H_0}{\overset{H_1}{\gtrless}} \lambda$$

Taking the natural logarithm leads to

$$-\frac{1}{2\sigma^2}\left(\sum_{n=1}^{N}[-2my_n] + Nm^2\right) \underset{H_0}{\overset{H_1}{\gtrless}} \ln\lambda$$

or

$$z = \sum_{n=1}^{N} y_n \underset{H_0}{\overset{H_1}{\gtrless}} \left(\frac{\sigma^2}{m}\ln\lambda + \frac{N}{2}m\right)$$

The optimal receiver performs a linear operation on the data (i.e., an FIR filter operation using unit weights). Figure 4.5 shows the likelihood functions for the variable z. If we use the familiar sample mean, we obtain

$$w = \frac{1}{N}\sum_{n=1}^{N} y_n \underset{H_0}{\overset{H_1}{\gtrless}} \left(\frac{\sigma^2}{Nm}\ln\lambda + \frac{m}{2}\right)$$

Note, both z and w are a one-dimensional test statistic (i.e., the sum of the samples or the sample mean versus a simple threshold). These new random variables are also called sufficient statistics. If the voltage m is negative (i.e., $m < 0$), then the inequalities in the last equation must be reversed.

Example 4.3 *Given N samples of $\mathbf{y}$ (i.e., $y_1, y_2, \cdots, y_N$), which are Gaussian, i.i.d., random variables with zero mean where the variance conveys the message. We want to detect which one of two messages is transmitted.*

$$\text{Under } H_0 : y_n \text{ has variance } \sigma_0^2 \; ; \qquad \text{for } n = 1, 2, \cdots, N$$

$$\text{Under } H_1 : y_n \text{ has variance } \sigma_1^2 \; ; \qquad \text{for } n = 1, 2, \cdots, N$$

where $\sigma_1^2 > \sigma_0^2$. The PDFs are given by

$$f_1(\mathbf{y}) = \frac{1}{\left(2\pi\sigma_1^2\right)^{N/2}} \exp -\frac{1}{2\sigma_1^2}\sum_{n=1}^{N} y_n^2$$

$$f_0(\mathbf{y}) = \frac{1}{\left(2\pi\sigma_0^2\right)^{N/2}} \exp -\frac{1}{2\sigma_0^2}\sum_{n=1}^{N} y_n^2$$

The LRT becomes

$$\Lambda(\mathbf{y}) = \left(\frac{\sigma_0}{\sigma_1}\right)^N \exp -\frac{1}{2}\sum_{n=1}^{N}\left(\left(\frac{y_n}{\sigma_1}\right)^2 - \left(\frac{y_n}{\sigma_0}\right)^2\right) \underset{H_0}{\overset{H_1}{\gtrless}} \lambda$$

where λ *is given by*

$$\lambda = P_0 \frac{(C_{10} - C_{00})}{(C_{01} - C_{11})}$$

Taking the natural logarithm leads to

$$\ln \Lambda(\mathbf{y}) = N \ln\left(\frac{\sigma_0}{\sigma_1}\right) + \frac{1}{2}\sum_{n=1}^{N} y_n^2 \left(\frac{1}{\sigma_0^2} - \frac{1}{\sigma_1^2}\right) \underset{H_0}{\overset{H_1}{\gtrless}} \ln \lambda \tag{4.9}$$

Note, the constant multiplier term (inside of the summation) is strictly positive for the given parameters; hence, one can bring it over to the other side of the inequality without any special consideration. Hence, the last expression can be rewritten as

$$\sum_{n=1}^{N} y_n^2 \underset{H_0}{\overset{H_1}{\gtrless}} \frac{2}{\left(\frac{1}{\sigma_0^2} - \frac{1}{\sigma_1^2}\right)}\left(\ln \lambda - N \ln\left(\frac{\sigma_0}{\sigma_1}\right)\right)$$

This can be simplified to

$$z = \sum_{n=1}^{N} y_n^2 \underset{H_0}{\overset{H_1}{\gtrless}} \frac{2\sigma_1^2\sigma_0^2}{(\sigma_1^2 - \sigma_0^2)}\left(\ln \lambda - N \ln\left(\frac{\sigma_0}{\sigma_1}\right)\right) = \gamma$$

The optimal detector consists of a non-linear operation on the data, (i.e., a square law type detection). Note there is only a single random variable (i.e., sufficient statistic) "z" rather than the N random variables of $\mathbf{y}$. *So the probabilities of interest are easily obtained using*

$$\begin{aligned} P_D &= \int_{\gamma}^{\infty} f_1(z)dz \\ P_M &= \int_{-\infty}^{\gamma} f_1(z)dz \\ P_{FA} &= \int_{\gamma}^{\infty} f_0(z)dz \end{aligned}$$

Note, that if the order of the magnitude of the variances is reversed (i.e., $\sigma_1^2 < \sigma_0^2$), then the inequalities in the Bayes' test are also reversed since the constant multiplier in (4.9) is negative. Hence, the Bayes' test becomes

$$z = \sum_{n=1}^{N} y_n^2 \underset{H_1}{\overset{H_0}{\gtrless}} \frac{2\sigma_1^2\sigma_0^2}{(\sigma_1^2 - \sigma_0^2)} \left(\ln \lambda - N \ln \left(\frac{\sigma_0}{\sigma_1} \right) \right) = \gamma$$

Example 4.4 *This is a numerical example using the results obtained in the last example. Only a single sample is received and based on this single sample, a decision is made. The prior probability for each symbol is identical (i.e., $P_0 = P_1 = 0.5$), the cost of a mistake of either type is equal, and the cost for proper decision is zero. The variances are given by $\sigma_1^2 = 4$ and $\sigma_0^2 = 1$. Hence, λ, the threshold, is unity.*

Under $H_0 : y \sim N(0, 1)$

$$f_0(y) = \frac{1}{(2\pi)^{1/2}} \exp -\frac{1}{2} y^2$$

Under $H_1 : y \sim N(0, 4)$

$$f_1(y) = \frac{1}{(2\pi 4)^{1/2}} \exp -\frac{1}{2 \cdot 4} y^2$$

The likelihood ratio test (and corresponding threshold) under the Bayes' detection philosophy becomes

$$\Lambda(y) = \frac{(2\pi)^{1/2} \exp -\frac{y^2}{8}}{(2\pi 4)^{1/2} \exp -\frac{y^2}{2}} \underset{H_0}{\overset{H_1}{\gtrless}} \lambda$$

This becomes

$$\frac{1}{2} \exp \frac{3y^2}{8} \underset{H_0}{\overset{H_1}{\gtrless}} \lambda = 1$$

Taking the logarithm on both sides and further simplifying, leads to a power or square law detector

$$y^2 \underset{H_0}{\overset{H_1}{\gtrless}} \frac{8}{3} \ln 2 = \frac{8}{3} \cdot 0.69 = 1.85$$

Equivalently, one can use an envelope detector of the form

$$|y| \underset{H_0}{\overset{H_1}{\gtrless}} \sqrt{1.85} = 1.36$$

Example 4.5 *A signal (5 volts) may or may not be transmitted and is observed in additive Gaussian noise. The noise has zero mean and unit variance.* $C_{00} = C_{11} = 0$, $C_{01} = C_{10} = k$, $P_0 = P_1 = 0.5$ *and* $k > 0$. *Hence, the threshold* $\lambda = 1$. N *equals 1, that is, one sample is available. Under*

$$\begin{aligned} H_0 &: \; y(t) = n(t) \\ H_1 &: \; y(t) = 5 + n(t) \end{aligned}$$

We sample the data at one point in time and work with one observation. The density function under each hypothesis becomes

$$\begin{aligned} f_0(y) &= \frac{1}{(2\pi)^{1/2}} \exp -\frac{1}{2}y^2 \\ f_1(y) &= \frac{1}{(2\pi)^{1/2}} \exp -\frac{1}{2}(y-5)^2 \end{aligned}$$

The likelihood ratio test becomes

$$\Lambda(y) = \exp -\frac{1}{2}(25 - 10y) \underset{H_0}{\overset{H_1}{\gtrless}} 1$$

Further simplifying it by taking the logarithm leads to

$$z = y \underset{H_0}{\overset{H_1}{\gtrless}} 2.5$$

Note, this result is somewhat expected. When the threshold λ *equals to unity, then we are just equating the two likelihood functions and test which one is the dominant one. Obviously, the point of dichotomy is exactly halfway between the means of the two possible densities, making the threshold selection equivalent to the crossover point of the two likelihood functions. The power of the test (i.e.,* P_D*) and the false alarm rate* P_{FA} *are easily computed. Also the probability of a miss (*P_M*), that is, the probability of saying* H_0 *is true while in reality the hypothesis* H_1 *is correct is easily obtained. The numerical* *values can be taken out of a table in Appendix D or be computed via one of the many numerical procedures available.*

$$\begin{aligned} P_D &= \int_{2.5}^{\infty} f_1(z)dz = 0.9938 \\ P_{FA} &= \int_{2.5}^{\infty} f_0(z)dz = 0.0062 \\ P_M &= \int_{-\infty}^{2.5} f_1(z)dz = 0.0062 \\ &= 1 - P_D = 1 - 0.9938 \end{aligned}$$

One procedure that is easily implemented on a calculator or a personal computer (see Appendix D) is advocated in Helstrom [4] and Abramowitz and Stegun [9]. The error is less than 7.5 10^{-8}*, limiting the usefulness somewhat. That is, for problems where the area under the tail section of a Gaussian is on the order of this error or smaller, more sophisticated procedures must be used.*

Example 4.6 *Given the same scenario as in the previous problem, except that now there is access to nine independent samples (i.e.,* $N = 9$*). Under*

$$\begin{aligned} H_0 &: \; y_n \sim N(0,1) \; ; \quad n = 1, 2, \cdots, 9 \\ H_1 &: \; y_n \sim N(5,1) \; ; \quad n = 1, 2, \cdots, 9 \end{aligned}$$

From Example 4.2 we have

$$z = \sum_{n=1}^{N} y_n \underset{H_0}{\overset{H_1}{\gtrless}} \frac{N}{2} m = \frac{9}{2} \cdot 5 = 22.5$$

Since y_n *is Gaussian under each hypothesis, it is obvious that* z *will be Gaussian under each hypothesis. To characterize* $f_i(z)$*, for* $i = 0, 1$*, all that is needed is the mean of* z *under each hypothesis and the variance of* z*. The means are*

$$\begin{aligned} E_0 \; z &= \sum_{n=1}^{9} E_0 \; y_n = 9 \cdot 0 = 0 \\ E_1 \; z &= \sum_{n=1}^{9} E_1 \; y_n = 9 \cdot 5 = 45 \end{aligned}$$

where the subscript of E *indicates which signal is assumed to be present (i.e., which hypothesis is true).*

We realize that the DC component has no effect on the variance computation; hence, we compute the variance for the zero hypothesis (i.e., no DC component). Therefore, the second moment equals the variance

$$\sigma_z^2 = E_0 \; z^2 \;\; = \;\; E_0 \sum_{n=1}^{9} \sum_{m=1}^{9} (y_n \; y_m)$$

$$= \;\; \sum_{n=1}^{9} \sigma_y^2 = 9 \cdot 1 = 9$$

So under

$$H_0 \;\; : \;\; z \sim N(0, 9)$$
$$H_1 \;\; : \;\; z \sim N(45, 9)$$

Hence, the test becomes

$$z \underset{H_0}{\overset{H_1}{\gtrless}} 22.5$$

and the detection performance becomes

$$P_D = \int_{22.5}^{\infty} f_1(z) \; dz = Q\left(\frac{22.5 - 45}{3}\right) = Q\left(-\frac{22.5}{3}\right) \approx 1.0$$

$$P_{FA} = \int_{22.5}^{\infty} f_0(z) \; dz = Q\left(\frac{22.5}{3}\right) = Q(7.5) = 3.19 \cdot 10^{-14}$$

Note that we could have used a normalized statistic that is the sample mean. Then the test and threshold would be as follows:

$$\text{Sample mean} = w = \frac{1}{N} \; z = \frac{1}{9} \sum_{n=1}^{9} y_n \underset{H_0}{\overset{H_1}{\gtrless}} \frac{22.5}{9}$$

so

$$w \underset{H_0}{\overset{H_1}{\gtrless}} 2.5$$

The sample mean will be 0 and 5 under the zero and one hypothesis, respectively. The variance under each hypothesis will be 1/9. So we have

$$H_0 \;\; : \;\; w \sim N(0, 1/9)$$
$$H_1 \;\; : \;\; w \sim N(5, 1/9)$$

Of course, the detection performance of the statistics of w relative to the z detection statistic will not change at all as the interested reader might quickly verify.

Example 4.7 *An experiment is performed and the number of events are counted. The number of events obey a Poisson law. Under*

$$
\begin{aligned}
H_0 &: \Pr\{n \text{ events}\} = (m_0^n)/n! \exp -(m_0) \\
H_1 &: \Pr\{n \text{ events}\} = (m_1^n)/n! \exp -(m_1) \ ; \quad \text{for } n = 0, 1, 2, 3, \cdots
\end{aligned}
$$

The Bayes' detector is given by the likelihood ratio test (LRT) which is compared to the threshold λ, which depends on prior probabilities and the cost associated with each decision. For this example we obtain

$$
\Lambda(n) = \frac{\Pr(n \text{ events}|H_1)}{\Pr(n \text{ events}|H_0)} = \left(\frac{m_1}{m_0}\right)^n e^{-m_1+m_0} \ \mathop{\gtrless}_{H_0}^{H_0} \ \lambda
$$

The log likelihood ratio test (LLRT) becomes

$$
n \ \ln\left(\frac{m_1}{m_0}\right) + m_0 - m_1 \ \mathop{\gtrless}_{H_0}^{H_1} \ \ln \lambda
$$

Depending on which parameter is the larger one, the multiplier associated with n is positive or negative. The detection scheme becomes

if $m_1 > m_0$, then

$$
n \ \mathop{\gtrless}_{H_0}^{H_1} \ \frac{\ln \lambda + m_1 - m_0}{\ln\left(\frac{m_1}{m_0}\right)}
$$

if $m_0 > m_1$, then

$$
n \ \mathop{\gtrless}_{H_1}^{H_0} \ \frac{\ln \lambda + m_1 - m_0}{\ln\left(\frac{m_1}{m_0}\right)}
$$

Example 4.8 *An event may or may not have occurred. The costs and prior probabilities are given by $C_{00} = C_{11} = 0$, $C_{01} = C_{10} = 1$, and $P_0 = P_1 = 0.5$, respectively. Hence, the threshold $\lambda = 1$. N equals 1, that is, one sample is available. The density functions, are two-sided exponential densities. The* *conditional PDFs are given by*

$$\begin{aligned} f_0(y) &= \frac{1}{2}\ \exp - \ |y| \\ f_1(y) &= \frac{1}{4}\ \exp - \frac{|y|}{2} \end{aligned}$$

The likelihood ratio test becomes

$$\Lambda(y) \;=\; \frac{1}{2}\exp(|y| - |y/2|) \begin{array}{c} H_1 \\ > \\ < \\ H_0 \end{array} T \;=\; 1$$

Further simplification (i.e., take natural logarithms) leads to

$$z \;=\; |y| \begin{array}{c} H_1 \\ > \\ < \\ H_0 \end{array} 2\ ln\ 2 \;=\; 1.3863$$

Equivalently, this can also be written as:

If $-1.3863 < y \ < 1.3863$ *then decide on* H_0*, otherwise decide on* H_1*. The numerical values for the performance can be obtained from the simple exponential integrals given below*

$$\begin{aligned} P_D &= \frac{1}{2}\int_{1.3863}^{\infty} \exp -(\frac{y}{2})\ dy \\ P_{FA} &= \int_{1.3863}^{\infty} \exp -\ (y)\ dy \\ P_M &= 1 - P_D \end{aligned}$$

Example 4.9 *An event may or may not have occurred. The costs and prior probabilities are given by* $C_{00} = C_{11} = 0$, $C_{01} = C_{10} = 1$*, and* $P_0 = P_1 = 0.5$*, respectively. Hence, the threshold* $\lambda = 1$*.* N *independent samples are available. The density functions, are one-sided exponential densities. The conditional PDFs, given that* H_1 *or* H_0 *are true, are given by*

$$\begin{aligned} f_0(y_n) &= \exp(-\ y_n)\ \ U(y_n) \\ f_1(y_n) &= \frac{1}{2}\ \exp(-\ \frac{y_n}{2})\ \ U(y_n) \end{aligned}$$

Accounting for all the observations, this becomes

$$f_0(\mathbf{y}) = \prod_{n=1}^{N} exp(-\ y_n)\ \ U(y_n)$$

$$f_1(\mathbf{y}) = \prod_{n=1}^{N} \frac{1}{2}\ \exp(-\ \frac{y_n}{2})\ \ U(y_n)$$

The likelihood ratio test (LRT), for positive y_n, becomes

$$\Lambda(\mathbf{y}) = \frac{\prod_{n=1}^{N} \frac{1}{2} \exp -(\frac{1}{2} y_n)}{\prod_{n=1}^{N} \exp -(y_n)}$$

$$= \exp \frac{1}{2} \sum_{n=1}^{N} y_n \underset{H_0}{\overset{H_1}{\gtrless}} \lambda = 2^N$$

Taking the logarithm leads to

$$z = \sum_{n=1}^{N} y_n \underset{H_0}{\overset{H_1}{\gtrless}} 2N\ ln\ 2$$

The optimal receiver performs a linear operation on the data (i.e., a FIR filter operation using unit weights).

Example 4.10 *This is a numerical example of a problem of the type shown in Example 4.7. A stock market analyst noticed that he could use the number of touchdowns scored by his favorite team as an indicator as to what the stock market will be doing the following week. In this model, the touchdowns follow a Poisson law and the threshold λ is set to one. We will notice later that this threshold selection corresponds to the maximum likelihood (ML) detection scheme. Under*

H_0 : bull market (i.e., expanding economy)

H_1 : bear market (i.e., stagnating or collapsing economy)

The probability law under each hypothesis is Poisson with parameter m_0 and m_1, respectively. From experimental data, the analyst had established the value of m_0 as 1 and of m_1 as 3. So he proposed to plot the number of

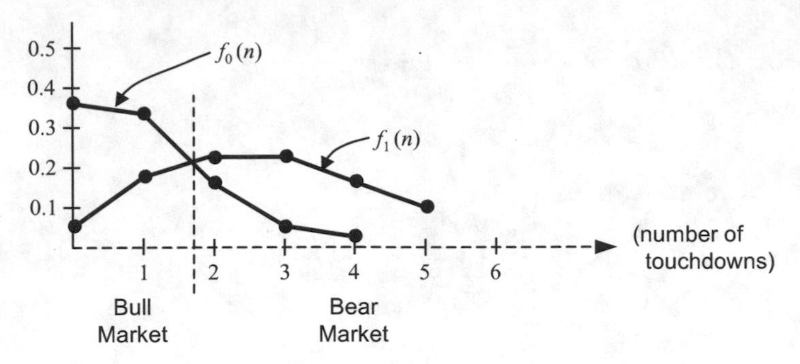

Figure 4.6: Likelihood functions of Example 4.10.

touchdowns on the horizontal axis (see the likelihood function plot in Figure 4.6). Then establish which of the two probability density functions is the dominant one, then detect the type of market valid for the next trading week. Of course this scheme seems to disregard cause and effect, but it seemed to work for this market analyst.

So far we have used the Bayes' test (detector, philosophy, approach, etc.), but many times not all the required information for the Bayes' test is available.

4.3 MAXIMUM *A POSTERIORI* (MAP) DETECTION

Suppose that we have information about the prior probabilities but either cannot, or do not want to, assign a cost to making the possible decisions. This approach is the maximum *a posteriori* (MAP) detection philosophy. To keep the notation simple, we assume in the derivation that the number of data samples is one. If one wants to derive the equivalent expressions for N data samples (i.e., $N > 1$), all that needs to be done is to replace the scalar quantities in 4.10–4.15 with the corresponding vector quantities. We start with a test that chooses the dominant (i.e., maximum) *a posteriori* probability, as given by

$$\Pr(H_1|y) \underset{H_0}{\overset{H_1}{\gtrless}} \Pr(H_0|y) \tag{4.10}$$

The left side of (4.10) is the *a posteriori* probability of hypothesis H_1 given the observation y, while the right side is the *a posteriori* probability of hypothesis H_0 given the observation y. In other words, given the data (y), which hypothesis has the larger probability (i.e., which one is more likely true, H_0

or H_1)? We usually want to work with probability density functions rather than with probabilities (i.e., distributions). Equation (4.10) can be written for an increment of y as follows:

$$\Pr(H_1 | \mathsf{y} < y \leq \mathsf{y} + d\mathsf{y}) \underset{H_0}{\overset{H_1}{\gtrless}} \Pr(H_0 | \mathsf{y} < y \leq \mathsf{y} + d\mathsf{y}) \tag{4.11}$$

The Bayes' theorem for conditional probabilities says

$$\Pr(H_i | \mathsf{y} < y \leq \mathsf{y} + d\mathsf{y}) = \frac{\Pr(\mathsf{y} < y \leq \mathsf{y} + d\mathsf{y} | H_i)}{\Pr(\mathsf{y} < y \leq \mathsf{y} + d\mathsf{y})} \Pr(H_i) \tag{4.12}$$

for $i = 0, 1$. The probability of an incremental area equals the PDF times $d\mathsf{y}$, given by

$$\begin{aligned} \Pr(\mathsf{y} < y \leq \mathsf{y} + d\mathsf{y} | H_i) &= f_i(y) d\mathsf{y} \\ \Pr(\mathsf{y} < y \leq \mathsf{y} + d\mathsf{y}) &= f(y) d\mathsf{y} \end{aligned} \tag{4.13}$$

As the vector $d\mathsf{y}$ becomes very small (i.e., a differential), the conditional probability (4.12) can be written as

$$\Pr(H_i | y) = \frac{f_i(y)}{f(y)} Pr(H_i) \tag{4.14}$$

for $i = 0, 1$. The detection inequality (4.11) becomes when using $\Pr(H_i) = P_i$

$$\frac{f_1(y)}{f_0(y)} \underset{H_0}{\overset{H_1}{\gtrless}} \frac{P_0}{P_1} \quad \text{(MAP detector)} \tag{4.15}$$

Note, no cost function is used, or equivalently $C_{10} - C_{00} = C_{01} - C_{11}$. We can also interpret the MAP detector as a Bayes' detector with a different threshold. The threshold for the MAP detector is just a function of the prior probabilities (i.e., P_0 and P_1). At this point, if it is desirable to work with a vector of data samples, all that needs to be done is to replace the scalar y with the vector $\mathbf{y}$ in (4.15).

4.4 MAXIMUM LIKELIHOOD (ML) CRITERION

Suppose no prior probability nor any cost information is available, but we still want to set up a meaningful detection scheme, we will see that comparing the likelihood function will us allow to do this. Our criterion is "Given the observation (data), which conditional PDF (i.e., $f_1(\mathbf{y})$ or $f_0(\mathbf{y})$) more likely generated the observation." We again prefer the scalar case in deriving the detector. Once the detector expression is obtained, it can easily be extended to the vector case. The detection scheme becomes

$$f_1(y) \underset{H_0}{\overset{H_1}{\gtrless}} f_0(y)$$

and for multiple observations

$$\frac{f_1(\mathbf{y})}{f_0(\mathbf{y})} \underset{H_0}{\overset{H_1}{\gtrless}} 1 \qquad \text{(ML detector)} \tag{4.16}$$

Note, one can interpret this as a Bayes' detector where the prior probabilities are equal and the cost functions are of the form $C_{10} - C_{00} = C_{01} - C_{11}$. The ML approach is frequently used since it does not require knowledge of either prior probabilities or of cost functions. This information is many times not available or sometimes it is not desirable to use it. At any rate, we see at this point in our venture that the difference between the three detection schemes (i.e., Bayes, MAP, and ML) is in the selection of the threshold but that the detector structure is identical. We will also see that this observation is general in that it also applies to the remaining detectors to be discussed in Sections 4.5–4.7.

4.5 MINIMUM PROBABILITY OF ERROR CRITERION

This detection criterion is used in binary communication problems, where the cost of making an error, that is calling a true zero a one or calling a true one a zero is the same and the cost of making a correct decision is zero. We do allow non-equal prior probabilities, which is a function of how the coding (i.e., source coding) is performed. This problem is in the same form as Example 4.1, which illustrated the conditional probabilities and the channel transition probabilities. We set up the average cost in the same fashion as was done in the Bayes' detector derivation. The average cost (which needs to be minimized) using (4.2) and the cost assignment $C_{00} = C_{11} = 0, C_{10} = C_{01} = 1$

$$C = P_0 \int_{R_1} f_0(\mathbf{y})\, d\mathbf{y} + P_1 \int_{R_0} f_1(\mathbf{y})\, d\mathbf{y}$$

which is the total probability of making an error. The Bayes' approach then leads directly to

$$\Lambda(\mathbf{y}) = \frac{f_1(\mathbf{y})}{f_0(\mathbf{y})} \underset{H_0}{\overset{H_1}{\gtrless}} \frac{P_0}{P_1} \tag{4.17}$$

Note, this is the same expression (including the threshold) as that of the MAP detector in vector form. In some literature, this particular test criterion is also called the "ideal observer criterion."

Example 4.11 *N independent observations of a Gaussian random process (with the signal embedded) are obtained (i.e., y_n for $n = 1, 2, \cdots, N$). Use the minimum probability of error criterion to obtain the detector and the associated probabilities. Under*

$$\begin{aligned} H_0 &: \; y_n \sim N(m_0, 1) \; ; \quad \text{for } n = 1, \cdots, N \; ; \quad \text{i.i.d.} \\ H_1 &: \; y_n \sim N(m_1, 1) \; ; \quad \text{for } n = 1, \cdots, N \; ; \quad \text{i.i.d.} \end{aligned}$$

with P_1 and P_0 given and $m_1 > m_0$. The LRT becomes

$$\Lambda(\mathbf{y}) = \frac{\prod_{n=1}^{N} \exp -\frac{1}{2}(y_n - m_1)^2}{\prod_{n=1}^{N} \exp -\frac{1}{2}(y_n - m_0)^2}$$

$$= \exp \frac{1}{2} \sum_{n=1}^{N} \left[(-2m_0 + 2m_1) y_n + m_0^2 - m_1^2 \right] \underset{H_0}{\overset{H_1}{\gtrless}} \frac{P_0}{P_1} = \lambda$$

This can be further reduced. As always, we try to get as close as possible to the raw data variables. This makes the receiver (detector) relatively simple and many times allows the computation of the detection quantities in terms of tractable probability density functions. Hence, we obtain

$$\sum_{n=1}^{N} y_n \underset{H_0}{\overset{H_1}{\gtrless}} \frac{\ln \lambda + \left(m_1^2 - m_0^2\right) N}{2(m_1 - m_0)} = \lambda_1$$

Or in terms of the sample mean (if it is more convenient)

$$z = \frac{1}{N} \sum_{n=1}^{N} y_n \underset{H_0}{\overset{H_1}{\gtrless}} \frac{\ln \lambda}{2N(m_1 - m_0)} + \frac{1}{2}(m_1 + m_0) = \eta$$

where the mean of the random variable z is given by

$$E_i\ z = \frac{1}{N}\sum_{n=1}^{N} E\ \mathbf{y}_n = m_i$$

and m_i is the transmitted symbol value $(i = 0, 1)$. Note, if we want to compute the performance of this detector using the original $\mathbf{y}$ variables, say we want to compute Pr(say $H_1|H_0$ is true) *or* Pr(say $H_1|H_1$ is true) *then we have to deal with N-dimensional integrals over an N-dimensional decision region. The problem becomes much simpler if we use the new sufficient statistic z. We recall that the sum of Gaussian random variables is Gaussian (the Gaussian family is closed under linear operations), so all we need to compute are the mean and variance of the new Gaussian random variable z.*

We see now that the detection statistic, the random variable z, is a simple one-dimensional Gaussian random variable.

$$\sigma_z^2 = \ = \frac{1}{N^2}\sum_{n=1}^{N}\sum_{m=1}^{N} E\left((y_n - m_{y_n})(y_m - m_{y_m})\right)$$

$$= \frac{1}{N^2}\sum_{n=1}^{N}\sigma_{\mathbf{y}}^2 = \frac{1}{N}$$

where m_{y_n} is the mean of the nth sample y_n. Any desired performance statistic, in contrast to the N-dimensional random variable y_n, is easily computed. For example, using $E_0\ z = m_0$ and $E_1\ z = m_1$, then $\Pr(H_1|H_0)$ *and* $\Pr(H_1|H_1)$ *are easily computed by*

$$\Pr(\text{say } H_1|H_0 \text{ is true}) =$$

$$\int_{\eta}^{\infty} \frac{1}{\left(2\pi\frac{1}{N}\right)^{1/2}} \exp[-\frac{(z-m_0)^2 N}{2}]\ dz = Q\left((\eta - m_0)\sqrt{N}\right)$$

$$\Pr(\text{say } H_1|H_1 \text{ is true}) =$$

$$\int_{\eta}^{\infty} \frac{1}{\left(2\pi\frac{1}{N}\right)^{1/2}} \exp[-\frac{(z-m_1)^2 N}{2}]\ dz = Q\left((\eta - m_1)\sqrt{N}\right)$$

$$\Pr(\text{say } H_0|H_1 \text{ is true}) =$$

$$= 1 - \Pr(\text{say } H_1|H_1 \text{ is true}) =$$

$$= \int_{-\infty}^{\eta} \frac{1}{\left(2\pi\frac{1}{N}\right)^{1/2}} \exp[-\frac{(z-m_1)^2 N}{2}]\, dz$$

$$= erf\left((\eta - m_1)\sqrt{N}\right)$$

$$= 1 - Q((n - m_1)\sqrt{N})$$

Example 4.12 *Given that N samples of* $\mathbf{y}$*, i.e.,* $y_1, y_2, \cdots, y_N$ *are Gaussian, i.i.d., random variables with zero mean and a variance which conveys the message, we want to detect which one of two messages was transmitted. Design the detector using the Bayes' error criterion. Under*

$$\begin{aligned} H_0 &: y_n \text{ has variance } \sigma_0^2\,; \quad \text{for } n = 1, 2, \cdots, N \\ H_1 &: y_n \text{ has variance } \sigma_1^2\,; \quad \text{for } n = 1, 2, \cdots, N \end{aligned}$$

where $\sigma_1^2 > \sigma_0^2$*,* $P_1 = P_0$*,* $C_{00} = C_{11} = 0$*,* $C_{01} = C_{10} = 1$*, and* $\lambda = 1$*; hence,* $ln\ \lambda = 0$*. Using (4.6) or equivalently Bayes' test from Example 4.3, we have*

$$z = \sum_{n=1}^{N} y_n^2 \underset{H_0}{\overset{H_1}{\gtrless}} \frac{2\sigma_1^2\sigma_0^2}{(\sigma_1^2 - \sigma_0^2)} N \ \ln\left(\frac{\sigma_1}{\sigma_0}\right) = z_0$$

Since the observations y_n *are i.i.d. Gaussian random variables, the random variable* z *is* χ^2 *(chi-squared) with N-degrees of freedom. The probability density function for* z *is (see Appendix A and [1])*

$$f_i(z) = \frac{z^{(N/2)-1}}{2^{N/2}\sigma_j^2\Gamma(N/2)} e^{-z/(2\sigma_i^2)}; \quad \text{for } i = 0, 1$$

The detection statistics are given by

$$\begin{aligned} P_{FA} &= \int_{z_0}^{\infty} f_0(z) dz \\ P_M &= \int_{-\infty}^{z_0} f_1(z) dz \\ P_D &= 1 - P_M \end{aligned}$$

To evaluate these expressions, we need to have tables of Pearson's incomplete gamma function [2] or use a numerical approximation technique that requires evaluation on a computer.

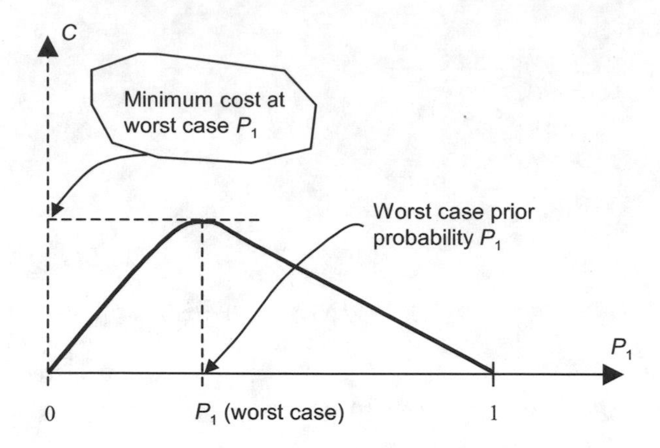

Figure 4.7: Average cost versus P_1.

4.6 MIN-MAX CRITERION

If the cost information (i.e., C_{ij} $i, j = 0, 1$) is available but not the "*a priori* probability" P_0 or P_1, then the best Bayes' related detection scheme (under the worst possible choice of H_i, $(i = 0, 1)$) is desirable. This approach assumes that nature plays the role of an adversary and moves or picks the worst possible value for P_0 and P_1 in order to make the average cost the largest one possible. We know this phenomena under a different name. Many times we call it Murphy's law, which says Mother Nature knows what hurts the most and uses it ruthlessly. We can plot the average cost versus the *a priori* probability, say P_1 as shown in Figure 4.7.

We assume without loss of generality that the costs C_{00}, C_{11} are smaller than the cost of making an error and the cost of any decision is some non-negative number. If we take Bayes' cost (rearranging (4.2)), we can replace the integrals using the symbols of P_{FA} and P_M.

$$
\begin{aligned}
C &= P_0\, C_{00} \int_{R_0} f_0(\mathbf{y})\, d\mathbf{y} + P_0\, C_{10} \int_{R_1} f_0(\mathbf{y})\, d\mathbf{y} \\
&\quad + P_1\, C_{01} \int_{R_0} f_1\, (\mathbf{y})\, d\mathbf{y} + P_1\, C_{11} \int_{R_1} f_1(\mathbf{y})\, d\mathbf{y}
\end{aligned}
\tag{4.18}
$$

The average cost (Bayes' approach) becomes

$$
\begin{aligned}
C &= P_0 C_{00} + P_1 C_{11} + P_1 (C_{01} - C_{11}) P_M \\
&\quad + P_0 (C_{10} - C_{00})\, P_{FA}
\end{aligned}
\tag{4.19}
$$

We note that at the end points we have the following conditions:

- When $P_0 = 1$ (i.e., certainty of event "zero"), then we must have $P_1 = 0$ and $P_{FA} = 0$. The average cost becomes $C = C_{00} + (C_{10} - C_{00})\ P_{FA} = C_{00}$.
- When $P_1 = 1$ (i.e., certainty of event "one"), then we must have $P_0 = 0$, $P_{FA} = 0$, and $P_M = 0$. The average cost becomes $C = C_{11} + (C_{01} - C_{11})\ P_M = C_{11}$.

If we assume that $C_{00} = C_{11} = 0$, as is the case in most real life situations of interest, then as we plot the average cost C as function of P_1 we obtain a graph of the form given in Figure 4.7. The maximum occurs where the slope of the curve is zero. The exception is when the average cost becomes a straight line, then the maximum occurs at one of the end points. To obtain the maximum of the curve we take a partial derivative of C with respect to P_1 and set it to zero (i.e., we find where the slope is zero). So, given the maximum average cost, we minimize it, hence the name Min-Max criterion. After we express every term in terms of P_1, taking a partial derivative leads to

$$\frac{\partial C}{\partial P_1} = \frac{\partial}{\partial P_1}$$
$$[(1 - P_1)C_{00} + P_1 C_{11} + P_1(C_{01} - C_{11})P_M + (1 - P_1)(C_{10} - C_{00})\ P_{FA}] \tag{4.20}$$

Setting the partial derivative to zero leads to

$$(C_{11} - C_{00}) + (C_{01} - C_{11})P_M - (C_{10} - C_{00})P_{FA} = 0 \tag{4.21}$$

If $C_{00} = C_{11} = 0$ we get

$$C_{01}P_M = C_{10}P_{FA} \tag{4.22}$$

The detector form using the Bayes' cost formula is the Bayes' detector with a threshold λ_0 that requires (4.21) to be true.

Example 4.13 *The cost for a correct decision is zero ($C_{ii} = 0$ for $i = 0, 1$). The number of samples is one ($N = 1$), $C_{01} = C_{10} = K > 0$. Under*

$$\begin{aligned} H_0 &: \ y = 1 + n\ ; \quad \text{where } n \sim N(0,1) \\ H_1 &: \ y = 2 + n \end{aligned}$$

The Min-Max test becomes (4.22) $P_{FA} = P_M$. Or in terms of integrals, we obtain

$$\frac{1}{\sqrt{(2\pi)}} \int_{-\infty}^{\lambda_0} \exp \frac{(y-2)^2}{2} dy = \frac{1}{\sqrt{(2\pi)}} \int_{\lambda_0}^{\infty} \exp \frac{(y-1)^2}{2} dy$$

With error function replacement, this becomes $erf(\lambda_0 - 2) = Q(\lambda_0 - 1) = 0.3085$. Using the error function tables, we see that the solution will be a λ_0

of 1.5. In general, if $C_{00} \neq C_{11}$ *and* $C_{01} \neq C_{10}$ *then we use* $(C_{01} - C_{11})P_M - (C_{10} - C_{00})P_{FA} = C_{00} - C_{11}$*; or in terms of the detection statistic*

$$P_{FA} = \int_{\lambda_0}^{\infty} f_0(y)\, dy$$

$$= \int_{-\infty}^{\lambda_0} f_1(y) dy \, \frac{C_{01} - C_{11}}{C_{10} - C_{00}} + \frac{C_{00} - C_{11}}{C_{00} - C_{10}}$$

4.7 NEYMAN-PEARSON CRITERION

4.7.1 Introduction

This criterion is typically used in radar and sonar applications since it allows us to hold the false alarm rate (P_{FA}) at a constant value. We cannot attach a cost to making mistakes in radar/sonar applications. This is particularly true since loss of life and loss of essential equipment can be involved. An alternative approach is to use the concept of a fixed false alarm rate. Furthermore, this is highly desirable when resource allocations are important and the maximum false alarm rate has to be kept at a tolerable level. This detector was the first in a class called CFAR (constant false alarm rate) detectors. In essence, the false alarm rate is fixed while the detection probability is maximized. This is best shown by using methods that maximize or minimize a cost term while meeting an equality constraint. To this end, we will introduce the method of Lagrange multipliers in Section 4.7.2.

4.7.2 Optimization and Lagrange Multipliers

The optimization of a function having a constraint may be done via two different ways:

(a) Direct approach (can be very cumbersome)

(b) Lagrange multipliers (adjustable multiplying parameters, usually a straightforward systematic method) [10]

These methods are best illustrated by working one of the classical problems in optimization when equality constraints are present.

Example 4.14 *Maximize the volume (V) of the cylindrical grocery can of Figure 4.8, for a given amount of sheet metal (i.e., for a fixed (constraint) area* (A_0) *maximize the volume).*

(a) **Direct approach:** *Maximize V such that* A_0 *is a fixed constant.*

(1) $V(r, \ell) = \pi r^2 \ell$

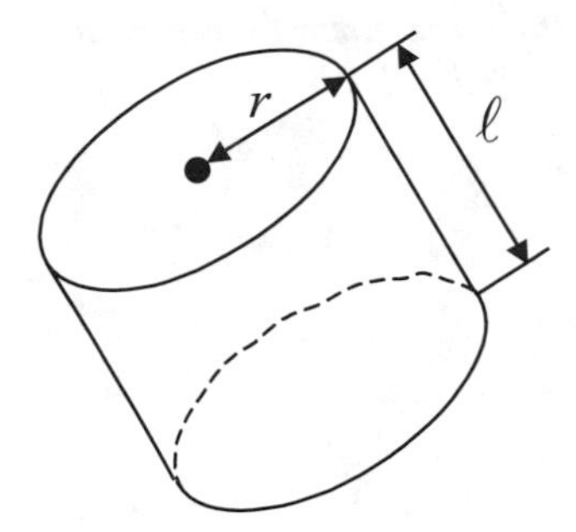

Figure 4.8: Geometry of Example 4.14.

(2) $A(r, \ell) = 2\pi r^2 + 2\pi r\ell = A_0 =$ *fixed constant*

Solution:

From (2) we solve for $\ell = (A_0 - 2\pi r^2)/(2\pi r)$ *and insert it into (1)*

$$V(r) = \pi r^2 \frac{A_0 - 2\pi r^2}{2\pi r} = \frac{A_0 r}{2} - \pi r^3$$

Since V *is now just a function of one parameter, the partial derivative with respect to* r *(i.e., the slope is zero therefore there is a maximum) leads to*

$$\begin{aligned} \frac{\partial V}{\partial r} &= \frac{A_0}{2} - 3\pi r^2 = 0 \\ \Longrightarrow r &= \sqrt{\frac{A_0}{6\pi}} \end{aligned}$$

and

$$\ell = \frac{A_0 - 2\pi r^2}{2\pi r} = \sqrt{\frac{2A_0}{3\pi}}$$

(b) **Via Lagrange multipliers:** *We define the appended cost function as shown by*

$$\begin{aligned} J(r, \ell) &= V(r, \ell) + \lambda\,(A(r, \ell) - A_0) \\ &= \pi r^2 \ell + \lambda\left(2\pi r^2 + 2\pi r\ell - A_0\right) \end{aligned}$$

The first term is the quantity to be optimized, while the second term is the equality constraint, weighted by the Lagrange multiplier λ. *We note that it is purely coincidental that the Lagrange multiplier is labeled with the Greek letter* λ, *which is also the symbol used to denote many* *of the thresholds in the detector implementations. Taking the partial derivatives w.r.t. ℓ, r, and λ leads to*

$$\frac{\partial J}{\partial \ell} = \pi r^2 + \lambda 2\pi r = 0 \implies r = -2\lambda$$

$$\frac{\partial J}{\partial r} = 2\pi r\ell + \lambda(4\pi r + 2\pi \ell) = 0 \implies \ell = -4\lambda$$

$$\frac{\partial J}{\partial \lambda} = 2\pi r^2 + 2\pi r\ell - A_0 \implies$$

$$A_0 = 2\pi\left(\underbrace{-2\lambda}_{r}\right)^2 + 2\pi\left(\underbrace{-2\lambda}_{r}\right)\left(\underbrace{-4\lambda}_{\ell}\right) = 24\pi\lambda^2$$

Hence

$$\lambda = \pm\left[\frac{A_0}{(24\pi)}\right]^{1/2}$$

Solving for r and ℓ leads to

$$r = 2\sqrt{\frac{A_0}{24\pi}} = \sqrt{\frac{A_0}{6\pi}}$$

$$\ell = 4\sqrt{\frac{A_0}{24\pi}} = \sqrt{\frac{2A_0}{3\pi}}$$

So we see this systematic way leads directly to the constraint solution.

4.7.3 Neyman-Pearson Approach

Let us now fix the false alarm rate P_{FA} at a value of α and optimize the probability of detection P_D, or equivalently minimize the probability of a miss P_M, since $P_M = 1 - P_D$. We define the appended cost function as

$$\begin{aligned} J &= \overbrace{\int_{R_0} f_1(\mathbf{y})d\mathbf{y}}^{\text{minimize}} + \lambda\left[\overbrace{\int_{R_1} f_0(\mathbf{y})d\mathbf{y} - \alpha}^{\text{equality constraint}}\right] \\ &= \int_{R_0} f_1(\mathbf{y})d\mathbf{y} + \lambda\left[1 - \int_{R_0} f_0(\mathbf{y})d\mathbf{y} - \alpha\right] \\ &= \lambda(1-\alpha) + \int_{R_0}\left(f_1(\mathbf{y}) - \lambda f_0(\mathbf{y})\right)d\mathbf{y} \end{aligned} \tag{4.23}$$

The first term in (4.23) represents a fixed positive cost, while the second term is an adjustable cost. The expression is set up similar to the Bayes' case

(4.5). When we examine the integral, we notice that we would like to select the region R_0 in such a fashion that the function J is minimized. There is no need to take any derivatives, we just used the Lagrange multiplier to set up the function J. Obviously, the term J is minimized when one uses the following philosophy: If the data $\mathbf{y}$ makes the second term in the integrand larger than the first term, then include the data as belonging to region R_0. Conversely, if the data makes the first term larger than the second one, then include this data as belonging to region R_1. In other words:

$$f_1(\mathbf{y}) \underset{H_0}{\overset{H_1}{\gtrless}} \lambda f_0(\mathbf{y})$$

The form of the detector is

$$\Lambda(\mathbf{y}) = \frac{f_1(\mathbf{y})}{f_0(\mathbf{y})} \underset{H_0}{\overset{H_1}{\gtrless}} \lambda \tag{4.24}$$

The threshold λ is chosen so that the P_{FA} constraint is met. The false alarm rate is given by

$$P_{FA} = \int_{\lambda}^{\infty} f_0(\mathbf{y})\, d\mathbf{y} \tag{4.25}$$

hence, in principle, one can solve for the threshold λ in a straightforward fashion.

Example 4.15 *Under the no signal condition only noise is received, while under the signal present condition 1 volt plus additive noise is received. The noise is Gaussian with zero mean and unit variance. We want to design a CFAR detector (i.e., Neyman-Pearson detector) with a fixed P_{FA} of 0.1. The noise is Gaussian as described by $n \sim N(0,1)$. Under*

$$\begin{aligned} H_0 &: \quad y = n \\ H_1 &: \quad y = 1 + n \end{aligned}$$

$$\Lambda(y) \quad = \quad \frac{f_1(y)}{f_0(y)} \quad = \quad \exp(y - 0.5) \underset{H_0}{\overset{H_1}{\gtrless}} \lambda$$

or

$$\ln \Lambda(y) \quad = \quad y - \frac{1}{2} \underset{H_0}{\overset{H_1}{\gtrless}} \ln \lambda$$

or

$$y \underset{H_0}{\overset{H_1}{\gtrless}} \ln \lambda + 0.5 = \gamma$$

The fixed alarm rate is given by

$$P_{FA} = \int_{\gamma}^{\infty} \frac{1}{2\pi} \exp - \left(\frac{y^2}{2}\right) dy = 0.1$$

We know the above expression is $Q(\gamma) = 0.1$. *From the tables (Appendix D), we can solve for* γ, *which is 1.29. The final test becomes:*

$$y \underset{H_0}{\overset{H_1}{\gtrless}} 1.29$$

Of course, we also want to compute the probability of detection. Using the tables, we find $P_D = Q(0.29) = 0.3859$, *which is not a particularly good detection performance, but it is the best one can obtain under the circumstances.*

Example 4.16 *Suppose that we have the same conditions as in Example 4.15, except that we have access to 100 samples rather than just one. Under*

$$\begin{aligned} H_0 &: \; y(n) = n(n) \; ; \quad \text{for } n = 0, 1, \cdots, 99 \quad \text{and} \\ H_1 &: \; y(n) = 1 + n(n) \; ; \quad \text{for } n = 0, 1, \cdots, 99 \end{aligned}$$

$$\Lambda(\mathbf{y}) = \frac{f_1(\mathbf{y})}{f_0(\mathbf{y})} = \exp -0.5 \sum_{n=0}^{99} (1 - 2y(n)) \underset{H_0}{\overset{H_1}{\gtrless}} \lambda$$

$$\sum_{n=0}^{99} y(n) \underset{H_0}{\overset{H_1}{\gtrless}} \ln \lambda + 50$$

Let us call the left-hand side of the inequality z *and the right-hand side* γ, *that is*

$$z = \sum_{n=0}^{99} y(n) \underset{H_0}{\overset{H_1}{\gtrless}} \gamma$$

We realize that z *is a Gaussian random variable (i.e., one-dimensional) with mean of 0 or 100 under* H_0 *or* H_1, *respectively. The variance is easily shown to be 100; hence, the standard deviation equals 10. To guarantee a* P_{FA} *of 0.1, we know* $0.1 = Q(\gamma/10)$ *or* $1.29 = \gamma/10$. *Hence,* $\gamma = 12.9$ *and the probability of detection* $P_D = Q(\{100 - 12.9\}/10) \approx 1.0$. *Compared to Example 4.15, the probability of detection has drastically improved. If desirable, one could increase* γ *to force a smaller probability of false alarm and yet still retain a satisfactory detection probability.*

As in all detection problems, a minute change in detection probability, say from 0.999 to 0.998, is too small a difference to be noted. But at the other end of the scale, at the probability of false alarm, a change by 0.001, say from 10^{-6} *to* $0.001001 \approx 0.001$, *is a change by three orders of magnitude, which constitutes a thousandfold increase in the false alarm rate.*

Note, in particular, in the last example the change from a 100-dimensional random variable space to a one-dimensional space. We always try to take advantage of a lower-dimensional equivalent random variable, if possible.

As a final binary hypothesis testing example, we shall use one in which the functional form of the probability density function depends on the hypothesis.

Example 4.17 *When the event "1" or "0" is true, the date will follow an exponential or uniform distribution, respectively.*

$$H_1 \ : \ y \text{ has a PDF given by } f_1(y) = e^{-y}U(y)$$

$$H_0 \ : \ y \text{ has a PDF given by } f_0(y) = \frac{1}{2}\left[U(y) - U(2-y)\right]$$

Clearly, negative valued data is impossible.

$$f_1(y) \underset{H_0}{\overset{H_1}{\gtrless}} f_0(y)$$

or

$$e^{-y} \gtrless \frac{1}{2}\left[U(y) - U(2-y)\right], \quad \text{for } y \geq 0$$

Clearly, the left-hand side (LHS) $>$ *right-hand side (RHS) if* $y > 2$ *(since* $f_0(y)$ *is zero for arguments greater than 2).*

If we restrict ourselves to the interval [0,2], then

$$-y \underset{H_0}{\overset{H_1}{\gtrless}} \ln\frac{1}{2}$$

or

$$y \underset{H_0}{\overset{H_1}{\gtrless}} \ln 2 = 0.6931$$

Hence, the decision regions are

If	$y < 0.6931$	then H_1 is true,
if	$y > 2$	then H_1 is true,
if	$0.6931 < y < 2$	then H_0 is true.

$$\Pr\ (D_1|H_1) = P_D = \int_0^{\ln\ 2} e^{-y} dy + \int_2^{\infty} e^{-y} dy = 0.6353$$

$$\Pr\ (D_1|H_0) = \int_0^{\ln\ 2} \frac{1}{2} dy = 0.3465$$

Suppose that we want to fix $\Pr\ (D_1|H_0)$ *at a level of 0.3, then we shift the original threshold* (ln2) *an increment to the left so that*

$$\int_0^{T} \frac{1}{2} dy = 0.3$$

Hence, $T = 0.6$. *The new threshold also changes* $\Pr\ (D_1|H_1)$ *to*

$$\Pr\ (D_1|H_1) = \int_0^{0.6} e^{-y} dy + \int_2^{\infty} e^{-y} dy = 0.5865$$

4.8 MULTIPLE HYPOTHESES

If there are more than two hypotheses (i.e., more than two signals), the approach used in the Bayes' detector derivation is still appropriate. There are, however, more hypotheses and more types of errors. We will derive the general M-dimensional case, and then via an example, detail the analysis for the case when $M = 3$. Suppose there are M different events (i.e., $H_0, H_1, \cdots, H_{M-1}$) and assume that we know

(a) $\Pr(H_0) = P_0$, $\Pr(H_1) = P_1, \cdots, \Pr(H_{M-1}) = P_{M-1}$, where the sum of these prior probabilities must equal one;

(b) C_{ij} that is the cost of choosing i when j is true;

(c) $f_i(\mathbf{y})$ for $i = 0, 1, \cdots, M-1$, the conditional density $f_{\mathbf{Y}|H_i}(\mathbf{y}|H_i)$, then the Bayes' formulation minimizes the average cost C given by

$$C = \sum_{i=0}^{M-1} \sum_{j=0}^{M-1} C_{ij} \Pr(\text{choose } H_i|H_j)\, P_j \qquad (4.26)$$

Note, the product of the last two terms in (4.26), Pr(choose $H_i|H_j)P_j$ is just Pr(choose H_i, H_j is true). Bayes' test will separate the observation space into M mutually disjointed and exhaustive regions $R_0, R_1, \cdots, R_{M-1}$ such that C is minimized. Mutually exhaustive and disjointed means:

$$R_i \bigcap R_j = \Phi \quad \text{(the empty set, the Null space) when } i \neq j$$

$$\bigcup_{i=0}^{M-1} R_i = R \quad \text{(the observation space)}$$

Now Pr(choose $H_i|H_j$) denoted by $\Pr(D_i|H_j)$ can be written as

$$\Pr(D_i|H_j) = \int \cdots \int_{R_i} f_j(\mathbf{y})\, d\mathbf{y}$$

The average cost can therefore be written as

$$C = \sum_{i=0}^{M-1} C_{ii} P_i \int_{R_i} f_i(\mathbf{y}) d\mathbf{y} + \sum_{i=0}^{M-1} \sum_{j=0, i\neq j}^{M-1} C_{ij} P_j \int_{R_i} f_j(\mathbf{y}) d\mathbf{y} \tag{4.27}$$

Using $R_i = R - \bigcup_{j=0,j\neq i}^{M-1} R_j$ and

$$\int_R f_j(\mathbf{y}) d\mathbf{y} = 1$$

we get

$$C = \overbrace{\sum_{i=0}^{M-1} C_{ii} P_i}^{\text{fixed cost}} + \sum_{i=0}^{M_1} \int_{R_i} \sum_{j=0,j\neq i}^{M-1} P_j (C_{ij} - C_{jj}) f_j(\mathbf{y}) d\mathbf{y}) \tag{4.28}$$

Define $I_i(\mathbf{y})$ as

$$I_i(\mathbf{y}) = \sum_{j=0, i\neq j}^{M-1} P_j (C_{ij} - C_{jj}) f_j(\mathbf{y}) \tag{4.29}$$

then C is given by

$$C = \text{fixed cost} + \sum_{i=0}^{M-1} \int_{R_i} I_i(\mathbf{y}) d\mathbf{y} \tag{4.30}$$

The first term in (4.30) is a fixed cost term, while the second term constitutes a variable cost. As in the binary decision case, we assign each observation $\mathbf{y}$ to the region which will make the variable cost the smallest. To do this, let us define our decision rule, that is, "choose that hypothesis that corresponds to the minimum of $I_i(\mathbf{y})$ as the correct one (i.e., say H_i is true)."

Dividing $I_i(\mathbf{y})$ of (4.29) by $f_0(\mathbf{y})$, we get the averaged (weighted by cost and prior probabilities) likelihood function $J_i(\mathbf{y})$ as follows

$$\begin{aligned} J_i(\mathbf{y}) &= \frac{I_i(\mathbf{y})}{f_0(\mathbf{y})} \\ &= \sum_{j=0, j\neq i}^{M-1} P_j(C_{ij} - C_{jj})\Lambda_j(\mathbf{y}) \end{aligned} \tag{4.31}$$

where

$$\Lambda_j(\mathbf{y}) = \frac{f_j(\mathbf{y})}{f_0(\mathbf{y})}$$

The decision (detection) criterion becomes: "choose the hypothesis for which $J_i(\mathbf{y})$ is minimum (i.e., say H_i is true)." For example, choose H_0 if all the following inequalities are true.

$$J_0 < J_1 \Rightarrow \sum_{\substack{j=0 \\ j\neq 0}}^{M-1} P_j(C_{0j} - C_{jj})\Lambda_j(\mathbf{y}) < \sum_{\substack{j=0 \\ j\neq 1}}^{M-1} P_j(C_{1j} - C_{jj})\Lambda_j(\mathbf{y})$$

$$J_0 < J_2 \Rightarrow \sum_{\substack{j=0 \\ j\neq 0}}^{M-1} P_j(C_{0j} - C_{jj})\Lambda_j(\mathbf{y}) < \sum_{\substack{j=0 \\ j\neq 2}}^{M-1} P_j(C_{2j} - C_{jj})\Lambda_j(\mathbf{y})$$

$$\vdots$$

$$J_0 < J_{M-1} \Rightarrow \sum_{\substack{j=0 \\ j\neq 0}}^{M-1} P_j(C_{0j} - C_{jj})\Lambda_j(\mathbf{y}) < \sum_{\substack{j=0 \\ j\neq M-1}}^{M-1} P_j(C_{(M-1)j} - C_{jj})\Lambda_j(\mathbf{y})$$

The decision will be made in a likelihood space of dimension M (i.e., an M dimensional hyper space).

Example 4.18 *Suppose* $M = 3$

If $J_0 < J_1 \bigcap J_0 < J_2$ then $\mathbf{y} \in R_0 \Rightarrow$ choose H_0

If $J_1 < J_0 \bigcap J_1 < J_2$ then $\mathbf{y} \in R_1 \Rightarrow$ choose H_1

If $J_2 < J_0 \bigcap J_2 < J_1$ then $\mathbf{y} \in R_2 \Rightarrow$ choose H_2

where $\bigcap$ *represents the logical "and" operation.*

Example 4.19 *Suppose one of three constant signals is present (m_0 or m_1 or m_2). The signal is embedded in Gaussian noise (i.i.d.) $\sim N(0, \sigma^2)$, where N data samples are available.*

$$\begin{aligned} H_0 &: y_n \sim N(m_0, \sigma^2) \\ H_1 &: y_n \sim N(m_1, \sigma^2) \\ H_2 &: y_n \sim N(m_2, \sigma^2) \end{aligned}$$

where n =0, 1,2, ..., N, $m_0 < m_1 < m_2$. $P_i = 1/3$, $C_{ii} = 0$, for $i = 0, 1, 2$. $C_{ij} = 1$ when $i \neq j$. Minimize the probability of error based on using all N samples. Under

$$H_i \;:\; f_i(\mathbf{y}) = \prod_{n=1}^{N} \frac{1}{\sqrt{2\pi\sigma^2}} \exp -\frac{1}{2\sigma^2}(y_n - m_i)^2 \;; \qquad i = 0, 1, 2$$

The $I_i(\mathbf{y})$, (4.29), becomes (for $i = 0, 1, 2$)

$$\begin{aligned} I_0(\mathbf{y}) &= \sum_{j=0, j\neq 0}^{2} P_j\, C_{0j}\, f_j(\mathbf{y}) \\ &= \frac{1}{3}\, f_1(\mathbf{y}) + \frac{1}{3}\, f_2(\mathbf{y}) \\ I_1(\mathbf{y}) &= \sum_{j=0, j\neq 1}^{2} P_j\, C_{1j}\, f_j(\mathbf{y}) \\ &= \frac{1}{3}\, f_0(\mathbf{y}) + \frac{1}{3}\, f_2(\mathbf{y}) \\ I_2(\mathbf{y}) &= \sum_{j=0, j\neq 2}^{2} P_j\, C_{2j}\, f_j(\mathbf{y}) \\ &= \frac{1}{3}\, f_0(\mathbf{y}) + \frac{1}{3}\, f_1(\mathbf{y}) \end{aligned}$$

In terms of the $J_i(\mathbf{y})$ (for $i = 0, 1, 2$) we have

$$\begin{aligned} J_0(\mathbf{y}) &= \frac{I_0(\mathbf{y})}{f_0(\mathbf{y})} = \frac{1}{3}\, \Lambda_1(\mathbf{y}) + \frac{1}{3}\, \Lambda_2(\mathbf{y}) \\ J_1(\mathbf{y}) &= \frac{1}{3}\, \frac{f_0(\mathbf{y})}{f_0(\mathbf{y})} + \frac{1}{3}\, \frac{f_2(\mathbf{y})}{f_0(\mathbf{y})} = \frac{1}{3} + \frac{1}{3}\, \Lambda_2(\mathbf{y}) \\ J_2(\mathbf{y}) &= \frac{1}{3}\, \frac{f_0(\mathbf{y})}{f_0(\mathbf{y})} + \frac{1}{3}\, \frac{f_1(\mathbf{y})}{f_0(\mathbf{y})} = \frac{1}{3} + \frac{1}{3}\, \Lambda_1(\mathbf{y}) \end{aligned}$$

Suppressing the argument in the likelihood functions, this says the following:

Choose H_0

if $J_0 < J_1$ *and* $J_0 < J_2$*, that is*

(*a*) $1/3\ (\Lambda_1 + \Lambda_2) < 1/3\ (1 + \Lambda_2) \quad \Rightarrow \quad \Lambda_1 < 1$

(*b*) $1/3\ (\Lambda_1 + \Lambda_2) < 1/3\ (1 + \Lambda_1) \quad \Rightarrow \quad \Lambda_2 < 1$

Choose H_1

if $J_1 < J_0$ *and* $J_1 < J_2$*, that is*

(*a*) $1/3\ (1 + \Lambda_2) < 1/3\ (\Lambda_1 + \Lambda_2) \quad \Rightarrow \quad 1 < \Lambda_1$

(*b*) $1/3\ (1 + \Lambda_2) < 1/3\ (1 + \Lambda_1) \quad \Rightarrow \quad \Lambda_2 < \Lambda_1$

Choose H_2

if $J_2 < J_0$ *and* $J_2 < J_1$*, that is*

(*a*) $1/3\ (1 + \Lambda_1) < 1/3\ (\Lambda_1 + \Lambda_2) \quad \Rightarrow \quad 1 < \Lambda_2$

(*b*) $1/3\ (1 + \Lambda_1) < 1/3\ (1 + \Lambda_2) \quad \Rightarrow \quad \Lambda_1 < \Lambda_2$

These decision regions are shown in Figure 4.9, which shows the regions in terms of the likelihood ratios Λ_1 *and* Λ_2*. The decision separating surfaces are* $\Lambda_1 = 1$*,* $\Lambda_2 = 1$*, and* $\Lambda_1 = \Lambda_2$*. As indicated in Figure 4.9, we decide on the event* i *if the event is located in region* i *(i.e., we decide on* H_i *in region* R_i*).*

This type of representation is not easily visualized, so rather than working with likelihood functions, we try to simplify the representation by taking logarithms and reducing the expressions to their simplest form. We know that the expression

$$\Lambda_j(\mathbf{y}) = \frac{f_j(\mathbf{y})}{f_0(\mathbf{y})} \underset{<}{\overset{>}{}} 1$$

by inserting the appropriate probability density function, the expression becomes for $j = 1, 2$

$$\exp -\frac{1}{2\sigma^2}\left[\sum_{n=1}^{N}(y_n - m_j)^2 - \sum_{n=1}^{N}(y_n - m_0)^2\right] \underset{<}{\overset{>}{}} 1$$

Taking logarithms and simplifying leads to

$$-\sum_{n=1}^{N}(m_j^2 - 2m_j y_n) + \sum_{n=1}^{N}(m_0^2 - 2m_0 y_n) \underset{<}{\overset{>}{}} 0$$

Simplifying these terms and using the sample mean as a convenient statistic leads to

$$z = \frac{1}{N}\sum_{n=1}^{N} y_n \underset{<}{\overset{>}{}} \frac{m_0 + m_j}{2}, \quad \text{for } j = 1, 2\,.$$

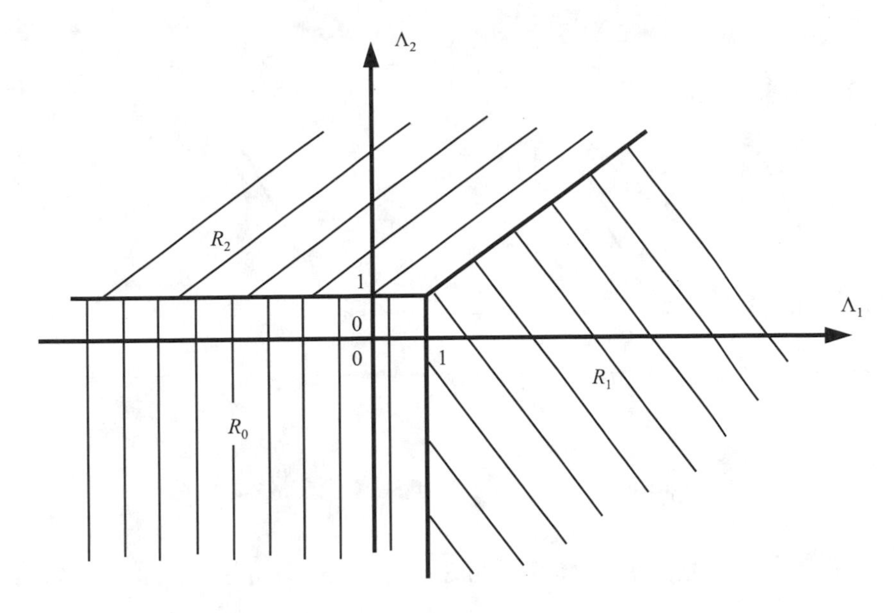

Figure 4.9: Decision regions for Example 4.19.

From $\Lambda_1 < 1$ *we obtain* $z < (m_0 + m_1)/2$.

From $\Lambda_2 < 1$ *we obtain* $z < (m_0 + m_2)/2$.

From $\Lambda_1 > 1$ *we obtain* $z > (m_0 + m_1)/2$.

From $\Lambda_2 > 1$ *we obtain* $z > (m_0 + m_2)/2$.

From $\Lambda_1 = \Lambda_2$, *we obtain*

$$f_1(\mathbf{y}) = f_2(\mathbf{y})$$

or

$$\sum_{n=1}^{N} (m_1^2 - 2m_1 y_n) = \sum_{n=1}^{N} (m_2^2 - 2m_2 y_n)$$

hence, $z = (m_1 + m_2)/2$. $\Lambda_2 > \Lambda_1$ *and* $\Lambda_2 < \Lambda_1$ *translate into* $z < 0.5(m_1 + m_2)$ *and* $z > 0.5(m_1 + m_2)$, *respectively. The decision regions are now very simple and intuitively pleasing:*

(a) decide on H_0 *when* $\Lambda_1 < 1$ *and* $\Lambda_2 < 1$, *becomes* $z < 0.5\ (m_1 + m_0)$ *and* $z\ < 0.5(m_0 + m_2)$,

(b) decide on H_1 *when* $1 < \Lambda_1$ *and* $\Lambda_2 < \Lambda_1$, *becomes* $z > 0.5\ (m_1 + m_0)$ *and* $z < 0.5\ (m_1 + m_2)$,

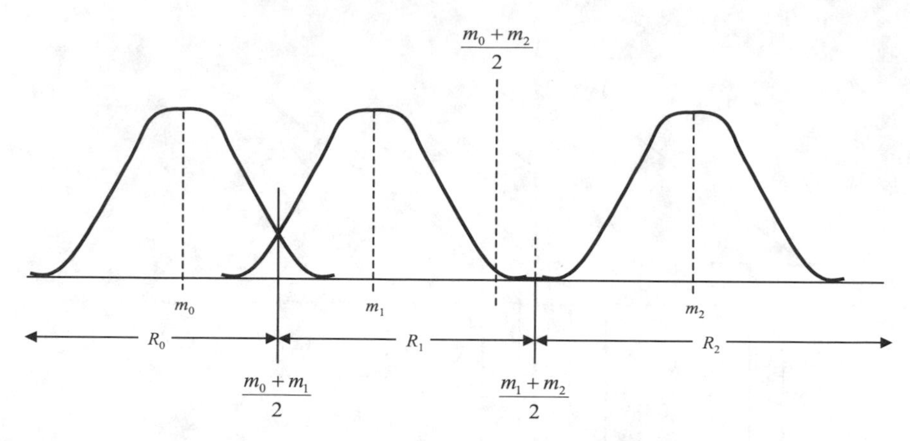

Figure 4.10: Decision surfaces for Example 4.19.

(c) *decide on* H_2 *when* $1 < \Lambda_2$ *and* $\Lambda_1 < \Lambda_2$, *becomes* $z > 0.5\ (m_2 + m_0)$ *and* $z > 0.5\ (m_1 + m_2)$.

These (one-dimensional) decision surfaces are shown in Figure 4.10.

4.9 COMPOSITE HYPOTHESIS TESTING

In the previous sections, we assumed that we know exactly how the quantities (i.e., signals) to be tested for look like. Usually, but not necessarily so, we have the statistical description of the noise. Many times the information about the signal component may be vague, in that we might not know all the parameters. We distinguish between three different types of scenarios, where we allow for a varying degree of uncertainty in one or several parameters. These three scenarios are

Type 1: The probabilistic description of some or all of the signal parameters (i.e., a known probability density function (PDF)) is available,

Type 2: A statistical description of the random signal or its parameters are not available. Or only a partial knowledge of the PDF is available, and

Type 3: The parameter(s) of the signal or signals are unknown; hence, a PDF description is not available.

We will only address the binary detection problem in this chapter knowing that an extension to the M-ary detection problem is straightforward but tedious. The unknown parameters are also called nuisance parameters. The nuisance parameters may be an issue appropriately addressed under either or both hypotheses. They allow us to incorporate randomness into the signals and uncertainties in the noise PDF. This may be used to model uncertainty in the arrival time, duration, carrier frequency, phase, amplitude, etc. If we denote the nuisance parameter by $\boldsymbol{\theta}$, where $\boldsymbol{\theta}$ may be one- or multi-dimensional, then under the two hypotheses we have:

$$
\begin{aligned}
H_0 &: \mathbf{y} \text{ is described by } f_0(\mathbf{y}|\boldsymbol{\theta}_0) \quad \text{and} \\
H_1 &: \mathbf{y} \text{ is described by } f_1(\mathbf{y}|\boldsymbol{\theta}_1)
\end{aligned}
$$

where $\boldsymbol{\theta}_i$, for $i = 0, 1$ is a vector of nuisance parameters (random or deterministic).

Before we deal with these types of problems, we will review the concept of removing (i.e., averaging out) a nuisance parameter. As we recall, we can average out, via the marginal approach, undesired quantities from a conditional probability density function. Suppose we have the conditional probability density function $f(x_1x_2|x_3x_4)$ and we want to remove (average out) the left-hand variable x_2. This is done by employing the so-called marginal property [1], as follows:

$$f(x_1|x_3x_4) = \int_{-\infty}^{\infty} f(x_1x_2|x_3x_4)\,dx_2 \tag{4.32}$$

To remove a right-hand variable, we multiply by the conditional density of the variable to be removed w.r.t. the remaining right-hand variables and integrate out the undesired nuisance parameter as follows:

$$f(x_1x_2|x_4) = \int_{-\infty}^{\infty} f(x_1x_2|x_3x_4)\,f(x_3|x_4)\,dx_3 \tag{4.33}$$

and

$$f(x_1x_2) = \int\int f(x_1x_2|x_3x_4)\;f(x_3|x_4)\,f(x_4)\;dx_3dx_4 \tag{4.34}$$

These identities can be verified very easily by using Bayes' rule. When nuisance parameters are involved, we call the hypothesis test a composite hypothesis test. If no unknown parameters are involved, we call the hypothesis test a simple hypothesis test. We note, when the limits of integration (or summation) are understood, then they are sometimes left off.

4.9.1 Nuisance Parameters with Known or Unknown Probability Density Function or with Unknown but Fixed Values

The likelihood ratio test is the optimum detection strategy, that is one computes $\Lambda(\mathbf{y})$ and compares it to the appropriate detection threshold (which is a function of the detector implementation, i.e., MAP, ML, etc.).

Type 1: We have

$$\Lambda(\mathbf{y}) = \frac{f_1(\mathbf{y})}{f_0(\mathbf{y})} = \frac{\int \cdots \int_{\boldsymbol{\theta}_1} f(\mathbf{y}|\boldsymbol{\theta}_1)\, f(\boldsymbol{\theta}_1)\, d\boldsymbol{\theta}_1}{\int \cdots \int_{\boldsymbol{\theta}_0} f(\mathbf{y}|\boldsymbol{\theta}_0)\, f(\boldsymbol{\theta}_0)\, d\boldsymbol{\theta}_0} \underset{H_0}{\overset{H_1}{\gtrless}} \lambda_0 \tag{4.35}$$

Type 2: When the vector $\boldsymbol{\theta}$ is random with unknown or partially known density, we approach the problem as follows:

> Design the detector as for Type 1 assuming the worst case probability density function (i.e., the uniform PDF reflects very little knowledge about the parameter density, except of course, the maximum and minimum values of the parameter(s)). If possible we use the partial knowledge about the PDF that may be available. Many times the decision rule turns out to be independent of the parameter.

Type 3: $\boldsymbol{\theta}$ is deterministic but unknown. Therefore, there is no PDF over which one can average. In principle, there are two approaches to this type of problem.

(a) We can use a Neyman-Pearson test. If the Neyman-Pearson test turns out to be independent of $\boldsymbol{\theta}$, then it is called a uniformly most powerful (UMP) test. The test must be independent of $\boldsymbol{\theta}$ in both threshold and in test decision variable(s) (the detector algorithm).

(b) If a UMP test does not exist, $\boldsymbol{\theta}$ can be estimated and these estimates are then used in a "generalized LRT"

$$\begin{aligned} \Lambda_g(\mathbf{y}) &= \frac{\max_{\boldsymbol{\theta}_1} f_1(\mathbf{y})}{\max_{\boldsymbol{\theta}_0} f_0(\mathbf{y})} \\ &= \frac{f_1(\mathbf{y}|\hat{\boldsymbol{\theta}}_1)}{f_0(\mathbf{y}|\hat{\boldsymbol{\theta}}_0)} \underset{H_0}{\overset{H_1}{\gtrless}} \lambda_0 \end{aligned} \tag{4.36}$$

where $\hat{\boldsymbol{\theta}}_i$ is the ML estimate of the parameter, for hypothesis $i = 0, 1$. These problems are best illustrated using examples. The examples use only a one-dimensional parameter. This serves well to illustrate the principle involved. The extension to more than one nuisance parameter is straightforward, i.e., replace the one-dimensional average with the corresponding multi-dimensional average.

Example 4.20 *Type 1, known nuisance PDF. Suppose that under the H_0 assumption we have a simple hypothesis, while under the H_1 assumption we have a composite hypothesis. This means that under H_0 we know the likelihood function for the zero event exactly, while under H_1 there is uncertainty and a statistical description of the parameter. In this example we have as a nuisance parameter the mean of the process, m. Under*

$$\begin{aligned} H_0 &: y \sim N(0, \sigma^2) \\ H_1 &: y \sim N(m, \sigma^2) \end{aligned}$$

where m has a Gaussian PDF given by $m \sim N(0, \sigma_m^2)$. The likelihood ratio is given by

$$\Lambda(y) = \frac{f_1(y)}{f_0(y)} = \frac{\int_{-\infty}^{\infty} f_1(y|m)\ f(m)\ dm}{f_0(y)} \begin{array}{c} H_1 \\ > \\ < \\ H_0 \end{array} \lambda$$

Using the available statistical information, this becomes

$$\Lambda(y) = \frac{\frac{1}{\sqrt{2\pi\sigma^2}} \int_{-\infty}^{\infty} \exp -\frac{(y-m)^2}{2\sigma^2}\ \frac{1}{\sqrt{2\pi\sigma_m^2}} \exp -\frac{m^2}{2\sigma_m^2}\ dm}{\frac{1}{\sqrt{2\pi\sigma^2}} \exp -\frac{y^2}{2\sigma^2}} \begin{array}{c} H_1 \\ > \\ < \\ H_0 \end{array} \lambda$$

This can be reduced to

$$y^2 \begin{array}{c} H_1 \\ > \\ < \\ H_0 \end{array} \frac{2\sigma^2(\sigma^2 + \sigma_m^2)}{\sigma_m^2} \left(\ln \lambda + 0.5 \ln \left(1 + \frac{\sigma_m^2}{\sigma^2} \right) \right)$$

Problems of type 2 are handled the same way as problems of type 1, that is, use whatever statistical information is at hand or assign a worst-case one. Other than that, the problem of type 2 is handled the same way as problems of type 1.

Example 4.21 *Type 3a, UMP type problem. Under*

$$\begin{aligned} H_0 &: y \sim N(0, 1) \\ H_1 &: y \sim N(m, 1) \end{aligned}$$

where m is an unknown positive quantity. The LRT is written as

$$\Lambda(y) = \exp\left(my - \frac{m^2}{2} \right) \begin{array}{c} H_1 \\ > \\ < \\ H_0 \end{array} \lambda$$

After some simplification this can be expressed as

$$y \underset{H_0}{\overset{H_1}{\gtrless}} \frac{m}{2} + \frac{1}{m} \ln \lambda = \lambda_0$$

Note, we could only reduce it to this simple form since we knew in advance that $m > 0$ *and the test left-hand side of the inequality can be made independent of the nuisance parameter. For the Neyman-Pearson test, the false alarm rate is fixed at some desired level of* P_{FA}. *Since our detection statistic is simply the observation, the expression for the* P_{FA} *becomes just the area under the integral of* $f_0(y)$ *from* λ_0 *to infinity. That is*

$$P_{FA} = \int_{\lambda_0}^{\infty} \frac{1}{\sqrt{2\pi}} \exp -\frac{y^2}{2} dy = Q(\lambda_0)$$

We see that the determination of the threshold λ_0 *is independent of the nuisance parameter* m *and that the Neyman-Pearson test (i.e., the left-hand side of the inequality) is also independent of the nuisance parameter* m. *This says that this test is a UMP test with respect to the parameter* m.

Example 4.22 *Type 3b, generalized LRT problem. A similar problem is given in Example 4.20 with two exceptions: (a) we do not know the polarity of the signal and (b) we have* N *samples (i.e.,* $N > 1$*) available. Under*

$$\begin{aligned} H_0 &: y_n \sim N(0, \sigma^2) \quad \text{for } n = 1, 2, \cdots, N \\ H_1 &: y_n \sim N(m, \sigma^2) \end{aligned}$$

where m *is a non-zero constant of unknown (polarity) sign. In terms of likelihood function we have*

$$f_0(\mathbf{y}) = \prod_{n=1}^{N} \frac{1}{\sigma\sqrt{2\pi}} \exp -\frac{1}{2\sigma^2} y_n^2 \qquad \text{(simple hypothesis)}$$

and for the composite hypothesis, assuming m *is known, we have*

$$f_1(\mathbf{y}|m) = \prod_{n=1}^{N} \frac{1}{\sigma\sqrt{2\pi}} \exp -\frac{1}{2\sigma^2} (y_n - m)^2$$

We use a maximum likelihood (ML)estimation procedure to obtain an estimate of m, *denoted by* $\hat{m}_{ML}$. *The derivation and idea behind the* ML *estimate will be discussed in Chapter 8. For now we use the sample mean as the* ML *estimate of the true mean. The expression for the sample mean is*

$$\hat{m}_{ML} = \frac{1}{N} \sum_{n=1}^{N} y_n$$

hence the LRT becomes

$$\Lambda(\mathbf{y}) = \frac{f_1(\mathbf{y}|\hat{m}_{ML})}{f_0(\mathbf{y})}$$

$$= \frac{\prod_{n=1}^{N} \exp -\frac{1}{2\sigma^2}\left(y_n - \frac{1}{N}\sum_{m=1}^{N} y_m\right)^2}{\prod_{n=1}^{N} \exp -\frac{1}{2\sigma^2} y_n^2} \begin{array}{c} H_1 \\ > \\ < \\ H_0 \end{array} \lambda$$

After some manipulation this can be simplified

$$\left|\frac{1}{\sqrt{N}}\sum_{n=1}^{N} y_n\right| \begin{array}{c} H_1 \\ > \\ < \\ H_0 \end{array} \gamma = \sqrt{2\sigma^2 \ln \lambda}$$

We introduced the idea of nuisance parameters using simple examples. We will return to the idea of averaging out undesired influences in Chapter 6.

4.10 RECEIVER OPERATOR CHARACTERISTIC CURVES AND PERFORMANCE

4.10.1 General Background

In all of the derivations and all of the examples we typically left out one important ingredient, namely the detector performance. This section will address this shortcoming. Usually, we want to know the probability of detection (P_D), the probability of a miss (P_M), and the probability of false alarm (P_{FA}). Of course, $P_D + P_M = 1$, hence knowing one of these two quantities provides the second one automatically. These terms have their origin in the detection literature where we describe the performance of binary decisions (target present or target absent) with these quantities. We interpret the term P_{FA} as the probability of saying the "1" event is true given that the "0" event actually occurred. P_M is the probability of saying the "0" event is true given that the "1" event actually occurred, while P_D is the probability of saying the "1" event is true given that the "1" event actually occurred. The majority of detection problems (i.e., radar, sonar, nuclear blast) consider as the "1" event the condition that a target or event to be detected is present. Without loss of generality, it is more convenient for us to describe the performance via the three earlier defined terms (P_{FA}, P_M, and P_D). Perhaps some better suited definitions are $\Pr(D_1|H_0)$ for P_{FA}, $\Pr(D_0|H_1)$ for P_M, and $\Pr(D_1|H_1)$ for P_D. But since the majority of problems deal with detection of targets, we use the radar/sonar definitions most of the time. Only when more clarity is provided by the general definition will we resort to it.

These quantities can be used to provide a graphical representation of the particular detection scheme. It seems a little strange to talk about different implementations of detectors at this time, but there are many reasons to compare detector performance. Four very typical reasons are listed below. They are:

(a) One could choose to use or disregard certain pieces of information when designing the detector. A typical scenario is not to use prior probabilities, even when they are available, and then obtain a measure of the trade-offs when ignoring that information.

(b) We may want to check out the design procedure by computing theoretical values of the detection performance and compare them with the results using the detector on simulated or real data set.

(c) Another frequent use of these quantities is to study the degradation of the performance as a function of signal-to-noise ratio (SNR) or as a function of the degree of straying away from the postulated ideal functional form of the detector. One can also display the degree of sensitivity to variations in PDF parameters.

(d) The performance of several detection schemes can be compared. We may have a choice of approaching the detection problem that is to say we may have to or want to select a sub-optimal design, which might lead to different detector implementations. In these cases, the trade-offs can be studied using the detection quantities.

4.10.2 ROC Curves

The basic receiver operating characteristic (ROC) curve is a two-dimensional graph of probability of detection (P_D) versus the probability of false alarm (P_{FA}). A typical example of the likelihood functions for the binary signal case is shown in Figure 4.11.

We note that one can slide the threshold away from the point as dictated by the detection philosophy (i.e., MAP, ML, Bayes, etc.) and examine the detection performance at each of these thresholds. To each unique threshold chosen, there corresponds a unique pair of points (P_D and P_{FA}) which describe the performance of the detector at that threshold. We typically plot these pairs of points against a vertical and horizontal scale of P_D and P_{FA}, respectively. A typical plot of these points for various SNRs is shown in Figure 4.12.

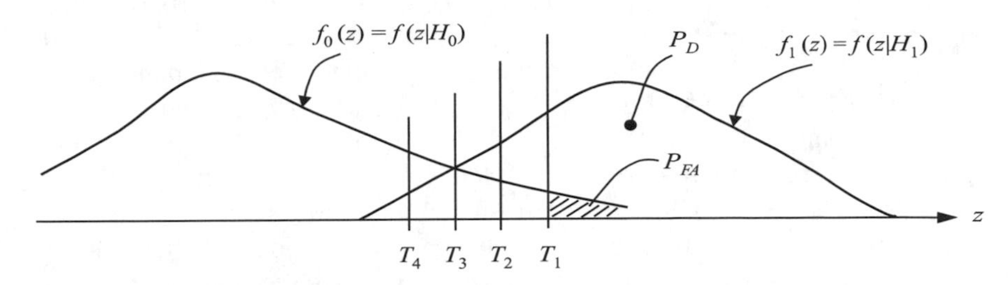

Figure 4.11: Probability density functions at the output of a detector.

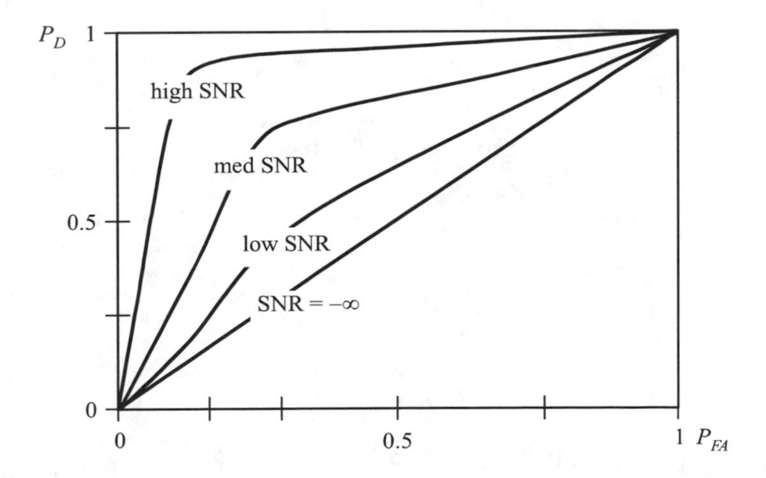

Figure 4.12: Typical ROC curve.

We note that if each event is equally likely and we randomly pick a choice, we would be right and wrong half of the time. This would trace a diagonal line from the lower left-hand corner to the upper right-hand corner in Figure 4.12. By the nature of designing a receiver (detector) we should do better than a random guess of the event. Hence, we always expect to operate with a higher probability of detection than the probability of false alarm. This ensures that all ROC curves will reside in the upper left-hand plane above the line based on a random guess. Figure 4.11 shows that if the threshold is reduced from T_1 to T_2 to T_3, etc., then the P_{FA} and the P_D will both increase as the threshold is decreased. Hence, we know that the curve must be concave. That is, if we lower the threshold and allow a higher P_{FA} at least as many or more events belonging to the $f_1(\mathbf{y})$ population are then allowed to pass through resulting in the same or higher P_D. We see that sliding along the ROC curve is equivalent to changing the threshold of the receiver. A low level of P_{FA} and P_D corresponds to a high threshold level, while a high level of P_{FA} and P_D corresponds to a low level of threshold. As a matter of fact, the numerical value of the slope of the ROC curve at any point corresponds to the numerical value of the threshold of the detector to provide the performance indicated by the coordinates of that point on the curve. A simple derivation of this result is given below. The detection statistic $\Lambda(\mathbf{y})$, which usually is a single random variable under each hypothesis, is compared to the threshold λ. The detector performance can then be described by:

$$P_D = \Pr(\Lambda > \lambda | H_1) = \int_{\lambda}^{\infty} f_{\Lambda|H_1}(\xi)\, d\xi \tag{4.37}$$

$$P_{FA} = \Pr(\Lambda > \lambda | H_0) = \int_{\lambda}^{\infty} f_{\Lambda|H_0}(\xi)\, d\xi \tag{4.38}$$

We can also rewrite P_D, the probability of detection using

$$f_{\Lambda|H_1}(\xi) = \frac{f_{\Lambda|H_1}(\xi)}{f_{\Lambda|H_0}(\xi)}\, f_{\Lambda|H_0}(\xi) = \Lambda(\xi)\ f_{\Lambda|H_0}(\xi)$$

as

$$\begin{aligned} P_D &= \int_{\lambda}^{\infty} f_{\Lambda|H_1}(\xi)\, d\xi \\ &= \int_{\lambda}^{\infty} \Lambda(\xi)\ f_{\Lambda|H_0}(\xi)\, d\xi \end{aligned} \tag{4.39}$$

Taking the derivatives of (4.38) and (4.39) with respect to λ (using Leibnitz's rule, Appendix D), leads to

$$\frac{\partial P_{FA}}{\partial \lambda} = f_{\Lambda|H_0}(\lambda)\ (-1) \tag{4.40}$$

$$\frac{\partial P_D}{\partial \lambda} = \Lambda(\lambda)\ f_{\Lambda|H_0}(\lambda)\ (-1) = -\lambda\ f_{\Lambda|H_0}(\lambda) \tag{4.41}$$

where $\Lambda(\lambda) = \lambda$ is substituted. The ratio of the two partial derivatives, that is, the slope of the ROC curve, becomes

$$\frac{\partial P_D}{\partial P_{FA}} = \lambda \tag{4.42}$$

We realize that the instantaneous slope of the ROC curve at any arbitrary point numerically corresponds to the threshold of the detector, which provides the probability of false alarm and probability of detection as indicated by the coordinate of the point under consideration. The detection performance display of the ROC curve is reasonable when considering a typical P_D of better than 0.5 at ranges of P_{FA} larger than 0.1. This is acceptable for our purposes, that is, the understanding of detectors and their performance. However, for many problems we would like to look at performance at probability of false alarm levels many orders of magnitude smaller than 0.1. A typical P_{FA} range might be 10^{-12} to 10^{-6}. In those cases, we will resort to other performance representations, i.e., use logarithmic values of the detection quantities, index the curves by SNR, etc. [3,5,6,7]. At this point, however, we recognize the relationship of threshold, performance, and the use of the ROC curve, since in any class project we would want to use manageable probabilities, hence a manageable number of simulation runs.

4.11 SUMMARY

The concept of hypothesis testing is introduced in Section 4.1. Bayes' cost function, criterion, and detectors are discussed in Section 4.2. Section 4.3 introduces the MAP, while Section 4.4 addresses the ML based detection. In Section 4.5 we investigate the concept of minimizing the probability of error, while in Section 4.6 we introduce the Min-Max criterion. The class of CFAR detector is introduced via the Neyman-Pearson technique in Section 4.7. Section 4.8 extends the binary detection scenario to an M-ary detection problem. An introduction to the removal of nuissance parameters (uncertainties) is given in Section 4.9. Section 4.10 investigates the concept and use of ROC curves.

4.12 PROBLEMS

1. When the zero hypothesis is true, a DC value of zero is transmitted. When the one hypothesis is true, a DC value of 0.75 is transmitted. The transmitted signal is embedded in uniformly distributed, zero mean, noise. The variance of the uniform noise is 0.5. Assume each event (i.e., signal) is equally likely, the cost for making no error is zero and the cost for making either one of the possible mistakes is unity. Only one single sample is available.

(a) Derive the optimal detector and state which criterion you are using.

(b) Derive the appropriate threshold.

(c) Compute the total probability of error.

2. Single sample scenario. Under the zero hypothesis, zero mean Gaussian noise is received. Under the one hypothesis, a DC value of 1 is transmitted and received embedded in zero mean Gaussian noise. The variance of the Gaussian noise is 1.0. Use the ML criterion to obtain the most convenient (most simple) implementation.

   (a) Set up the detector and threshold. Compute the probability of $Pr(D_1|H_0)$ based on this information.

   (b) Go back to the likelihood ratio test (an exponential); call that random variable L. Compute the density of L under the zero and the one hypothesis. Integrate the appropriate density function over the appropriate range of L (i.e., one limit is the value of λ obtained in the original setup of the LRT) to obtain $Pr(D_1|H_0)$. *HINT:* One would expect the values from part (a) and (b) to agree.

3. Under

$$\begin{aligned} H_0 \ : \ y \text{ s.t. } f(y) &= \exp(-y)\, U(y) \\ H_1 \ : \ y \text{ s.t. } f(y) &= 2\exp(-2y)\, U(y) \end{aligned}$$

   where $U(y)$ is the unit step.

   (a) Derive the appropriate detector and corresponding threshold (simplest form).

   (b) Sketch and label the detector output densities. Show threshold and decision regions.

   (c) Compute $Pr(D_1|H_0)$.

4. For $i = 1, 2$, x_i are independent random variables uniformly distributed with $x_i \sim U(-2/3, 2/3)$. Under the zero hypothesis, only noise is received. Under the one hypothesis, a DC value of 1/2 is embedded in the noise (for the duration of 2 samples). The receiver inadvertently adds the two input samples. That is, $y = x_1 + x_2$ under either one of the two hypotheses.

   (a) Based on this single piece of information y, design the optimal detector and threshold.

   (b) Compute $Pr(D_1|H_0)$.

5. Computer exercise. Under

$$\begin{aligned} H_0 \ : \ r(n) &= n(n)\ ; \qquad 0 \le n \le 63 \\ H_1 \ : \ r(n) &= s(n) + n(n)\ ; \qquad 0 \le n \le 63 \\ s(n) &= \cos\left\{ \frac{2\pi(k_0 n^2 + k_1 n)}{64} \right\} \end{aligned}$$

$n(n)$ is white Gaussian noise with unit variance.

$k_0 = 0.15, \quad k_1 = 7.0$ (known signal in white Gaussian noise)

(a) Design the optimal receiver for these conditions for a P_{FA} of 10^{-3}.

(b) Show all work in sufficient detail. Provide a diagram of the detector and the numerical value of the threshold.

(c) Verify results using a MATLAB simulation.

6. Under H_1 :

$$x \sim N(0,4) \implies f_x(x) = \frac{1}{\sqrt{2\pi 4}} \exp - \frac{x^2}{8}$$

Under H_0 :

$$f_x(x) = \frac{1}{2} \exp - |x|$$

Single sample scenario: Set up L.R.T. Determine threshold, decision regions, and detector form.

7. Under

$$\begin{aligned} H_0 \ : \ y &= \text{exponential with parameter } a \\ H_1 \ : \ y &= \text{exponential with parameter } b \end{aligned}$$

where $a > b > 0$. See table in Appendix D for the exponential PDF. Find the optimal detector, its threshold, the P_{FA} and P_M.

8. Under

$H_0 \ : \ y = x$

where x is rectified Gaussian noise n, with the Gaussian $n \sim N(0,1)$.

Under

$H_1 \ : \ y = z;$

where

$$f_z(z) = \exp - z\,U(z)$$

(a) Find the detector to minimize the probability error (detector and threshold).

(b) Compute P_e.

9. Under

$$\begin{aligned} H_0 : y &= \text{chi-squared with 2 degrees of freedom } (\sigma = 1) \\ H_1 : y &= \text{chi-squared with 2 degrees of freedom } (\sigma = 4) \end{aligned}$$

(a) Find the optimal detector (say also which detection criterion is used and why).

(b) What is the detection threshold?

(c) Find $P(D_1|H_0)$.

(d) Find $P(D_0|H_1)$.

10. Under

$$\begin{aligned} H_0 : y &= a \\ H_1 : y &= c \end{aligned}$$

where $c = a + b$

$$\begin{aligned} f_A(a) &= 0.5 \exp - (0.5a)\, U(a) \\ f_B(b) &= 0.25 \exp - (0.25b)\, U(b) \end{aligned}$$

A and B are statistically independent and $U(\ )$ are unit step functions.

(a) Determine the detector (in its simplest form).

(b) Determine the threshold(s) and sketch the decision regions for your detection statistic.

(c) Compute $Pr(D_1|H_0)$.

(d) Compute $Pr(D_1|H_1)$.

11. Under

$$\begin{aligned} H_0 : y &= x \ ; \text{ where } f_x(x) = 2\,(2\pi)^{-1/2} \exp - (x^2/2)\, U(x) \\ H_1 : y &= z \ ; \text{ where } f_z(z) = \exp - (z)\, U(z) \end{aligned}$$

x and z are statistically independent, and $U(\ )$ are unit step functions.

(a) Determine the detector (in its simplest form).

(b) Determine the threshold(s) and sketch the decision regions.

12. Prove that

$$\frac{\int_{-\infty}^{\infty} \frac{1}{\sqrt{2\pi}\sigma} \exp\left(-\frac{(y-m)^2}{2\sigma^2}\right) \frac{1}{\sqrt{2\pi}\sigma_m} \exp\left(-\frac{m^2}{2\sigma_m^2}\right) dm}{\frac{1}{\sqrt{2\pi}\sigma} \exp\left(-\frac{y^2}{2\sigma^2}\right)} \begin{array}{c} H_1 \\ > \\ < \\ H_o \end{array} \eta$$

can be reduced to

$$y^2 \begin{array}{c} H_1 \\ > \\ < \\ H_0 \end{array} 2\sigma^2 \frac{\sigma^2 + \sigma_m^2}{\sigma_m^2} \left(\ln(\eta) + \frac{1}{2} \ln\left(1 + \frac{\sigma_m^2}{\sigma^2}\right)\right)$$

13. Show that

$$\Lambda(y) = \frac{\prod_{i=1}^{N} \frac{1}{\sqrt{2\pi}\sigma} \exp\left(-\frac{1}{2\sigma^2}\left(y_i - \frac{1}{N}\sum_{i=1}^{N} y_i\right)^2\right)}{\prod_{i=1}^{N} \frac{1}{\sqrt{2\pi}\sigma} \exp\left(-\frac{y_i^2}{2\sigma^2}\right)} \begin{array}{c} H_1 \\ > \\ < \\ H_0 \end{array} \eta$$

can be reduced to

$$\left| \frac{1}{\sqrt{N}} \sum_{i=1}^{N} y_i \right| \begin{array}{c} H_i \\ > \\ < \\ H_0 \end{array} \gamma = \sqrt{2\sigma^2 \ln(\eta)}$$

References

[1] Papoulis, A., *Probability, Random Variables, and Stochastic Processes*, New York: McGraw-Hill, 1965.

[2] Pearson, K., *Tables of Incomplete* Γ*-Functions*, London: Cambridge University Press, 1965.

[3] Whalen, A., *Detection of Signals in Noise*, Orlando, FL: Academic Press, 1971.

[4] Helstrom, C.W., *Elements of Signal Detection and Estimation*, Englewood Cliffs, NJ: Prentice-Hall, 1995.

[5] Gradshteyn, I.S., and Ryzhik, I.M., *Table of Integrals, Series, and Products*, fifth ed., San Diego, CA: Academic Press, 1994.

[6] Melsa, J.L., and Cohn, D.L., *Decision and Estimation Theory*, New York: McGraw-Hill, 1978.

[7] Srinath, M.D., Rajasekaran, P.K., and Viswanathan, R., *Introduction to Statistical Signal Processing with Applications*, Englewood Cliffs, NJ: Prentice-Hall, 1996.

[8] Barkat, M., *Signal Detection and Estimation*, London: Artech House, 1991.

[9] Abramowitz, M., and Stegun, I.A., *Handbook of Mathematical Functions*, New York: Dover Publications, 1970.

[10] Pierre, D.A., *Optimization Theory with Applications*, New York: Dover Publications, 1986.

Chapter 5

Non-Parametric and Sequential Likelihood Ratio Detectors

5.1 INTRODUCTION

This chapter deals with non-parametric detection, a topic that is more difficult to analyze than the detection problems discussed in Chapter 4. We shall introduce the reader to the topic in a simple fashion. This chapter serves as an introduction to non-parametric and sequential likelihood ratio detection and is meant to make the reader aware of these topics as well as to provide some references. Much information about these topics can be found in the *IEEE Transactions on Aerospace*, and some detailed information is also available in Helstrom [1], in Kazakos and Kazakos [2], and in [3–5,7,8].

5.2 NON-PARAMETRIC DETECTION

Non-parametric detection can be used when the density function of the noise is not known or is known only approximately. We usually assume that the noise PDF is an even function or that the noise, on the average, takes on a positive value just as often as a negative value. Typical book references discussing this material are the texts by Gibson and Melsa [3] and Gibbons [4]. Also, the book by Melsa and Cohn [5] has a chapter dedicated to this material.

The statistical test is usually easily implemented, but the resulting density functions are somewhat difficult to deal with. We shall look only at the two simplest variations of this type of detector, but generalizations to more complicated weighting schemes can easily be visualized. The tests are based on the Neyman-Pearson criterion, which is familiar to the reader (see Chapter 4, Section 4.7). Many times we refer to these types of detectors as constant false alarm rate (CFAR) detectors, since the false alarm rate is kept at a constant level. We are using the terminology "non-parametric" in the engineering sense referring to non-parametric and distribution free methods. As the name implies, non-parametric refers to a class of distributions where the number of parameters is so large that one cannot use just a few of them. The distribution free method refers to a test statistic distribution which tends to be insensitive to having exact knowledge of the data distribution [7].

5.2.1 Sign Detector

The optimal detector for a fixed voltage level (i.e., a positive constant) in the presence of zero mean additive Gaussian noise is (see Example 4.6)

$$\sum_{n=1}^{N} y_n \overset{H_1}{\underset{H_0}{\gtrless}} T$$

We could also count the number of times the samples exceed zero, which requires N to be greater than one.

$$\sum_{n=1}^{N} U(y_n) \overset{H_1}{\underset{H_0}{\gtrless}} T_u$$

where $U()$ denotes the standard unit step function

$$U(y_n) = \begin{cases} 1\,; & y_n > 0 \\ 0\,; & y_n < 0 \end{cases}$$

and T_u denotes an appropriately chosen threshold. Basically, this detector is a counter that counts the number of times the observation takes on positive values. It does so on a sample by sample basis employing a uniform weight of one. That is, no special emphasis (i.e., weight) is given to the magnitude value of the data, i.e., the distance from zero. Since the number of positive observations is counted, the detector is essentially counting the number of positive signs (i.e., it is a sign detector). The implementation is very simple, typically consisting of a hard limiter and an adder, but the performance analysis is somewhat tedious. To provide some insight into the performance the following analysis is undertaken.

5.2.2 Performance Analysis of the Sign Detector

Given the observation (data), under

$$\begin{array}{lll} H_0 : & \Pr\{y_n \geq 0|m_0\} = 1/2 \\ H_1 : & \Pr\{y_n \geq 0|m_1\} = p > 1/2 \end{array} ; \quad \text{for } n = 1, 2, \cdots, J$$

Let us denote the sign of y_n by

$$s_n = \begin{cases} 1 ; & y_n > 0 \\ 0 ; & y_n < 0 \end{cases}$$

and

$$\Pr(s_n|m_0) = \begin{cases} 1/2 ; & y_n = 0 \\ 1/2 ; & y_n = 1 \end{cases}$$

$$\Pr(s_n|m_1) = \begin{cases} 1-p ; & y_n = 0 \\ p ; & y_n = 1 \end{cases}$$

where p is the probability of the data being positive, given that the fixed voltage level m_1 is transmitted. This leads to the likelihood ratio test (LRT) as follows

$$\Lambda(\mathbf{s}) = \frac{p^{\sum_{n=1}^{J} s_n} (1-p)^{J-\sum_{n=1}^{J} s_n}}{(1/2)^J} \begin{array}{c} H_1 \\ > \\ < \\ H_0 \end{array} \Lambda_0$$

Let J^+ denote the number of positive observations, then the LRT can be written as

$$\begin{aligned} \Lambda(\mathbf{s}) &= \frac{p^{J^+} (1-p)^{J-J^+}}{(1/2)^J} \\ &= [2(1-p)]^J \left(\frac{p}{1-p}\right)^{J^+} \begin{array}{c} H_1 \\ > \\ < \\ H_0 \end{array} \Lambda_0 \end{aligned}$$

To simplify it to the point where it is in its most useful form, we take the logarithm to the base $(p/(1-p))$. The expression can then be reduced to

$$J^+ \begin{array}{c} H_1 \\ > \\ < \\ H_0 \end{array} \log_{p/(1-p)} \Lambda_0 - J \log_{p/(1-p)} [2(1-p)] = \Lambda_1$$

J^+ is the sum of Bernoulli distributed random variables. If the event with parameter m_0 is true, then the binomial distribution J^+ has parameters J and 1/2. That is,

$$\begin{aligned}
\Pr\left(J^+ = n|m_0\right) &= \begin{pmatrix} J \\ n \end{pmatrix} \left(\frac{1}{2}\right)^n \left(1 - \frac{1}{2}\right)^{J-n} \\
&= \begin{pmatrix} J \\ n \end{pmatrix} \left(\frac{1}{2}\right)^J
\end{aligned}$$

The probability of saying H_1, given that H_0 is true, which in radar/sonar terms is called the P_{FA}, is given by

$$P_{FA} = \sum_{n=\Lambda_1+1}^{J} \begin{pmatrix} J \\ n \end{pmatrix} \left(\frac{1}{2}\right)^J$$

If the event with parameter m_1 is true, then the binomial distribution J^+ has parameters J and p

$$\Pr\left(J^+ = n|m_1\right) = \begin{pmatrix} J \\ n \end{pmatrix} p^n \left(1 - p\right)^{J-n}$$

and the probability of saying H_1, given that H_1 is true, which in radar/sonar terms is called the P_D, is given by

$$P_D = \sum_{n=\Lambda_1+1}^{J} \begin{pmatrix} J \\ n \end{pmatrix} p^n (1 - p)^{J-n}$$

These concepts are best illustrated with an example.

Example 5.1 *Given that eight samples $(y_1, y_2, \cdots, y_8)$ are used, and insisting on a P_{FA} of 0.1, derive all pertinent information that governs a sign detector and analyze the performance. Under*

$$\begin{aligned}
H_0 &: \; y_n = n_n \qquad\qquad n = 1, 2, \cdots 8 \\
H_1 &: \; y_n = m + n_n
\end{aligned}$$

where n_n are i.i.d. zero mean Gaussian random variables $n_n \sim N(0, \sigma^2)$ and the signal is a positive constant.

Solution: $J = 8$, $P_{FA} = 0.1$.

$$P_{FA} = \sum_{n=\Lambda_1+1}^{8} \begin{pmatrix} 8 \\ n \end{pmatrix} \left(\frac{1}{2}\right)^8 \leq 0.1$$

or

$$\sum_{n=\Lambda_1+1}^{8} \binom{8}{n} \leq 25.6$$

Now,

$$\binom{i}{k} = \frac{i!}{k!(i-k)!}$$

hence,

$$\binom{8}{8} = 1\,; \qquad \binom{8}{7} = 8\,; \qquad \binom{8}{6} = 28\,; \qquad \binom{8}{5} = 56$$

and

$$\sum_{n=7}^{8} \binom{8}{n} = 1 + 8 \leq 25$$

while

$$\sum_{n=6}^{8} \binom{8}{n} = 1 + 8 + 28 > 25$$

This makes $\Lambda_1 = 6$. Hence, the LRT becomes

$$J^{+} = \sum_{n=1}^{8} U(y_n) \underset{H_0}{\overset{H_1}{\gtrless}} 6$$

$$P_{FA} = \sum_{n=7}^{8} \binom{8}{n} \left(\frac{1}{2}\right)^{8} = \frac{8+1}{256} = 0.035 \leq 0.1$$

$$P_D = \sum_{n=7}^{8} \binom{8}{n} p^{n}(1-p)^{8-n}$$

To obtain a feel for how the probability of detection varies with the probability p under the H_1 hypothesis (i.e., probability of the random variable y_n, for $n = 0, 1, \cdots, J$, to be positive), we compute the detection probability for different values of p.

Suppose $p = 3/4$, then

$$\Pr(\text{choose } H_1 | H_1 \text{ is true})$$

is given by

$$P_D = 8(3/4)^7(1/4)^1 + (3/4)^8 = 0.367$$

Suppose $p = 0.99$, then

$$\Pr(\text{choose } H_1 | H_1 \text{ is true})$$

is given by

$$P_D = 8(0.99)^7(0.01)^1 + (0.99)^8 = 0.997$$

Suppose $p = 1.0$, then

$$\Pr(\text{choose } H_1 | H_1 \text{ is true})$$

would result in

$$P_D = 8(1.0)^7(0.0)^1 + (1.0)^8(1)^0 = 1.0$$

5.3 WILCOXON DETECTOR

In the previous section we developed a counting procedure to find the number of times the observation is positive. This approach also points out how the detection scheme can be modified. This modification is a ranking or weighting according to the distance from the line of dichotomy (i.e., zero in this case) under the H_0 hypothesis. Again the detector is used with a threshold guaranteeing a CFAR type performance. The detector is modified to count according to the rank of the observation. By this we mean

$$\sum_{n=1}^{J} d_n \underset{H_0}{\overset{H_1}{\gtrless}} T \; ; \qquad \text{where} \quad |y_{k_1}| \leq |y_{k_2}| \leq |\cdots \leq |y_{k_J}|$$

$$\text{where} \quad d_n = \begin{cases} 0 & \text{for } y_{k_n} < 0 \\ n & \text{for } y_{k_n} \geq 0 \end{cases}$$

$$\text{for n} = 1, 2, \cdots, \text{J}$$

To illustrate the ranking, we show a typical data set and extract the detection statistic.

Example 5.2 *Given the data set, where* $J = 7$

$$\{y_i\}_{i=1}^{7} = \{0.11, \; -4.0, \; 9.12, \; 0.12, -4.1, \; 6.0, \; 0.99\}$$

the magnitude ordered set is given by

$$\{0.11, \; 0.12, \; 0.99, \; |-4|, \; |-4.1|, \; 6, \; 9.12\}$$

$$\{d_n\} = \{1, \; 2, \; 3, \; 0, \; 0, \; 6, \; 7\}$$

$$z = \sum_{n=1}^{7} d_n = 19 \underset{H_0}{\overset{H_1}{\gtrless}} T_Z$$

where T_Z still has to be determined.

Note this ranking is directly proportional to the position of the positive values while accounting for the magnitude of the negative values. This technique can be modified to allow other rankings (weightings). Analysis of the rank detector is not as simple as that of the sign detector. For more information regarding non-parametric detection schemes, we refer the reader to the references quoted at the end of the chapter.

A new, exciting area of signal processing deals with higher order moments and cumulants [9]. In some applications, one can use the moment (or cumulant) to detect the absence or presence of a signal or component. For example, Aktas and Hippenstiel [6] used thresholds based on fourth order moments to reject noise in the reconstruction of a waveform in a wavelet band processing scheme.

5.4 SEQUENTIAL DETECTION

The sequential likelihood ratio detection criterion was introduced by A. Wald (1959), whose name it carries (i.e., Wald's sequential LRT). One can interpret this technique as a modified Neyman-Pearson test in which two thresholds are established. Testing is done until one of the two thresholds is crossed. This modified Neyman-Pearson test fixes the probability of a miss in addition to the probability of a false alarm. In the classical Neyman-Pearson test, we compare the LR with a threshold which in turn is governed by the P_{FA}, a given constant. In the modified Neyman-Pearson test (also called the sequential probability ratio test (SPRT)), we use the likelihood ratio (LR) and compare it at every update time (i.e., at each new sequential observation point) with two thresholds. These thresholds are denoted by η_0 and η_1. The thresholds are determined by specifying a P_{FA} fixed at a value of α and a P_M fixed at a value of β. If the LR is larger than η_1, we decide on H_1, while if the LR is smaller than η_0, we decide on H_0. If either threshold is crossed, the test is terminated and the appropriate decision is declared. If the LR falls between the two thresholds then we defer the decision, that is we take another sample and repeat the test. Let $y_1, y_2, \cdots, y_I$, represent the observations (assumed to be independent and identically distributed) denoted by $\mathbf{y}_I = (y_1, y_2, \cdots, y_I)$. Then the LRT becomes

$$\Lambda(\mathbf{y}_I) = \frac{f_1(\mathbf{y}_I)}{f_0(\mathbf{y}_I)} = \prod_{n=1}^{I} \frac{f_1(y_n)}{f_0(y_n)} = \prod_{n=1}^{I-1} \frac{f_1(y_n)}{f_0(y_n)} \frac{f_1(y_I)}{f_0(y_I)} \tag{5.1}$$

or

$$\Lambda(\mathbf{y}_I) = \Lambda(\mathbf{y}_{I-1})\Lambda(y_I)$$

where the boldface quantities are vectors and the non-boldface quantities are scalars. The vectors have an index that is used to indicate that they change in the size of their dimension (i.e., increase as the test progresses). Hence, we have a recursive arrangement for the LRT with initial condition

$$\Lambda(\mathbf{y}_1) = \Lambda(y_1)$$

where $\Lambda(\mathbf{y}_2)$ is given by

$$\Lambda(\mathbf{y}_2) = \Lambda(\mathbf{y}_1)\Lambda(y_2) = \Lambda(y_1)\Lambda(y_2)$$

For a specified (fixed) P_{FA} and P_M, we need to derive thresholds η_0 and η_1 so that we meet the constraints:

$$\begin{aligned}
\alpha &= P_{FA} = \Pr\left(\text{choose } H_1 | H_0 \text{ is true}\right) \\
&= \int \cdots \int_{R_1} f_0(\mathbf{y}_n)\, d\mathbf{y}_n \;; \quad \text{for } n = 1, 2, \cdots, I \qquad (5.2) \\
\beta &= P_M = 1 - P_D \\
&\quad \text{where} \quad P_D = \Pr\left(\text{choose } H_1 | H_1 \text{ is true}\right) \\
P_D &= \int \cdots \int_{R_1} f_1(\mathbf{y}_n)\, d\mathbf{y}_n \;; \quad \text{for } n = 1, 2, \cdots, I \qquad (5.3)
\end{aligned}$$

If we multiply and divide the integrand of (5.3) by $f_0(\mathbf{y}_n)$, we obtain

$$\begin{aligned}
P_D &= \int \cdots \int_{R_1} \frac{f_1(\mathbf{y}_n)}{f_0(\mathbf{y}_n)}\, f_0(\mathbf{y}_n)\, d\mathbf{y}_n \\
&= \int \cdots \int_{R_1} \Lambda(\mathbf{y}_n)\, f_0(\mathbf{y}_n)\, d\mathbf{y}_n \quad \text{for } n = 1, 2, \cdots, I
\end{aligned}$$

To declare detection, the sequential arrangement of the LRT is denoted by SLRT

$$\Lambda(\mathbf{y}_n) = \Lambda(\mathbf{y}_{n-1})\ \Lambda(\mathbf{y}_{n-2}) \cdots \Lambda(\mathbf{y}_1) \geq\ \eta_1$$

must be true. Hence,

$$\begin{aligned}
P_D &= \int \cdots \int_{R_1} \Lambda(\mathbf{y}_n)\, f_0(\mathbf{y}_n)\, d\mathbf{y}_n \geq\ \eta_1 \int \cdots \int_{R_1} f_0(\mathbf{y}_n)\, d\mathbf{y}_n \\
&= \eta_1 P_{FA} = \eta_1 \alpha
\end{aligned}$$

We also know that $P_D = 1 - P_M = 1 - \beta$ resulting in

$$P_D = 1 - \beta \geq\ \eta_1\ \alpha$$

hence,

$$\eta_1 \leq \frac{1-\beta}{\alpha}$$

Similarly,

$$P_M = \beta = \int \cdots \int_{R_0} f_1(\mathbf{y}_n)\, d\mathbf{y}_n = \int \cdots \int_{R_0} \Lambda(\mathbf{y}_n)\, f_0(\mathbf{y}_n)\, d\mathbf{y}_n$$

To declare a miss

$$\Lambda(\mathbf{y}_n) \leq \eta_0$$

hence,

$$\beta \leq \eta_0 \int \cdots \int_{R_0} f_0(\mathbf{y}_n)\, d\mathbf{y}_n \;=\; \eta_0\,(1-\alpha)$$

This leads to a bound on η_0

$$\eta_0 \geq \frac{\beta}{1-\alpha}$$

The SLRT, at step I, becomes

$$\Lambda(\mathbf{y}_I) \quad \begin{matrix} H_1 \\ > \eta_1 \\ \text{take another sample} \\ < \eta_0 \\ H_0 \end{matrix}$$

This is interpreted as if $\Lambda(\mathbf{y}_I) > \eta_1$, we declare H_1 to be true, and if $\Lambda(\mathbf{y}_I) < \eta_0$, we declare H_0 to be true. If $\Lambda(\mathbf{y}_I) > \eta_0$, but smaller than η_1, we take another sample, going to $\Lambda(\mathbf{y}_{I+1})$ and repeat the test. Note, we assumed in all the derivations that the samples are i.i.d. and that the probabilities $P_M = \beta$ and $P_{FA} = \alpha$ remain constant throughout the whole test.

This type of test allows the user to use computational resources more effectively, since the test typically could be terminated earlier (with the modified Neyman-Pearson constraints). In other words, once the presence or absence of a target has been determined with an acceptable level of error (P_{FA} or P_M), other possible targets can be interrogated while data related to the just declared detection or non-detection is delegated to a tracking or classification network.

Example 5.3

$$\begin{array}{lll} \text{Under} \;\; H_0 : & y(n) = n(n) & \\ \text{Under} \;\; H_1 : & y(n) = m + n(n) & \end{array} \quad \text{for } n = 1, 2, \cdots$$

The additive noise is Gaussian (i.i.d.) with zero mean and variance σ^2. We want to terminate the test (SPRT) when $P_M \leq \alpha_1$ or $P_{FA} \leq \alpha_2$.

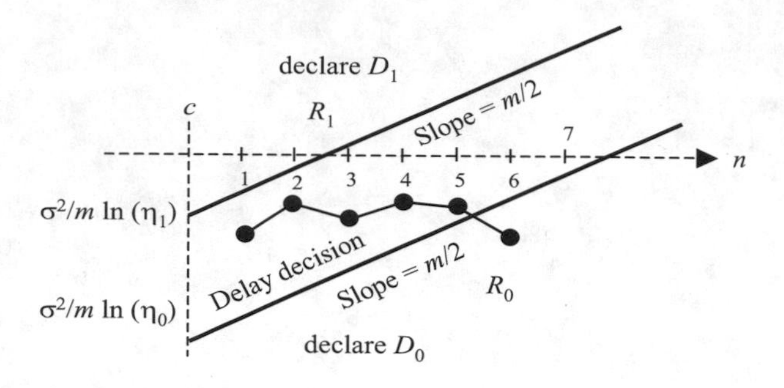

Figure 5.1: LRT versus the number of samples (test terminated at $n = 6$).

We know that

$$\eta_1 \leq \frac{1 - P_M}{P_{FA}} = \frac{1 - \alpha_1}{\alpha_2}$$

and that

$$\eta_0 \geq \frac{P_M}{1 - P_{FA}} = \frac{\alpha_1}{1 - \alpha_2}$$

The SLRT, at step I, becomes

$$\Lambda(\mathbf{y}_I) = e^{1/(2\sigma^2)\left(\sum_{n=1}^{I} y_n^2 - \sum_{n=1}^{I}(y_n - m)^2\right)} \begin{array}{c} H_1 \\ > \eta_1 \\ \\ < \eta_0 \\ H_0 \end{array} \quad \text{take another sample}$$

Simplifying the above expression leads to

$$z = \sum_{n=1}^{I} y_n \begin{array}{c} H_1 \\ > \\ \\ < \\ H_0 \end{array} \begin{array}{c} \frac{\sigma^2}{m}\ln(\eta_1) + \frac{mn}{2} \\ \\ \text{take another sample} \\ \\ \frac{\sigma^2}{m}\ln(\eta_0) + \frac{mn}{2} \end{array}$$

Figure 5.1 demonstrates the outcome of a particular test where for $I = 1, 2, 3, 4,$ *and* 5, *another sample needs to be taken since neither of the two thresholds are crossed. When* $I = 6$, *the lower threshold is crossed, allowing the declaration of event hypothesis zero to be true.*

5.5 SUMMARY

The general concept of non-parametric detection is introduced in Section 5.1. The concept is illustrated with the sign detector in Section 5.2, whose characteristics are also derived. Section 5.3 introduces the Wilcoxon detector, which ranks the observations. The concept of sequential detection is introduced in Section 5.4. Sequential detection permits termination of the detection test when a prescribed level of false alarm or probability of a miss is achieved.

5.6 PROBLEMS

1. Computer Exercise.

   Under

$$
\begin{aligned}
H_0 &: r(i) = s_0(i) + n_r(i) \quad \text{, where } \mathrm{s}_0(\mathrm{i}) = 0 \\
H_1 &: r(i) = s_1(i) + n_r(i) \quad \text{, where } \mathrm{s}_1(\mathrm{i}) = 1 \qquad \mathrm{i} = 0, \cdots, 9
\end{aligned}
$$

   Let $n(i)$ be Gaussian, i.i.d. $\sim N(0, 16)$. For this simulation exercise, we define the received noise $n_r(i)$ having three different PDFs (i.e., Gaussian, product of Gaussians, and the ratio of Gaussian that is a Cauchy):

$$
\begin{aligned}
\text{(i)} \quad \mathrm{n_r(i)} &\equiv n(i) \\
\text{(ii)} \quad \mathrm{n_r(i)} &\equiv \frac{n(i)}{n_1(i)} \\
\text{(iii)} \quad \mathrm{n_r(i)} &\equiv n(i) \cdot n_1(i)
\end{aligned}
$$

   where $n_1(i)$ is a Gaussian r.v. zero mean, variance $= 16$ and statistically independent of $n(i)$. Use 1000 realizations.

   (a) Design optimum detector for noise defined in (i). Plot simulation based ROC curves for optimum, sign, and Wilcoxon detector.

   (b) Repeat the experiments for noise defined in (ii). Use the optimum detector based on assuming the noise is Gaussian, the sign and Wilcoxon detector. Plot their experimental ROC curves.

   (c) Repeat the experiments for noise defined in (iii). Use the optimum detector based on assuming the noise is Gaussian, the sign and Wilcoxon detector. Plot their experimental ROC curves.

Comment on your results.

References

[1] Helstrom, C.W., *Elements of Signal Detection and Estimation*, Englewood Cliffs, NJ: Prentice-Hall, 1995.

[2] Kazakos, D., and Kazakos, P., *Detection and Estimation*, New York: Computer Science Press, 1990.

[3] Gibson, J.D., and Melsa, J.L., *Introduction to Non-Parametric Detection with Applications*, New York: Academic Press, 1975.

[4] Gibbons, J.D., *Non-Parametric Statistical Inference*, New York: McGraw-Hill, 1978.

[5] Melsa, J.L., and Cohn, D.L., *Decision and Estimation Theory*, New York: McGraw-Hill, 1978.

[6] Aktas, U., and Hippenstiel, R., "Localization of GSM Signals Using Wavelet Denoising Based on the Fourth Order Moment," *33rd Asilomar Conf. on Signals, Systems, and Computers*, Pacific Grove, CA, October 1999.

[7] Srinath, M.D., Rajasekaran, P.K., and Viswanathan, R., *Introduction to Statistical Signal Processing with Applications*, Englewood Cliffs, NJ: Prentice-Hall, 1996.

[8] Wald, A., *Sequential Analysis*, New York: Wiley, 1959.

[9] Nikias, C.L., and Petropulu, A.P., *Higher-Order Spectra Analysis — A Nonlinear Processing Framework* , Englewood Cliffs, NJ: Prentice-Hall, 1993.

Chapter 6

Detection of Dynamic Signals in White Gaussian Noise

6.1 INTRODUCTION

Initially, we will cover the detection of known signals embedded in white Gaussian noise. That is, we assume that we know all of the signal and noise parameters. This is an ideal case. Later in this chapter, we will remove some of the limitations in that we will allow the parameters to be random or unknown. In Chapter 7, we will allow the Gaussian noise to be colored (i.e., the noise process is not delta correlated). The bilateral white noise spectral density is given by $S_N(\omega) = N_0/2$ and the corresponding correlation function is given by $R_N(\tau) = N_0/2\ \delta(\tau)$. Note, as long as the noise is spectrally flat over the region of interest it can be interpreted as bandlimited white noise. We minimize the use of complex valued functions when possible and point out the changes when dealing with complex valued functions. Typically, the extensions to complex valued functions require the complex conjugation of the second term in the definition of the inner product. The second edition of *Detection of Signals in Noise* by McDonough and Whalen [10], has ample examples using complex valued entities.

We will let the signals be continuous functions of time and sample them properly, rather than starting with fixed quantities as in Chapter 4. One could have introduced signals with a time varying nature in Chapter 4, but for clearness of presentation this was not attempted. This particular approach introduces the processing of data in a natural fashion. That is, the

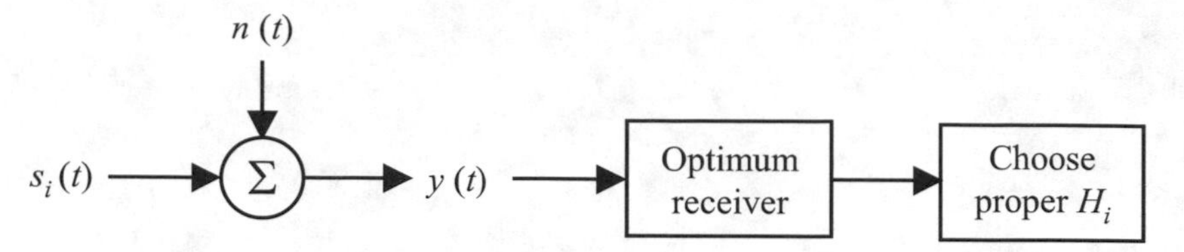

Figure 6.1: Binary detection.

setting up of the continuous time problem as a discrete time problem allows the application of the techniques learned in Chapter 4. It also allows the direct derivation of the discrete time version of the appropriate detectors. This chapter derives the discrete time correlator and hence, the discrete time matched filter. In addition, we are exposed to the projection of the data onto a set of basis functions. The data is projected onto a replica of the signal (i.e., the signals serve as the basis function). This is our first exposure to the concept of an inner product in a Hilbert space setting.

Section 6.2 will address the digital and analog correlator solution, while Sections 6.3 and 6.4 address the matched filter approach. In Section 6.5 we will extend the binary case to the more general M-ary case. The notion of parameter uncertainty will be taken up in Section 6.6. In Sections 6.7 and 6.8 we will examine the multi-pulse scenario as well as the notion of coherent and non-coherent averaging.

6.2 THE BINARY DETECTION PROBLEM

Suppose that one of two possible communication signals is transmitted. That is, when hypothesis H_0 is true signal $s_0(t)$ and when hypothesis H_1 is true, signal $s_1(t)$ is sent. This is a binary hypothesis problem where the additive noise (as specified above) makes the detection difficult. The binary detection scenario is shown in Figure 6.1. Since at this time, we only know an approach based on samples (i.e., data points), we will address this problem in a way very similar to the hypothesis testing procedure in Chapter 4. To analyze the situation, we conduct a Gedanken experiment (thought experiment) which allows us to use the tools that we have already acquired. During this derivation, we assume that the noise is bandlimited. This translates into knowledge about the zero crossing of the noise correlation function and allows, in theory at least, a sampling procedure where the sampling interval corresponds to the distance from the origin to the first zero crossing of the noise correlation function. Since the noise is Gaussian and the samples due to the sampling approach are uncorrelated, the samples are also independent under each hypothesis. The continuous waveform $y(t)$ is sampled and once a description in terms of a likelihood ratio of the data sequence is obtained, a limiting argument allows the sampling interval go to zero which in turn allows the description of the optimum receiver in terms of time continuous variables. This approach is advocated in Whalen [1] and helps to relate to the concept of representing signals and/or noise using weighted basis functions. In this context, we realize that a data sequence can be thought of to be an arrangement of weighted Kronecker delta functions, where the Kronecker delta functions form a complete ortho-normal basis set [5–7].

The known analog signals of duration T have the following representation, under

$$\begin{aligned} H_0 &: \quad y(t) = s_0(t) + n(t) \\ H_1 &: \quad y(t) = s_1(t) + n(t) \end{aligned} \quad ; \quad \text{for } 0 \leq t \leq T$$

Sampling these components (we assume that the signal is sampled at least at a rate corresponding to the Nyquist rate) leads to the description

$$\begin{aligned} y(t_m) \rightarrow y_m &= s_{im} + n_m \\ &\text{for } 1 \leq m \leq k \quad \text{and} \quad i = 0, 1 \end{aligned} \tag{6.1}$$

that is, under each hypothesis k samples are available. Using the expression from Chapter 4, we can easily write an expression for the LRT as

$$\Lambda(\mathbf{y}) = \frac{f_1(y_1, y_2, \cdots, y_k)}{f_0(y_1, y_2, \cdots, y_k)} \underset{H_0}{\overset{H_1}{\gtrless}} \lambda \tag{6.2}$$

where λ depends on the criterion that is being employed. For the time being and without loss of generality, we leave it as λ, knowing that at any time we can replace this symbol with the appropriate numerical quantity if that is desired. We notice that the left-hand side of expression (6.2), is just the ratio of two likelihood functions (or, if interpreted correctly, the ratio of two conditional probability density functions).

Bandlimited white Gaussian noise with spectral density as shown in Figure 6.2 has the corresponding correlation function shown in Figure 6.3. We recall that the correlation and spectral density function are related by a Fourier transform (i.e., the Wiener-Khintchine theorem, also known as the Wiener-Khintchine-Kolmogorov-Einstein theorem).

In equation form, the spectral density and corresponding correlation function are given by

$$S_N(\omega) = \begin{cases} N_0/2 & \text{for } |\omega| < \Omega \\ 0 & \text{else} \end{cases} \tag{6.3}$$

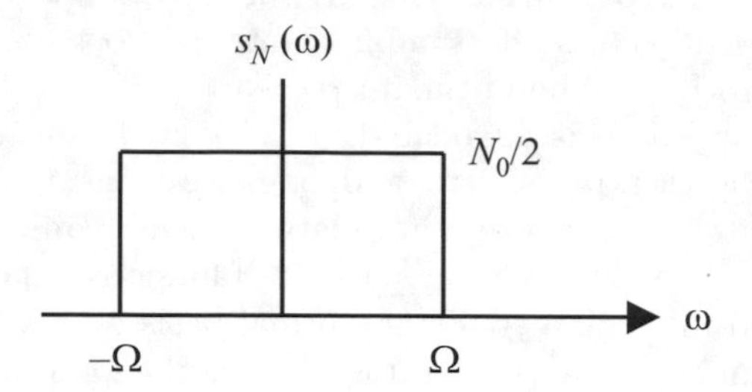

Figure 6.2: Bandlimited white noise: spectral density.

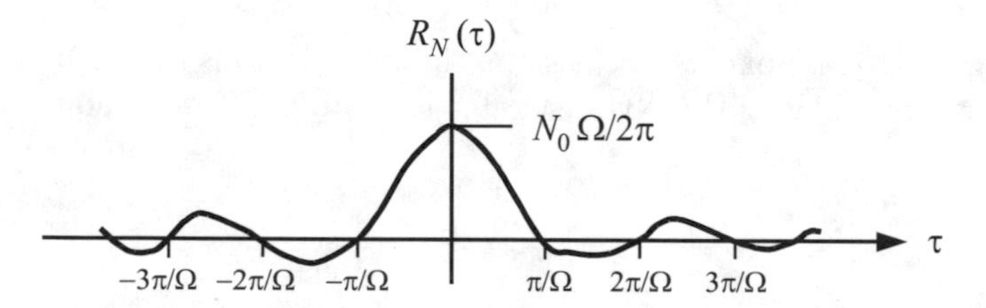

Figure 6.3: Bandlimited white noise: correlation function.

and

$$\begin{aligned} R_N(\tau) &= \frac{N_0}{2\pi}\,\Omega\,\frac{\sin(\Omega\tau)}{(\Omega\tau)} \\ &= \frac{N_0}{2\pi}\,\Omega\,\text{sinc}\,(\Omega\tau) \end{aligned} \tag{6.4}$$

The first zero crossing of the correlation function is at the correlation lag where $\tau = \Delta t = \pi/\Omega$. If we sample using this sampling interval, then all noise samples are uncorrelated and the joint density (likelihood) function can be expressed as the product of independent terms. This makes the derivation of the optimal detector very simple. To have a proper LR function expression, all we need to find is the mean under each hypothesis and the variance. The variance in this case, is independent of which hypothesis is true. The mean under each hypothesis is

$$\begin{aligned} E[y_m] = E[s_{im} + n_m] &= s_{im}\,; \quad \text{for } i = 0, 1 \\ & \qquad m = 1, \cdots, k \end{aligned} \tag{6.5}$$

The computation of the variance is just slightly more complicated. For example, under H_i we have

$$\sigma_{y_m}^2 = E\left[y_m - Ey_m\right]^2$$

$$
\begin{aligned}
&= E\left[s_{im} + n_m - s_{im}\right]^2 \\
&= E\left[n_m^2\right] \\
&= \sigma_N^2 = R_N(0) = \frac{N_0}{2\pi}\,\Omega
\end{aligned} \tag{6.6}
$$

which is independent of the hypothesis i. Using k independent samples (Δt units apart, taken over the time segment $(0, T)$), we obtain the likelihood functions as given by

$$
f_1(\mathbf{y}) = \frac{1}{(2\pi\sigma_N^2)^{k/2}} \exp - \frac{1}{2\sigma_N^2} \sum_{m=1}^{k} (y_m - s_{1m})^2 \tag{6.7}
$$

$$
f_0(\mathbf{y}) = \frac{1}{(2\pi\sigma_N^2)^{k/2}} \exp - \frac{1}{2\sigma_N^2} \sum_{m=1}^{k} (y_m - s_{0m})^2 \tag{6.8}
$$

The LRT (6.2) becomes

$$
\Lambda(\mathbf{y}) = \exp - \frac{1}{2\sigma_N^2} \sum_{m=1}^{k} \left(2y_m s_{0m} - 2y_m s_{1m} - \left[s_{0m}^2 - s_{1m}^2\right]\right) \underset{H_0}{\overset{H_1}{\gtrless}} \lambda \tag{6.9}
$$

As always, these types of expressions are reduced to the simplest form. Taking the natural logarithm and simplifying, leads to

$$
- \sum_{m=1}^{k} \frac{y_m s_{0m}}{\sigma_N^2} + \sum_{m=1}^{k} \frac{y_m s_{1m}}{\sigma_N^2} \underset{H_0}{\overset{H_1}{\gtrless}} \ln(\lambda) - \frac{1}{2\sigma_N^2} \sum_{m=1}^{k} \left(s_{0m}^2 - s_{1m}^2\right) \tag{6.10}
$$

If we multiply through by the variance of the noise, we obtain the discrete (also called digital) correlator form. So, if we are dealing with discrete samples, we would stop here and implement the digital correlator. This correlator is of the form

$$
\sum_{m=1}^{k} y_m\ s_{1m} - \sum_{m=1}^{k} y_m\ s_{0m} \underset{H_0}{\overset{H_1}{\gtrless}} \eta \tag{6.11}
$$

The threshold is easily obtained from the discussion leading to the digital correlator form. The discrete (digital) correlator is shown in Figure 6.4.

The final part of this section deals with the analog version of the detector implementation, so we need to apply some limiting arguments. To do this, consider the expression for the variance of the noise samples. For a bandlimited white noise process with spectral height $N_0/2$, the variance is given by

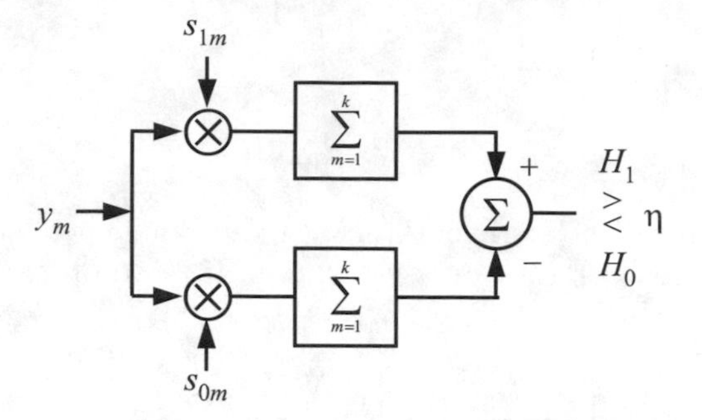

Figure 6.4: Discrete time correlator.

$\sigma_N^2 = N_0\Omega/(2\pi)$. A bandlimited process needs only to be sampled at twice the highest spectral frequency. In this case, the sampling frequency in radians/sec is $f_s = 2\Omega = 2\pi/\Delta t$, hence $\Omega = \pi/\Delta t$. The noise variance becomes $\sigma_N^2 = N_0\Omega/(2\pi) = N_0/(2\Delta t)$. This allows (6.10) to be written as

$$\lim_{\substack{k\to\infty \\ \Delta t\to 0}} {}_{\ni \Delta t\ k\ =\ T} \left\{ -\sum_{m=1}^{k} \frac{2y_m s_{0m}}{N_0}\Delta t + \sum_{m=1}^{k} \frac{2y_m s_{1m}}{N_0}\Delta t \underset{H_0}{\overset{H_1}{\gtrless}} \ln\lambda \right.$$

$$\left. -\frac{1}{N_0}\sum_{m=1}^{k}\left(s_{0m}^2 - s_{1m}^2\right)\Delta t \right\} \tag{6.12}$$

In the limit, the summations become integrals so that we obtain the following expression

$$\int_0^T \frac{2}{N_0}\, y(t)\, s_1(t)\, dt - \int_0^T \frac{2}{N_0}\, y(t)\, s_0(t)\, dt \underset{H_0}{\overset{H_1}{\gtrless}} \ln\lambda + \int_0^T \frac{1}{N_0}\left(s_1^2(t) - s_0^2(t)\right) dt \tag{6.13}$$

One can normalize this expression to obtain

$$\int_0^T y(t)\, s_1(t)\, dt - \int_0^T y(t)\, s_0(t)\, dt \underset{H_0}{\overset{H_1}{\gtrless}} \frac{N_0}{2}\ln\lambda + \frac{1}{2}\int_0^T \left(s_1^2(t) - s_0^2(t)\right) dt = T_0 \tag{6.14}$$

This detector is shown in Figure 6.5. In the right-hand side, the threshold has been simplified to be represented by the single symbol T_0.

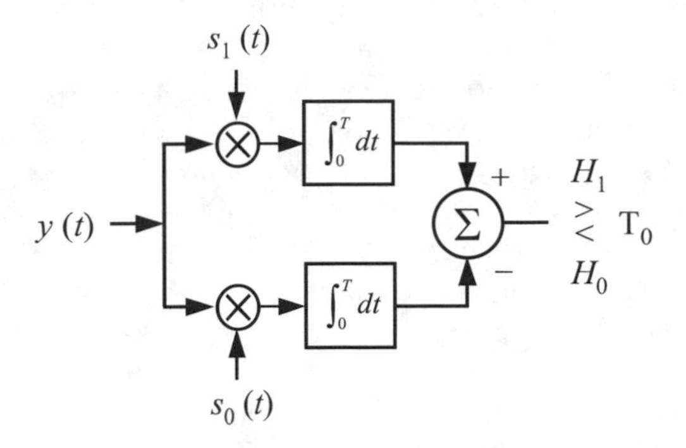

Figure 6.5: Correlation receiver.

We note that this is a correlation receiver. That is, the input is correlated with the stored replicas of the signals $s_0(t)$ and $s_1(t)$ for a time delay of zero. If a maximum likelihood approach is used and both signals have the same energy, then the test reduces to the determination of which correlator has the dominant output. For the Neyman-Pearson type criterion, we will disregard the value of T_0 altogether. The appropriate threshold is obtained by solving for the value under the zero hypothesis, to guarantee the desired P_{FA}. We also note that if our problem calls for the implementation as a communication receiver, then substituting the known values into the expression for λ (i.e., $P_1 = P_0$, $C_{00} = C_{11} = 0$, $C_{01} = C_{10} = 1$) leads to a λ of one. Let us now define a convenient test statistic for the receiver and obtain the performance of the receiver.

The desired test quantity is G, and with $\ln \lambda = 0$, we obtain the following expression

$$G = \int_0^T y(t)\left(s_1(t) - s_0(t)\right) dt + \frac{1}{2}\int_0^T \left(s_0^2(t) - s_1^2(t)\right) dt \underset{H_0}{\overset{H_1}{\gtrless}} 0 \qquad (6.15)$$

Since $y(t)$ has a Gaussian PDF, a scaled and integrated version of it also has a Gaussian PDF. Hence, G as defined in (6.15) is normally distributed. To obtain values for the performance of the detector, only the means and the variance of the Gaussian random variable G have to be obtained.

6.2.1 Performance Analysis

This is also a very convenient place to evaluate the performance of some standard digital communication modulation techniques. We use some simplifying assumptions to enhance the comparison. We assume that $P_0 = P_1$,

$C_{00} = C_{11} = 0$, and $C_{01} = C_{10} = 1$. This is the minimum probability of error criterion introduced in Chapter 4, resulting into a threshold of $\lambda = 1$. Hence, the receiver (detector) and threshold can be made as shown in (6.15).

To obtain the optimal receiver (detector) form and threshold is not sufficient. One also needs to determine the performance. This is most easily done using the probability of error, which, due to the costs and prior probabilities is numerically the same as Pr(choose $H_1|H_0$ is true). We recall that a scaled Gaussian random variable is still a Gaussian random variable. Hence, the random variable G is Gaussian under each hypothesis. To compute any of the performance indicators we need to find the means of G under each hypothesis and its variance. Under each hypothesis (may it be H_0 or H_1), the variance is identical for this type of detector. The mean of G, under the zero hypothesis, is denoted by

$$E(G|H_0) = E_0(G) \tag{6.16}$$

Disregarding the noise dependent terms that equate to zero, i.e., the first order moments of the noise are zero, we obtain

$$\begin{aligned} E_0(G) &= E\left[\int_0^T s_0(t)\,(s_1(t) - s_0(t))\,dt - \frac{1}{2}\int_0^T \left(s_1^2(t) - s_0^2(t)\right)dt\right] \\ &= E\left[\int_0^T \left(s_0(t)s_1(t) - s_0^2(t) - 0.5s_1^2(t) + 0.5s_0^2(t)\right)dt\right] \end{aligned} \tag{6.17}$$

This can be simplified to

$$E_0(G) = -\frac{1}{2}\int_0^T (s_1(t) - s_0(t))^2\,dt \tag{6.18}$$

The mean under the one hypothesis is computed similarly and is given by

$$\begin{aligned} E(G|H_1) &= E_1(G) = \frac{1}{2}\int_0^T (s_1(t) - s_0(t))^2\,dt \\ &= -E_0(G) \end{aligned} \tag{6.19}$$

The variance is a little more difficult to compute. We recall that the variance corresponds to the AC power in electronic circuits or in terms of statistics, it describes the spread of the probability density function about its mean. That is, it is independent of the mean (i.e., the DC term) and in this case can be computed either under the zero or the one hypothesis. Usually, we check which way the result can be obtained more easily. In many cases, the zero hypothesis results in a zero mean. Under this condition we prefer to compute the variance under the H_0 hypothesis since the second moment corresponds to the variance and no special consideration for any mean value has to be taken. In our case, we arbitrarily choose to work with the zero hypothesis (the one hypothesis requires the same amount of computation). We define the variance of the Gaussian test statistic as

$$\text{var}_{G_0} = E\left[(G|H_0) - E(G|H_0)\right]^2 = E\left[(G|H_1) - E(G|H_1)\right]^2 = \text{var}_{G_1} \quad (6.20)$$

Now, we first examine the expression

$$\begin{aligned} G|H_0 - E_0(G) &= \int_0^T (s_0(t) + n(t))\,(s_1(t) - s_0(t))\,dt \\ &\quad + \frac{1}{2}\int_0^T \left(s_0^2(t) - s_1^2(t)\right) dt + \frac{1}{2}\int_0^T \left[s_1(t) - s_0(t)\right]^2 dt \\ &= \int_0^T n(t)\left[s_1(t) - s_0(t)\right] dt \end{aligned} \quad (6.21)$$

Using (6.20), we note that the variance expression becomes

$$\begin{aligned} \text{var}_{G_0} &= E\left(\int_0^T n(t)\left[s_1(t) - s_0(t)\right] dt\right)^2 \\ &= \int_0^T \int_0^T E\left[n(t)n(\sigma)\right]\left[s_1(t) - s_0(t)\right]\left[s_1(\sigma) - s_0(\sigma)\right] dt\, d\sigma \\ &= \frac{N_0}{2}\int_0^T \left[s_1(t) - s_0(t)\right]^2 dt = \text{var}_{G_1} \end{aligned} \quad (6.22)$$

Rather than carrying this somewhat cumbersome expression further along, we use some simplifying terms. Let us define the following terms

$$\varepsilon = \frac{1}{2}\int_0^T \left[s_1^2(t) + s_0^2(t)\right] dt \quad (6.23)$$

$$\rho = \frac{1}{\varepsilon}\int_0^T s_0(t)s_1(t)dt \quad (6.24)$$

We note that ε and ρ represent the average energy and the normalized cross-correlation coefficient, respectively. This means that ρ is bounded above by 1 and below by -1. We can now express the following integrals in a more compact form

$$\begin{aligned} \int_0^T \left[s_1(t) \pm s_0(t)\right]^2 dt &= \int_0^T \left[s_0^2(t) + s_1^2(t)\right] dt \pm 2\int_0^T s_0(t)s_1(t)dt \\ &= 2\varepsilon \pm 2\varepsilon\rho \\ &= 2\varepsilon(1 \pm \rho) \geq 0 \end{aligned} \quad (6.25)$$

Using these expressions, the means and the variance of the Gaussian random variable G become

$$\begin{aligned} E_0(G) &= -\varepsilon(1-\rho) && (6.26) \\ E_1(G) &= \varepsilon(1-\rho) && (6.27) \\ \mathrm{var}_{G_0} = \mathrm{var}_{G_1} &= N_0(1-\rho)\varepsilon && (6.28) \end{aligned}$$

The PDFs of G under the zero and one hypothesis are $\sim N(-\varepsilon(1-\rho), N_0\varepsilon(1-\rho)$, and $\sim N(\varepsilon(1-\rho), N_0\varepsilon(1-\rho))$, respectively. The threshold is given by 0, the halfway mark between the means where either H_0 or H_1 is free. The probability of choosing H_1 while H_0 is true (i.e., P_{FA} in the radar and sonar terminology) is given by

$$\begin{aligned} P_{FA} &= \int_0^{\infty} f_0(g)dg \\ &= \int_0^{\infty} \frac{e^{-(g+\varepsilon(1-\rho))^2/(2N_0\varepsilon(1-\rho))}}{\sqrt{2\pi N_0\varepsilon(1-\rho)}} dg \\ &= Q\left(\frac{\varepsilon(1-\rho)}{\sqrt{N_0\varepsilon(1-\rho)}}\right) \\ &= Q\left(\sqrt{\frac{\varepsilon(1-\rho)}{N_0}}\right) && (6.29) \end{aligned}$$

The probability of error is

$$\begin{aligned} P_E &= \frac{1}{2}P_{FA} + \frac{1}{2}P_M \\ &= P_{FA} && (6.30) \end{aligned}$$

hence,

$$P_E = Q\left(\sqrt{(1-\rho)\frac{\varepsilon}{N_0}}\right) \qquad (6.31)$$

We notice that the probability of error depends only on ρ, ε, and N_0 and is therefore independent of the signal shape (i.e., only moments related to the second order are important).

We can now evaluate this expression for various values of ρ. Especially, the values 0, -1, and $+1$ are the most interesting ones. If $\rho = 0$, then

$$P_E = \int_0^{\infty} f_0(g)\, dg$$

$$= \int_0^\infty \frac{1}{(2\pi N_0\varepsilon)^{1/2}} e^{-(y+\varepsilon)^2/(2N_0\varepsilon)}\, dy$$
$$= Q\left(\left[\frac{\varepsilon}{N_0}\right]^{1/2}\right)$$

If $\rho = -1$, then

$$P_E = \int_0^\infty \frac{1}{\sqrt{2\pi 2\varepsilon N_0}} e^{-(g+\varepsilon 2)^2/(2N_0 2\varepsilon)} dg$$
$$= Q\left(\frac{2\varepsilon}{\sqrt{N_0 2\varepsilon}}\right) = Q\left(\sqrt{\frac{2\varepsilon}{N_0}}\right)$$

If we compare the complementary error function integrals for ρ equal to 0 and -1, we note that the latter expression has half the variance of the first one. In other words, the noise power has been halved or the effective average signal energy has doubled. This shows that if signals having a normalized cross-correlation coefficient of -1 are used, then this modulation scheme has a 3 dB advantage over the one that uses signals with a normalized cross-correlation coefficient of 0.

If we let $\rho = +1$, then $s_1(t)$ equals $s_0(t)$ and

$$P_E \equiv \int_0^\infty \frac{e^{-g^2/2}}{\sqrt{2\pi}} dg = \frac{1}{2}$$

This indicates that only half of the time we would make the correct decision. This constitutes very poor reception, but that should be obvious when we recognize that we would use the same symbol to represent two different messages.

Example 6.1 *Coherent Phase Shift Keying (CPSK). Under*

$$\begin{array}{lll} H_0 : & s_0(t) = A\sin(\omega_0 t) & \\ H_1 : & s_1(t) = -A\sin(\omega_0 t) & \end{array} \quad \text{for } 0 \le t \le T$$

and ω_0 is such that in T seconds an integer number of periods are generated. The two messages are two sinusoids 180 degrees out of phase. During transmission, white Gaussian noise with variance equal to $N_0/2$ is added to the signal. Hence,

$$\varepsilon = \frac{1}{2}\int_0^T \left(A^2 \sin^2 \omega_0 t + A^2 \sin^2 \omega_0 t\right) dt = \frac{A^2 T}{2}$$

and

$$\rho = \frac{1}{\varepsilon}\int_0^T -A^2 \sin^2 \omega_0 t dt$$
$$= -1$$

The probability of error is given by

$$\begin{aligned} P_E &\equiv \int_0^\infty \frac{e^{-(y+2\varepsilon)^2/(2N_0 2\varepsilon)}}{\sqrt{2\pi 2\varepsilon N_0}} dy \\ &= Q\left(\sqrt{\frac{2\ \varepsilon}{N_0}}\right) \end{aligned}$$

This is the optimum system since the normalized cross-correlation coefficient equals -1. *Typical error probabilities are shown in Figure 6.6.*

Example 6.2 *Coherent Frequency Shift Keying (CFSK). The two messages are transmitted at two different frequencies.*

$$\begin{array}{ll} H_0 : & s_0(t) = A\sin(\omega_0 t) \\ H_1 : & s_1(t) = A\sin(\omega_1 t) \end{array} \quad \text{for } 0 \le t \le T$$

We assume that ω_0 *and* ω_1 *are orthogonal, that is, over* T *seconds of observation the projection of one of the sinusoids onto the other is zero (i.e.,* $\rho = 0$*). The communication channel adds white Gaussian noise with variance equal to* $N_0/2$. *Since both signals have the same amplitude (i.e., amplitude* $= A$*), they have the same energy* ($A^2T/2$). *Hence, the average energy*

$$\begin{aligned} \epsilon &= \frac{1}{2}\int_0^T \left(s_1^2(t) + s_0^2(t)\right) dt \\ &= \frac{1}{2}\int_0^T \left(A^2 \sin^2 \omega_0 t + A^2 \sin^2 \omega_1 t\right) dt = \frac{A^2 T}{2} \end{aligned}$$

is the same as in Example 6.1. Using (6.29) the probability of error is given by

$$P_E = Q\left(\sqrt{\frac{\varepsilon}{N_0}}\right)$$

This performance is 3 dB worse than the one for CPSK. A typical performance curve is shown in Figure 6.6.

Example 6.3 *On-Off Keying (OOK, or amplitude shift keying (ASK)).*

$$\begin{array}{ll} H_0 : & s_0(t) = 0 \\ H_1 : & s_1(t) = B\cos(\omega_1 t) \end{array} \quad \text{for } 0 \le t \le T$$

The normalized cross-correlation coefficient is, by inspection, zero ($\rho = 0$). *The communication channel adds white Gaussian noise with variance equal to* $N_0/2$. *Only one correlator is needed to account for the signal* $s_1(t)$. *By definition, since there is no energy used for signal* $s_0(t)$, *the average energy*

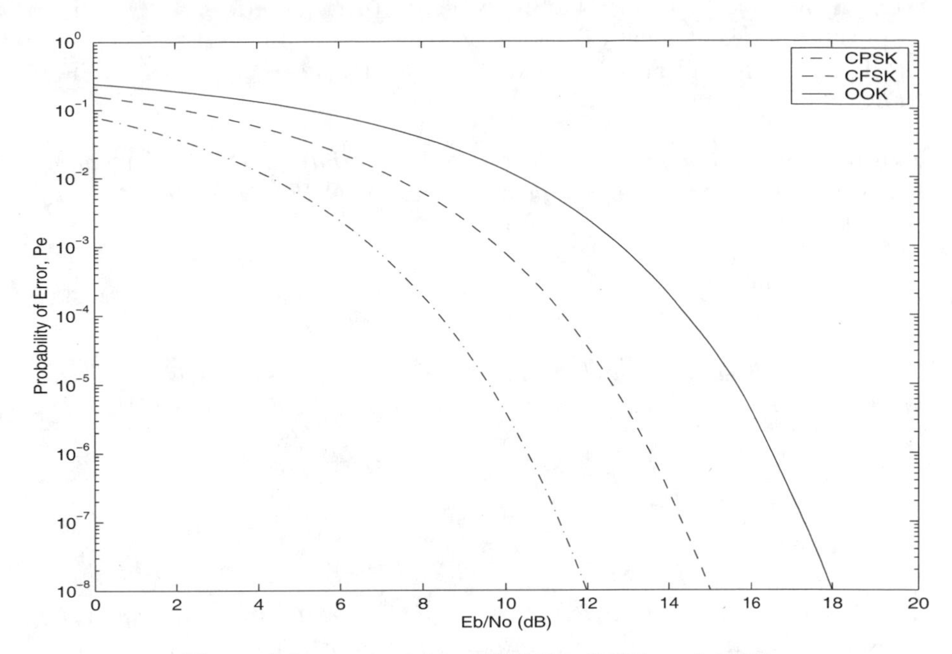

Figure 6.6: Detectability of CPSK, CFSK, and OOK.

is half of the energy of signal $s_1(t)$. Sometimes we refer to this type of modulation as amplitude shift keying (ASK). The energy of signal $s_1(t)$ is given by

$$\varepsilon_1 = \frac{1}{2}\int_0^T B^2 \cos^2 \omega_1 t dt = \frac{B^2 T}{4}$$

We denote this average energy by ε_1, where ε_1 represents half of the energy in signal $s_1(t)$. The probability of error becomes

$$P_E \equiv Q\left(\sqrt{\frac{\varepsilon_1}{N_0}}\right)$$

If the average energy ε_1 can be made numerically equal to the average energy ε transmitted in the continuous frequence shift keying (CFSK) scheme (i.e., let $B = \sqrt{2}A$), then OOK will have the same performance as CFSK. This of course would require a higher modulation amplitude (by 41.4%) which may be a serious consideration in some applications. A larger peak power, higher voltages, etc., would be needed. If the amplitude of signal $s_1(t)$ equals the amplitude of the CFSK (i.e., $B = A$), then there is a 3 dB loss relative to CFSK performance. Typical performance results are shown in Figure 6.6.

In summary, these three examples show the performance of three very typical digital modulation schemes. If the signal amplitudes are identical then CPSK will outperform CFSK, which in turn outperforms OOK. If the amplitudes are kept identical, the processing loss is 3 dB when going from CPSK to CFSK and an additional loss of 3 dB when going from CFSK to OOK.

Example 6.4 *Radar/Sonar Reception. We can interpret the radar and active sonar detection scheme as a realization of OOK. The received data is given by*

$$\begin{array}{lrcl} H_0 : & y(t) & = & n(t) \\ H_1 : & y(t) & = & A\cos(\omega_1 t) + n(t) \end{array} \qquad \text{for } 0 \le t \le T$$

The channel (medium and electronics) adds white Gaussian noise with variance equal to $N_0/2$. *The optimal detection statistic, modifying (6.14), is given by*

$$L = \int_0^T y(t)\ s(t)dt \ \underset{H_0}{\overset{H_1}{\gtrless}}\ \eta$$

where $s(t)$ *is the signal under the* H_1 *hypothesis and* η *is the appropriate threshold. We assume at this time (known signal) that we have an accurate description for the signal (i.e., we disregard Doppler, dispersion, attenuation, etc.). For the engineer studying detection theory, this particular example is a very important one that helps to form an understanding of the core ideas. The majority of details are worked out earlier in this section and will be used to minimize the work.*

The mean of L under the zero assumption is zero. The mean of L under the one assumption is $\varepsilon_1 = (A^2T)/2$. *The likelihood functions are given by*

$$f_0(\ell) \sim N(0, \sigma_1^2)$$

and

$$f_1(\ell) \sim N(\varepsilon_1, \sigma_1^2)$$

The variance (under the "zero" as well as under the "one" assumption) is $\sigma_1^2 = (N_0\varepsilon_1)/2$. *The probability of false alarm* P_{FA} *is given by*

$$\begin{aligned} P_{FA} &\equiv \int_\eta^\infty f_0(\ell)\, d\ell = \int_\eta^\infty \frac{e^{-\ell^2/(N_0\varepsilon_1)}}{\sqrt{(2\pi N_0 \varepsilon_1/2)}} d\ell \\ &= Q\left(\eta\sqrt{\frac{2}{\varepsilon_1 N_0}}\right) \end{aligned}$$

In the radar/sonar problem we use the Neyman-Pearson criterion which fixes the constant false alarm rate (CFAR). The actual threshold η *is obtained by*

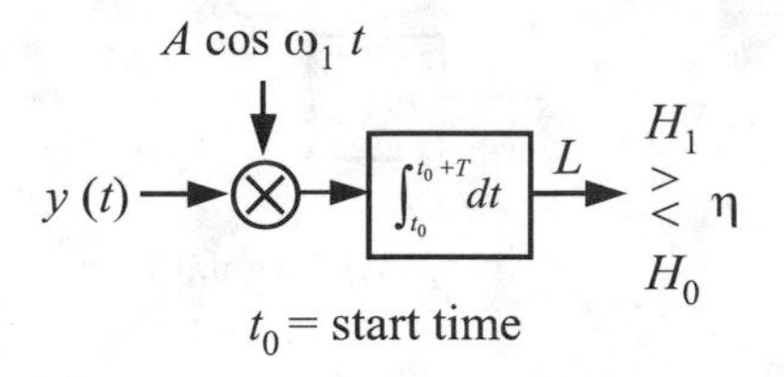

Figure 6.7: Radar/sonar detector.

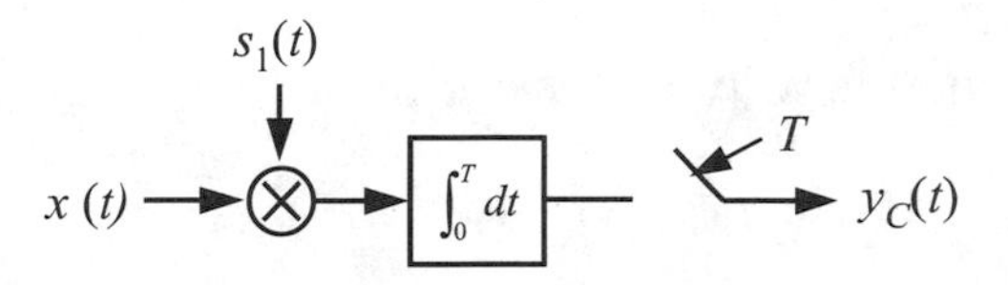

Figure 6.8: Correlator.

examining the test statistic (i.e., detector output) under the H_0 hypothesis and computing the threshold η from the statistic and the fixed alarm rate value. The probability of detection is given by

$$
\begin{aligned}
P_D &\equiv \int_{\eta}^{\infty} \frac{e^{-(\ell-\varepsilon_1)^2/(2\sigma_1^2)}}{\sqrt{2\pi\sigma_1^2}} d\ell \\
&= Q\left((\eta - \varepsilon_1)\sqrt{\frac{2}{\varepsilon_1 N_0}}\right)
\end{aligned}
$$

The final detector and its threshold, for an arbitrary start time t_0, are given in Figure 6.7. Typically, the false alarm rate is kept at a very small level, in the order of 10^{-12} *to* 10^{-6}.

6.3 MATCHED FILTERS

In Sections 6.3 and 6.4, to make the notation simple, we re-label the input as $x(t)$. This follows the general systems notation, labeling the inputs as $x(t)$ and the output as $y(t)$. If we re-examine the detector of the last section (the correlator), we realize that the multiplication by the replica and the follow-on integration can be replaced by a standard convolutional filter. The correlator is reproduced in Figure 6.8 while the convolutional filter is shown in Figure 6.9.

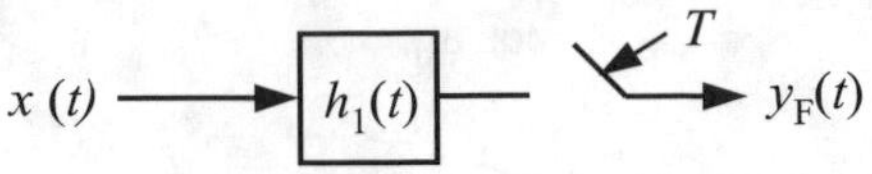

$x(t) \longrightarrow \boxed{h_1(t)} \longrightarrow y_F(t)$ (sampled at T)

Figure 6.9: Convolutional filter.

The correlator output, for an arbitrary input $x(t)$, at time T is given by

$$y_C(T) = \int_0^T x(t)\ s_1(t)\ dt \tag{6.32}$$

while the filter output at time T is described by

$$y_F(T) = \int_0^T x(T-\tau)\ h_1(\tau)\ d\tau \tag{6.33}$$

We note that the filter output is written as being produced by a causal input operating on a causal system. Suppose $h_1(t)$ is related to $s_1(t)$ by

$$h_1(t) = s_1(T-t)\ ; \quad \text{for } 0 \le t \le T \tag{6.34}$$

where $h_1(t)$ is a time reversed (i.e., a time mirrored) version of $s_1(t)$. Then the output of the filter is given by

$$y_F(T) = \int_0^T x(T-\tau)\ s_1(T-\tau)d\tau \tag{6.35}$$

With the change in variables $T - \tau = \sigma$ and $-d\tau = d\sigma$, the output becomes

$$\begin{aligned} y_F(T) &= \int_T^0 x(\sigma)\ s_1(\sigma)\ (-1)\ d\sigma \\ &= \int_0^T x(\sigma)\ s_1(\sigma)\ d\sigma \end{aligned} \tag{6.36}$$

We note this equals the output of the correlator if it is evaluated at time T. Since the response of the filter is matched to the signal to be detected, this technique is called the "matched filter."

For binary hypothesis testing, we have two channels as shown in Figure 6.10. Note the concept easily extends to the many (M-ary) hypothesis by allowing a matched filter for each signal to be detected. A typical set up is shown in Figure 6.11. This parallel filter arrangement is also known as a matched filter bank. We can also express the response of the matched filter in the frequency domain. Let

$$S(j\omega) = \int_0^T s(t)\ e^{-j\omega t}\ dt \tag{6.37}$$

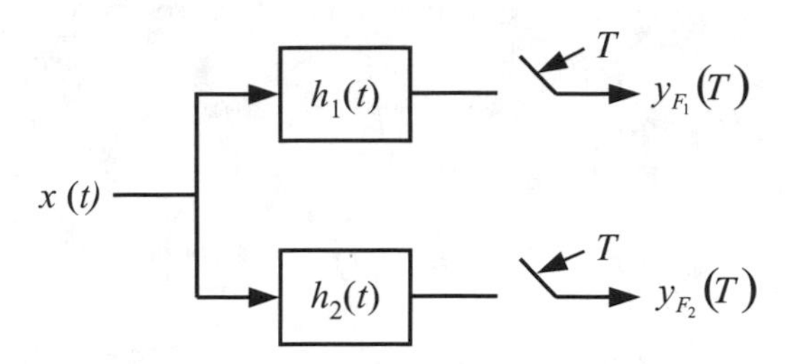

Figure 6.10: Two channel detection.

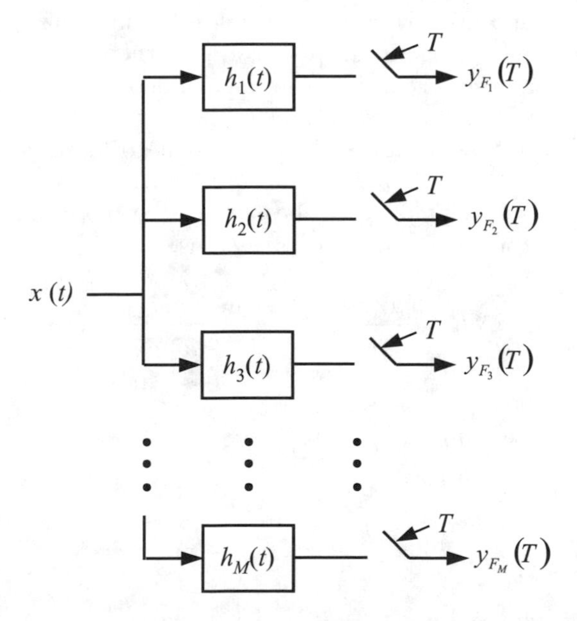

Figure 6.11: Matched filter bank.

and

$$H(j\omega) = \int_0^T h(t)\, e^{-j\omega t}\, dt \tag{6.38}$$

Now, $h_{\text{opt}}(t) = s(T - t)$ and assuming that $s(t)$ is real valued, hence

$$\begin{aligned} H_{\text{opt}}(j\omega) &= \int_0^T s(T-t)\, e^{-j\omega t}\, dt \\ &= \left(\int_0^T s(t)\, e^{-j\omega t}\, dt \right)^* e^{-j\omega T} \end{aligned}$$

or

$$H_{\text{opt}}(j\omega) = e^{-j\omega T} S^*(j\omega) \tag{6.39}$$

6.4 MATCHED FILTER APPROACH (maximizing the output SNR)

When a known signal might be embedded in white Gaussian noise, the matched filter can be derived using a criterion where one maximizes the output SNR. This obtains an optimal receiver structure by maximizing the output SNR, rather than minimizing the Bayes' cost. For the Gaussian noise case, the optimal detector (receiver) will be the same, independent of the criterion used. A brief description follows. Suppose there is a deterministic signal embedded in white Gaussian noise, then the input to filter can be written as $x(t) = s(t) + n(t)$. The output of the filter can be described as $y(t) = y_s(t) + y_n(t)$. The subscript denotes the output due to the signal and the noise input. The output SNR can be expressed as

$$SNR_0 \hat{=} \frac{[y_s(T)]^2}{\text{var}_{y_N(T)}} = \frac{\text{signal energy}}{\text{noise power}} \tag{6.40}$$

where

$$y_s(T) = \int_0^T h(\tau)\, s(T-\tau)\, d\tau$$

and

$$y_N(T) = \int_0^T h(\tau)\, n(T-\tau)\, d\tau$$

The noise correlation function is given by $E\{n(t_1)n(t_2)\} = (N_0/2)\delta(t_1 - t_2)$. The noise power at time T is obtained by first computing

$$[y_N(T)]^2 = \int_0^T \int_0^T h(\tau_1)\, h(\tau_2)\, n(T-\tau_1)\, n(T-\tau_2)\, d\tau_1\, d\tau_2 \tag{6.41}$$

and then applying the expectation operator

$$\begin{aligned} E\left[y_N(T)\right]^2 &= \int_0^T \int_0^T h(\tau_1)\, h(\tau_2)\, \frac{N_0}{2}\, \delta(\tau_2 - \tau_1)\, d\tau_2\, d\tau_1 \\ &= \frac{N_0}{2} \int_0^T h^2(\tau)\, d\tau \end{aligned} \tag{6.42}$$

The second moment is numerically equal to the variance when no DC component is present. In our case then, the last expression represents the output noise variance at time T. The output SNR becomes

$$SNR_0 \hat{=} \frac{2}{N_0} \frac{\left[\int_0^T h(\tau)\, s(T-\tau)\, d\tau\right]^2}{\int_0^T h^2(\tau)\, d\tau}$$

Using the Schwarz inequality, the last expression becomes

$$SNR_0 \hat{=} \frac{2}{N_0} \frac{\left[\int_0^T h(\tau)\, s(T-\tau)\, d\tau\right]^2}{\int_0^T h^2(\tau)\, d\tau}] \leq \frac{2}{N_0} \int_0^T s^2(T-\tau)\, d\tau \tag{6.43}$$

The equality only occurs when the impulse response $h(t)$ is proportional to the time mirrored signal $s(T-t)$. Under these conditions, the theoretical maximum of the output SNR is achieved, that is

$$SNR_0 = \frac{2}{N_0} \int_0^T s^2(T-\tau) d\tau \tag{6.44}$$

The optimum detector, using the criterion that maximize the output SNR, is called the matched filter for the obvious reason that the impulse response and the time signal are related (ie., matched). Note we obtain for this scenario (known signal in white Gaussian noise) the same solution as if we minimized a cost function as defined in Chapter 4.

6.5 M-ARY COMMUNICATION SYSTEMS

Suppose there are M possible signals which are orthogonal over the signal interval and that they have equal energy. Under

$$\begin{aligned} H_1 &: \quad y(t) = s_1(t) + n(t) \\ H_2 &: \quad y(t) = s_2(t) + n(t) \\ &\vdots \qquad\qquad \vdots \\ H_M &: \quad y(t) = s_M(t) + n(t) \end{aligned}$$

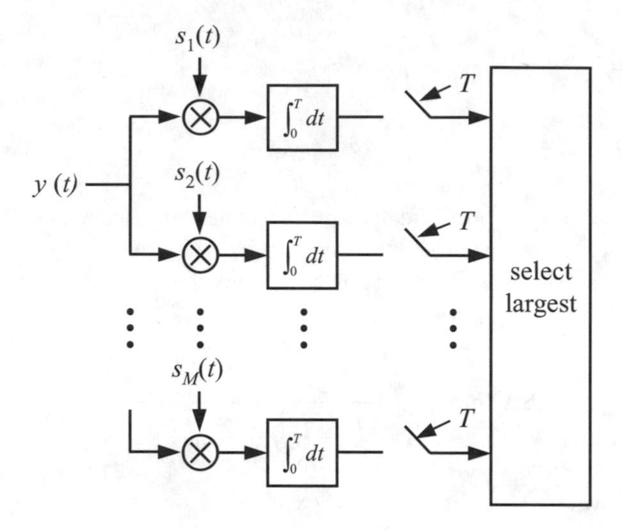

Figure 6.12: Correlator bank (M-ary detection).

Since the signals are orthogonal and have equal energy, we have

$$\rho_{ij} = \int_0^T s_i(t)s_j(t)dt = \begin{cases} E\,; & i = j \\ 0\,; & i \neq j \end{cases}$$

where the detector is of the form

$$G_i = \int_0^T y(t)s_i(t)dt$$

The detector consists of M parallel correlators (see Figure 6.12), which are compared at time T to choose the dominant output. This equates to choosing the proper hypothesis (i.e., signal). Figure 6.13 shows the equivalent implementation using matched filters.

6.6 DETECTION OF SIGNALS WITH RANDOM PARAMETERS

In this section, we will allow some uncertainty in some of the parameters of the signals that are to be detected. This is a continuation of the material introduced as the nuisance topic in Chapter 4, Section 4.9. Again, by using Bayes' rule of conditional densities and the application of marginal densities, we are able to average out the random fluctuations of the parameter(s) if the situation warrants it. To simplify the presentation, we will only address and

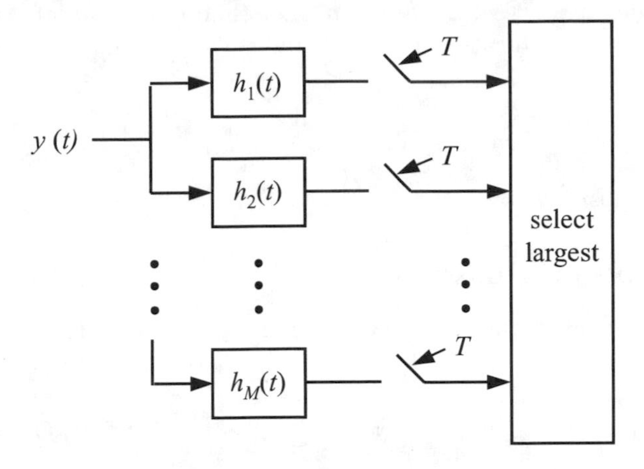

Figure 6.13: Matched filter bank (M-ary detection).

illustrate the simple uncertainties that can occur in the reception of a signal. In particular, the random phase, random amplitude, random frequency, and random arrival time will be addressed. But, this suffices to illustrate the idea behind the removal of the nuisance parameters. Since we work mainly with analog (continuous time) functions, we need to address the continuous time version of the likelihood function first.

6.6.1 Likelihood Functions

In Section 6.2, the continuous time correlator was derived. This was accomplished by performing a Gedanken (thought) experiment. It was assumed that the bandlimited noise is white, having a sinc-type correlation function. The sinc function was examined for the time lag of its first zero crossing which, if the process is sampled at multiples of this time increment, results in uncorrelated, hence independent noise samples. A limiting argument was then applied to obtain the continuous version of the likelihood ratio. The discrete form of the correlator was derived since it paves the way for decompositions in which orthogonal representations, using a variety of series expansions, will be used (see Chapter 7). The likelihood function development is introduced here since we need the likelihood function in its analog (i.e., continuous) form in this section.

For the ith hypothesis we have the discrete version of the likelihood function

$$f_i(\mathbf{y}) = \left(\frac{\Delta t}{\pi N_0}\right)^{m/2} e^{-(1/N_0)\sum_{k=0}^{m}(y_k - s_{ik})^2 \Delta t} \tag{6.45}$$

where $\mathbf{y}$ denotes a vector of size m. Let time $m \to \infty$ and $\Delta t \to 0$ such that $m\Delta t = T$, the duration of the signal. The continuous version of the likelihood function becomes

$$f_i(y) = K\, e^{-(1/N_0)\int_0^T (y(t)-s_i(t))^2 dt} \tag{6.46}$$

where K is a positive constant.

6.6.2 Random Phase

To examine the technique on signals having a phase uncertainty, we assume that under

$$\begin{array}{lll} H_0 : & s_0(t) = 0 & \\ H_1 : & s_1(t) = A\sin(\omega_c t + \theta) & \text{for } 0 \le t \le T \end{array}$$

The PDF for θ is given by $f_\Theta(\theta) = 1/(2\pi)$, for $-\pi < \theta \le \pi$. The noise is additive white Gaussian noise with spectral density $N_0/2$. The received signal is given by

$$\begin{array}{ll} H_0 : & y(t) = n(t) \\ H_1 : & y(t) = A\sin(\omega_c t + \theta) + n(t) \end{array}$$

As shown in the composite hypothesis testing section, we use

$$\Lambda(y) = \frac{\displaystyle\int f_1(y|\theta) f_1(\theta) d\theta}{f_0(y)} \underset{H_0}{\overset{H_1}{\gtrless}} \lambda_0 \tag{6.47}$$

Now, using (6.46) for the zero hypothesis, we obtain

$$f_0(y) = K e^{-(1/N_0)\int_0^T [y(t)]^2 dt}$$

while the one hypothesis leads to

$$f_1(y|\theta) = K e^{-(1/N_0)\int_0^T [y(t) - A\sin(\omega_c t+\theta)]^2 dt}$$

which becomes

$$f_1(y|\theta) = K e^{-(1/N_0)\int_0^T \left[y^2(t) \;-\; 2A\, y(t)\, \sin(\omega_c t+\theta) \; + A^2 \sin^2(\omega_c t+\theta)\right]\, dt}$$

Expanding the second quadratic term of the exponential in the last expression leads to

$$\int_0^T A^2 \sin^2(\omega_c t + \theta) dt = \int_0^T \frac{A^2}{2} dt - \int_0^T \frac{A^2}{2} \cos(2\omega_c t + 2\theta)\, dt$$

If the integration time $T \gg 2\pi/\omega_c$ is much longer than the period of one cycle of the sinusoid, then the second term in this equation can be neglected and the integral is approximately equal to $A^2T/2$. Of course, if the period of the sinusoid is such that in T seconds an integer number of half cycles are described, then the second term is automatically equal to zero. Hence, the LRT becomes

$$\Lambda(y) = \frac{e^{-(A^2T)/(2N_0)} \int_0^{2\pi} e^{-(1/N_0)\int_0^T \left(y^2(t) - 2y(t)A\sin(\omega_c t + \theta)\right)dt} d\theta/(2\pi)}{e^{-(1/N_0)\int_0^T y^2(t)\ dt}}$$

This can be further reduced using the trigonometric identity $\sin(\omega_c t + \theta) = \sin(\omega_c t)\cos(\theta) + \cos(\omega_c t)\sin(\theta)$ to

$$\Lambda(y) = e^{-A^2T/(2N_0)} \int_0^{2\pi} e^{(2Aq/N_0)\cos(\theta - \theta_0)} \frac{d\theta}{2\pi} \tag{6.48}$$

Here a simplification of the form

$$\begin{aligned} \int_0^T y(t)\sin(\omega_c t)dt &= q\cos(\theta_0) \\ \int_0^T y(t)\cos(\omega_c t)dt &= q\sin(\theta_0) \end{aligned}$$

and the trigonometric identity $\cos(\theta_0)cos(\theta) + \sin(\theta_0)\sin(\theta) = \cos(\theta_0 - \theta)$ were used. Now, the following integral will simplify the expression for the LRT

$$\frac{1}{2\pi}\int_0^{2\pi} e^{x\cos(\theta+\alpha)} d\theta = I_0(x)$$

where α can be any real number and $I_0(\ )$ is the modified zero-order Bessel function of the first kind. Note, the parameters can also be translated into the lower and upper limits of the integral (i.e., change of variables). What matters is that the support of the integral is exactly over 2π. The LRT becomes

$$\Lambda(y) = e^{-\ (A^2T)/(2N_0)}\ I_0\left(\frac{2Aq}{N_0}\right) \underset{H_0}{\overset{H_1}{\gtrless}} \lambda_0 \tag{6.49}$$

This can be further simplified

$$I_0\left(\frac{2Aq}{N_0}\right) \underset{H_0}{\overset{H_1}{\gtrless}} \lambda_0\ e^{A^2T/(2N_0)} \tag{6.50}$$

The modified Bessel function $I_0(2Aq/N_0)$ is a monotonically increasing function of q (see Appendix D). Therefore, either q itself or q^2 can be used as a decision rule. The detection rule becomes

$$q \underset{H_0}{\overset{H_1}{\gtrless}} \eta_0 \tag{6.51}$$

or

$$q^2 \underset{H_0}{\overset{H_1}{\gtrless}} \eta_0^2 = \eta_1 \tag{6.52}$$

For a specified λ_0, the threshold η_0 for the detection rule in (6.51) is obtained by solving

$$\lambda_0 = e^{-\,(A^2T)/(2N_0)} I_0\left(2\frac{A\eta_0}{N_0}\right) \tag{6.53}$$

The correlator and matched filter based detectors are shown in Figures 6.14 and 6.15, respectively. One can use q or equivalently, q^2 as the detection statistic. The quantity q^2 is proportional to the power spectral density at frequency ω_c. A little thought will reveal that the implementation in Figure 6.14 corresponds to a power spectrum analyzer. Based on the discussions in Chapters 2 and 3, we realize that under the noise only condition, the quantity q^2 will have an exponential PDF (i.e., chi-squared with two degrees of freedom). If the signal is present (i.e., H_1 is true), the random variable q^2 will have a non-central chi-squared PDF. We notice that the output of the correlators of Figure 6.14 correspond to a finite time Fourier transform of $y(t)$, where the top leg and bottom legs correspond to the imaginary and real part of the transform, respectively.

The quantity q corresponds to the envelope. When the hypothesis H_0 is true, q will have a Rayleigh PDF. Conversely, if the hypothesis H_1 is true, q will be a Rician variate.

6.6.3 Random Amplitude

The same procedure as for the random phase case is followed. That is, an LRT of the form

$$\Lambda(y) = \frac{\int_A f_1(y|a)\, f_1(a)\, da}{f_0(y)} \underset{H_0}{\overset{H_1}{\gtrless}} \lambda_0 \tag{6.54}$$

is formed. From physical insight or a worst-case scenario, the probability density of the random variable a can be estimated if the PDF is not available.

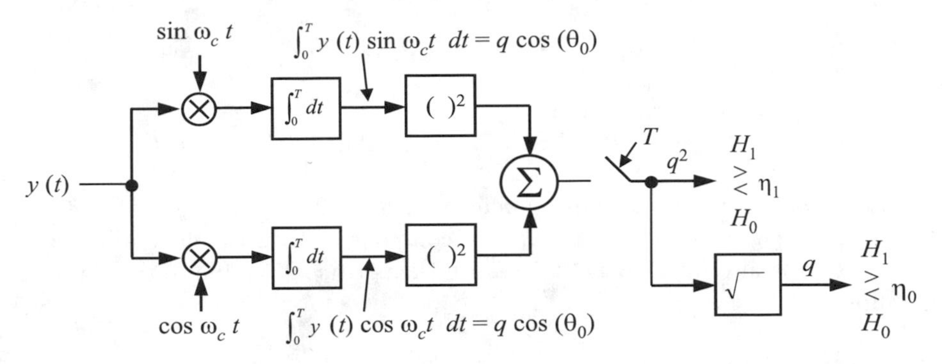

Figure 6.14: Correlator implementation.

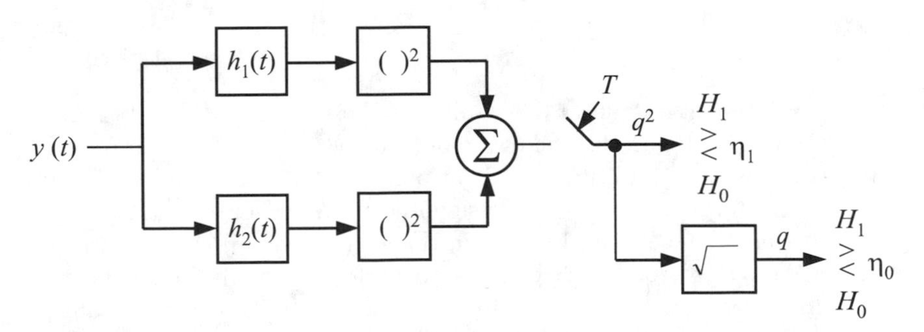

Figure 6.15: Matched filter implementation.

6.6.4 Random Amplitude and Phase

The LRT is of the form

$$\Lambda(y) = \frac{\int_{\boldsymbol{\Theta}} f_1(y|\boldsymbol{\theta})\ f_1(\boldsymbol{\theta})\ d\,\boldsymbol{\theta}}{f_0(y)} \overset{H_1}{\underset{H_0}{\gtrless}} \lambda_0 \tag{6.55}$$

where

$$\int_{\boldsymbol{\Theta}} f_1(y|\boldsymbol{\theta})\ f_1(\boldsymbol{\theta})\ d\,\boldsymbol{\theta} = \int_A \int_{\Theta} f_1(y|a,\theta)\ f_1(a,\theta)\ da\, d\theta$$

If a and θ are statistically independent, then $f_1(a,\theta) = f_1(a)\ f_1(\theta)$. Note, if the zero hypothesis contains a random component, then the denominator of the LRT will also be averaged to remove the perturbation due to the random component.

6.6.5 Random Frequency

Again the LRT is set up to allow the averaging out of the random component. The LRT is of the form

$$\Lambda(y) = \frac{\int_{\Omega} f_1(y|\omega)\ f_1(\omega)\ d\omega}{f_0(y)} \overset{H_1}{\underset{H_0}{\gtrless}} \lambda_0 \tag{6.56}$$

The PDF $f_1(\omega)$ can be approximated as

$$f_1(\omega) = \sum_{i=1}^{M} \Pr(\omega_i)\ \delta(\omega - \omega_i) \tag{6.57}$$

The spacing between ω_i and ω_{i+1} is suitably chosen (i.e., M evenly spaced spectral locations). Figure 6.16 shows a detector (matched filter) implementation for a constant amplitude, uniform phase distribution, and a uniform frequency distribution. The block designated by $h_i(t)$ for $i = 1, 2, \cdots, M$, and the envelope detector produces q_i in the same way as illustrated in Figure 6.15. We realize that each channel to perform the envelope detection properly must have an $h_I(t)$ and a corresponding $h_Q(t)$ (I and Q channel if the phase is unknown).

6.6.6 Random Arrival Time

The basic detector (i.e., correlator or matched filter) can be used to examine adjacent contiguous or overlapping data segments in time. If one looks for

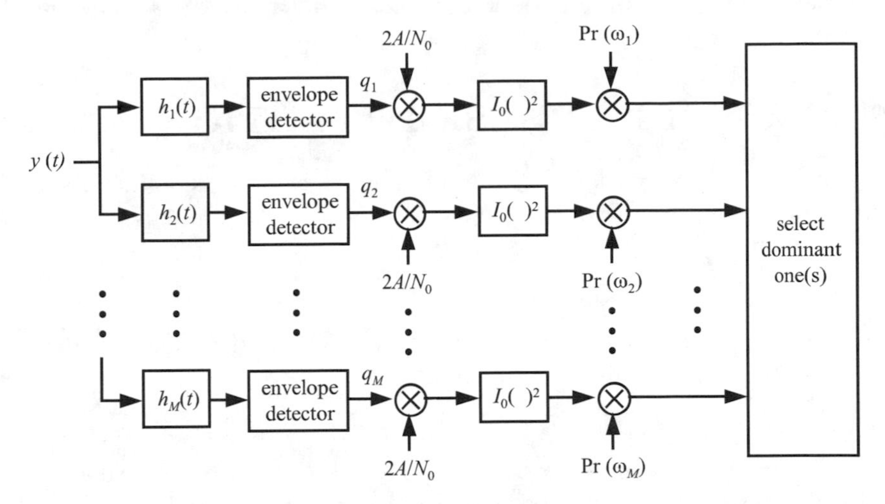

Figure 6.16: Matched filter (known amplitude, random uniform phase and frequency) embedded in additive white Gaussian noise.

a signal of length T, while the segments are contiguous and of size T, a mismatch of 3 dB relative to a perfect line up is possible in the worst-case scenario. This is because half of the energy would fall into each one of two contiguous segments.

A simple method to reduce the loss is to use overlap processing, that is, select data segments of duration T that overlap. At 50% overlap, also called 2:1 overlap, the worst-case processing loss is limited to 1.25 dB (i.e., the worst-loss is 1/4 of the energy). If one goes to a 75% overlap, also called a 4:1 overlap, the worst-case processing loss is limited to 0.58 dB (i.e., the worst loss is 1/8 of the energy). In the extreme case, one can go to an arbitrarily dense overlap, which is what wavelet processing (see Chapter 3) achieves at the highest spectral band prior to resampling. However, in conventional processing (i.e., correlator or matched filters), the processing load may become too large while not providing a measurable gain in performance.

6.6.7 Summary

It is obvious that the optimum receiver (in the Bayes' sense or by maximizing the output SNR) for narrowband signals (tonals, sinusoids) with unknown frequency, unknown phase, and unknown arrival time is the power spectral density. Usually, this is implemented using the periodogram, which in turn is the scaled magnitude squared output of the FFT. As we recall from the discussions in Chapter 3, the FFT is an extremely efficient processing tool.

Clever choice of the window function, number of padded on zeros, and overlap factor will provide a good match in frequency and time [3,4].

6.7 MULTIPLE PULSE DETECTION

When multiple pulses are available, two processing techniques lend themselves for use when trying to use the information from more than one pulse. The two choices are

(a) Incoherent processing or integration, also called power averaging.

This usually reduces the variance of the output statistic and enhances the processing gain by combining the projections.

(b) Coherent processing or integration.

This usually takes the inner product obtained by performing the correlation or matched filtering and provides the processing gain by extending the effective integration time.

6.7.1 Incoherent Averaging

We recall that the exact expression for a single known pulse is given by (6.50). The Bessel function $I_0(x)$ has expansions that are valid for small arguments (i.e., low SNRs) and for large arguments (i.e., large SNR). The approximation of the zero-order modified Bessel function for small arguments is given by

$$I_0(x) \approx 1 + \frac{x^2}{4}$$

hence,

$$I_0\left(2\frac{Aq}{N_0}\right) \approx 1 + \frac{A^2q^2}{N_0^2} \tag{6.58}$$

Taking the natural log of $(1+x)$, for small x, leads to

$$\ln(1+x) = x - \frac{x^2}{2!} + \frac{x^3}{3!} - \frac{x^4}{4!} \cdots$$

Hence, the natural log of the $I_0(2Aq/N_0)$ at small SNRs is approximated by

$$\ln I_0\left(2\frac{Aq}{N_0}\right) \approx \ln\left(1 + \frac{A^2q^2}{N_0^2}\right) \approx \frac{A^2q^2}{N_0^2} \tag{6.59}$$

Figure 6.17 shows the general (i.e., the exact) radar/sonar detector implementation for M pulses, while Figure 6.18 shows the receiver based on a low SNR approximation.

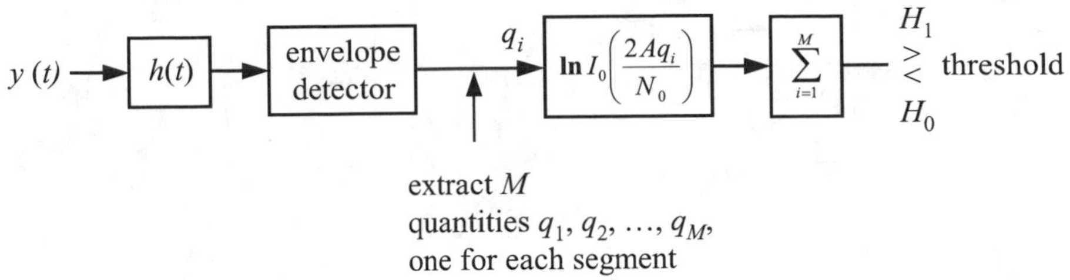

y (t)
h(t)
envelope detector
q_i
$\ln I_0\left(\frac{2Aq_i}{N_0}\right)$
$\sum_{i=1}^{M}$
H_1
$\gtrless$ threshold
H_0
extract M
quantities $q_1, q_2, \ldots, q_M$,
one for each segment

Figure 6.17: Radar/sonar M-pulse averaging (exact) scheme.

y (t)
h(t)
quadr. detector
q_i^2
$\sum_{i=1}^{M} q_i^2$
H_1
$\gtrless$ threshold
H_0
(M segments)
extract M quantities
$q_1^2, q_2^2, \ldots, q_M^2$,
one for each segment

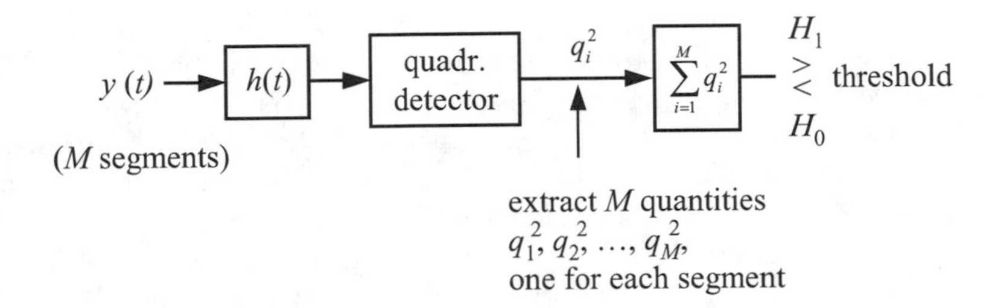

Figure 6.18: Radar/sonar M-pulse approximation at low SNR averaging scheme.

At a high SNR we can use a different approximation. The approximation of the zero-order modified Bessel function is given by

$$I_0(x) \approx \frac{e^x}{\sqrt{2\pi x}}$$

In our case, the approximation of the Bessel function (high SNR) becomes

$$I_0\left(2\frac{Aq}{N_0}\right) \approx \frac{e^{2(Aq/N_0)}}{\sqrt{4\pi\frac{Aq}{N_0}}} \tag{6.60}$$

The natural log becomes

$$\begin{aligned} \ln\left(I_0\left(2\frac{Aq}{N_0}\right)\right) &\approx 2\frac{Aq}{N_0} - \frac{1}{2}\ln\left(4\pi\frac{Aq}{N_0}\right) \\ &\approx 2\frac{Aq}{N_0} \end{aligned} \tag{6.61}$$

The last approximation takes advantage of the fact that the logarithm of a large positive number is much smaller than the positive number itself. An implementation using the high SNR approximation is shown in Figure 6.19.

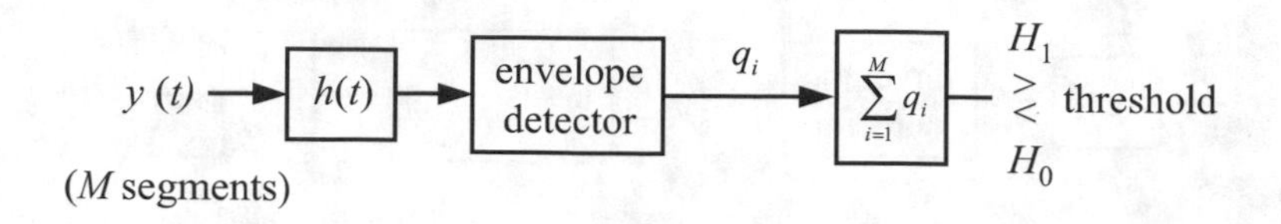

Figure 6.19: High SNR approximation.

The operation of summing M-successive q or q^2 outputs is also called post detection integration [8]. From an analysis point of view, square law detection (i.e., q^2) is preferred since the mathematical expressions are relatively simple. This is due to the fact that averaged chi-square probability densities are mathematically more tractable than averaged Rayleigh densities. See for example, the section on chi-squared densities in Chapter 2 or Appendix A. Chi-squared probability density functions, similar to the Gaussian density functions, are closed under addition. That is, when adding i.i.d. chi-squared random variables, the resultant is chi-squared. Of course the degrees of freedom will change according to the number of i.i.d. chi-squared random variables that are added. The performance of the linear sum of post detection outputs (i.e., sum of envelopes) and that of the sum of post detection outputs power wise (i.e., q^2) differ by less than 0.2 dB at a typical P_{FA} value of 10^{-6}, over a P_D range of 0.5 to 0.995, for values of M from 2 to 2^{14} [1,2].

This is why the performance of an envelope based system can be approximated using envelope squared quantities.

6.7.2 Coherent Versus Incoherent Integration (Averaging)

Sometimes the operation of coherent or incoherent averaging follows the basic detector implementation. By this we mean that the outputs of the detector (as a function of time) are averaged in a power or in a phase preserving sense. An example of a coherent integration scheme is shown in Figure 6.20. We note that a quantity, accessed and averaged, takes on other than just positive values. The quantity to be averaged can also be complex valued. Coherent processing (averaging or integration) can improve the processing gain by 3 dB per doubling of the number of terms involved (see also Chapter 9, Section 9.2).

Incoherent averaging asymptotically gains 1.5 dB per doubling (for a large number of terms in the average) [1,8,9]. A typical example for coherent averaging is the addition of sequential FFT outputs prior to magnitude squaring. If we magnitude squares first (spectral density estimate) and then add the

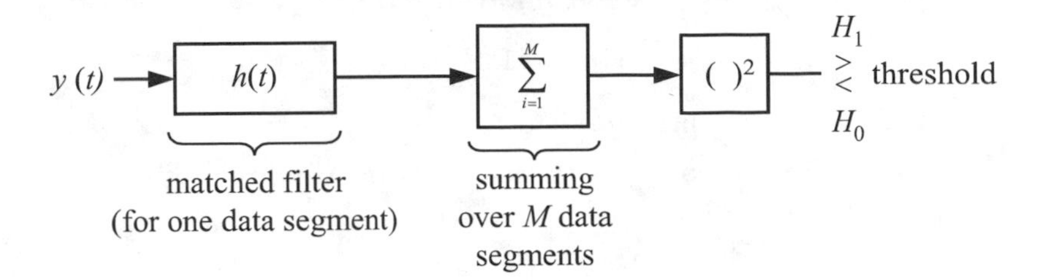

Figure 6.20: Coherent averaging example.

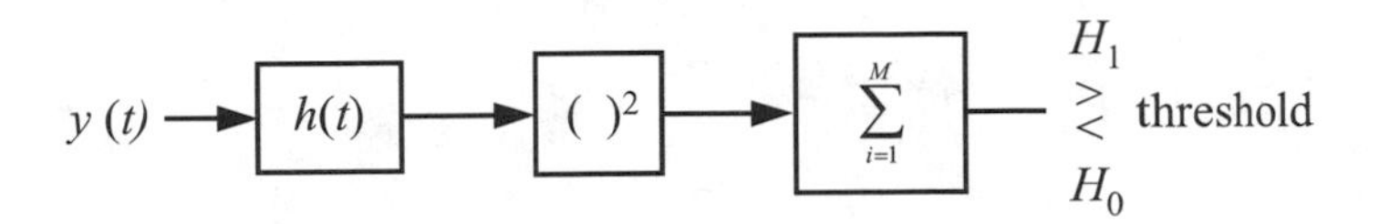

Figure 6.21: Incoherent averaging example.

components, we have implemented power (or incoherent) averaging. See Figure 6.21 for an example [3,4].

6.8 SUMMARY

Chapter 6 extends the basic detection technique, introduced in Chapter 4, to the more general case allowing the signal to be functions of time, rather than a constant. That is $s = s(t)$ or $s(n)$ in the continuous and discrete time case, respectively. With the exception of Section 6.6, the signals are assumed to be known and embedded in white Gaussian noise. Section 6.6 allows for randomness or uncertainties in the signal and/or noise parameters, but still assumes a Gaussian delta correlated (i.e., white) noise environment. Section 6.2 establishes the correlator structure of the optimal receiver (detector) and evaluates the performance. Sections 6.3 and 6.4 examine the matched filter minimizing the cost and maximizing the output SNR, respectively. Section 6.5 extends the binary to the M-ary detection case. In Section 6.7, multi-pulse processing is introduced. Coherent and incoherent averaging techniques are also addressed.

6.9 PROBLEMS

1. Under

$$H_0 : y_i = n_i \qquad i = 1, 2, \cdots, 10 \qquad , n_i \sim N(0, 1) \text{ i.i.d.}$$

$$\begin{aligned} H_1 \ : \ y_i \ &= \ 1 + n_i \quad \text{for} \quad i = 1, 2, \cdots, 5 \\ & \quad -1 + n_i \quad \text{for} \quad i = 6, 7, \cdots, 10 \end{aligned}$$

(a) Derive the optimum detector.

(b) Derive the optimum threshold for the conditions given (also state what design philosophy you used).

(c) Compute the probability $Pr(D_1|H_1)$.

(d) Compute the probability $Pr(D_1|H_0)$.

2. Under

$$\begin{aligned} H_0 \ : \ y_i \ &= \ n_i \ ; \qquad n_i \sim N(0, 1), \text{ i.i.d.} \\ H_1 \ : \ y_i \ &= \ s_i + n_i \ ; \qquad i = 1, 2, \cdots, 4 \end{aligned}$$

$s_i = (-1)^i$ for $i = 1, 2, 3, 4$,

$c_{01} = 2$; $c_{10} = 1$; $P_0 = 0.25$.

(a) Which criterion is to be used?

(b) Derive the optimum detector and threshold.

(c) What is $Pr(D_1|H_0)$? (obtain a numerical value)

3. Given the optimal detector for a pulsed sinusoid of known duration (unknown: frequency, phase, arrival time) in white Gaussian noise at low SNR as shown below:

$$r(n) \longrightarrow \boxed{\text{FFT}} \longrightarrow R(k) \longrightarrow \boxed{\text{magnitude squarer}} \longrightarrow |R(k)|^2$$

$0 \le n,\ k \le 511$, FFT-size $= N = 512$, no overlap, no window.

$$R(k) = \sum_{n=0}^{511} r(n) \exp\left(-j\frac{2\pi kn}{512}\right)$$

(a) How do you establish the threshold to guarantee an overall P_{FA}? Elaborate.

(b) Assume that the variance of $n(n)$ is unity, $P_{FA} = 10^{-6}$, 512 spectral bins, disregard any window losses, compute the appropriate threshold.

(c) For an 8:1 overlap (i.e., shift 1/8 of the transform length, say 32 points) assuming an unknown arrival time, what is the worst case processing loss due to misalignment in time?

(d) Computer Exercise:

Verify (b) using a MATLAB simulation (don't forget to explain your results).

4. Computer Exercise:

$$H_0 \;:\; y \sim N(0, \sigma^2) \qquad \text{(single sample)}$$
$$H_1 \;:\; y \sim N(5, \sigma^2) \qquad \text{(single sample)}$$

Using MATLAB:

(a) Derive the optimal detector. Show all work.

(b) Plot theoretical ROC curves for $\sigma^2 = 0.5, 1, 2, 4,$ and 8 (one plot with five curves).

(c) Pick the worst detection case of (b) above. Increase the number of data samples from 1 to 8 and plot the corresponding ROC curve superimposed on the corresponding worst-case ROC curve. Repeat the computation using $N = 64$ samples and superimpose the result on the two ROC curves. Of course, the noise will be statistically independent for all and any sample. This part of the exercise demonstrates the performance enhancement as we are allowed to see more and more samples in the detection process. Use sufficiently many computation points to obtain a fairly smooth set of curves. Give a brief explanation of your results (one plot with three ROC curves).

(d) Perform a numerical simulation. Use the theoretical plots from part (b) for $\sigma^2 = 0.5$ and for $\sigma^2 = 8$ and compare them to experimental results obtained from a simulation. The number of experiments should be sufficiently large to illustrate that the theoretical and experimental ROC curves (for both SNRs) agree.

Comment on your results.

5. Under

$$H_0 \;:\; y_i \;=\; -1 + n_i \;; \qquad i = 0, 1, \cdots, I-1$$
$$H_1 \;:\; y_i \;=\; +1 + n_i \;; \qquad i = 0, 1, \cdots, I-1$$

where n_i i.i.d. $\sim N(0, 1)$.

(a) Find the optimal detector and its threshold.

(b) Compute the P_{FA} when $I = 9$.

(c) If we fix the P_{FA} at 10^{-6} (i.e., Neyman-Pearson criterion): find the threshold.

(d) For the problem in part (c), compute the P_D.

6. Under

$$H_0 \; : \; r_i = s_{0i} + n_i \; ; \qquad i = 0, 1, 2, 3, 4, 5$$

$$H_1 \; : \; r_i = s_{1i} + n_i \; ; \qquad n_i \text{ i.i.d. } \sim N(0, 1)$$

$$\mathbf{s_0}^T = [1, 1, 1, -1, -1, -1] \qquad \mathbf{s_1}^T = [-1, -1, -1, 1, 1, 1]$$

$$Pr(H_0) = Pr(H_1) \qquad c_{00} = c_{11} = 0 \qquad c_{10} = c_{01} = 1$$

   (a) Derive the optimal detector by minimizing the probability of error.

   (b) What is the P_e, the probability of error?

7. The binary signal 0 and 1, represented by -1 and 1 volt, respectively, are transmitted over a symmetric channel. The prior probabilities are the same. There is zero cost for correct decisions and a given constant cost, C_c, for making a mistake. The channel adds zero mean, white Gaussian noise with a prescribed variance σ^2. So, we have:

$$H_0 \; : \; y \sim N(-1, \sigma^2) \qquad \text{(single or multiple samples)}$$

$$H_1 \; : \; y \sim N(1, \sigma^2) \qquad \text{(single or multiple samples)}$$

   (a) Design the appropriate detectors and provide for each one the appropriate detector form, threshold, and P_{FA} and P_D. *Note:* ROC curves display P_D versus P_{FA}, where P_D and P_{FA} are plotted on the vertical and horizontal axis, respectively.

      (i) Computer Exercise: Using one data sample, obtain and plot theoretical ROC curves for $\sigma^2 = 0.5, 1, 2, 4$, and 8. This part shows the theoretical degradation as a function of SNR (one plot with five theoretical ROC curves).

      (ii) Pick the worst detection scenario of (i) above, increase the number of data samples from 1 to 8 and plot the corresponding ROC curve superimposed on the respective worst-case detector curve (which is used on sample). Repeat the computation using $N = 64$ samples and superimpose on the other two curves. Of course, the noise will be statistically independent for all and any sample. This part of the exercise demonstrates the performance enhancement, as we are allowed to use more and more samples in the detection process. Use many computation points sufficiently to obtain a fairly smooth set of curves. Give a brief explanation of your results (one plot with three theoretical ROC curves).

      (iii) Perform a numerical simulation. Use the theoretical plots from part (i) for $\sigma^2 = 0.5$ and for $\sigma^2 = 8$ and compare to experimental results obtained from a simulation. The number of experiments in the simulation should be sufficiently large to illustrate that the theoretical and experimental ROC curves agree (two plots: one for each σ selected, each plot displaying a theoretical and one experimental ROC curve).

Comment on your results.

8. Under

$$H_0 \; : \; r(t) = n(t)$$

where $n(t)$ is a white Gaussian noise process with variance of 0.3.

$$H_1 \; : \; r(t) = s(t) + n(t)$$

$$s(t) = t, \quad \text{for } 0 \leq t \leq 1$$
$$\Pr(H_0) = \Pr(H_1)$$
$$C_{00} = C_{11}, \; C_{10} = C_{01} = 1$$

(a) Derive the optimal detector by minimizing the probability of error.

(b) What is the threshold?

(c) What is P_e, the probability of error?

(d) Give the expression and draw the impulse response of the appropriate matched filter.

9. Under

$$H_0 \; : \; r(t) = n(t); \; 0 \leq t \leq T$$
$$H_1 \; : \; r(t) = s(t) + n(t); \; 0 \leq t \leq T$$

where $n(t)$ is a white Gaussian noise process with bi-spectral height $N_0/2 = 1$, $\omega_0 = 2\pi/T$, $T = 1$ millisecond, and $s(t) = \sin \omega_0 \, t$.

(a) Design the optimal detector (receiver) for these conditions. Provide a diagram of the detector and the numerical value of the threshold.

(b) What is $Pr(D_1|H_0)$ and $Pr(D_1|H_1)$? Show all work in sufficient detail.

(c) Suppose instead of the required replica, we can only use a hard clipped version of the sinusoid (i.e., a square wave with the same period). What is the change in $Pr(D_1|H_1)$ if we adjust the threshold to provide a $Pr(D_1|H_0)$ as obtained in the ML approach in part (a) above?

(d) What is the loss (or gain) comparing the results in part (a) and (c)?

References

[1] Whalen, A., *Detection of Signals in Noise*, New York: Academic Press, 1971.

[2] Robertson, G.M., "Performance Degradation by Post Detector Nonlinearities," *Bell Syst. Tech. J.*, Vol. 47, No. 3, 1968, pp. 407–414.

[3] Kay, S.M., *Modern Spectral Estimation, Theory and Application*, Englewood Cliffs, NJ: Prentice-Hall, 1988.

[4] Marple, S.L., *Digital Spectral Analysis*, Englewood Cliffs, NJ: Prentice-Hall, 1987.

[5] Franks, L.E., *Signal Theory*, Englewood Cliffs, NJ: Prentice-Hall, 1969.

[6] Noble, B., and Daniel, J.W., *Applied Linear Algebra*, Englewood Cliffs, NJ: Prentice-Hall, 1988.

[7] Strang, G., *Linear Algebra and Its Applications*, 2nd ed., New York: Academic Press, 1980.

[8] Robertson, G.M., "Operating Characteristics for a Linear Detector of CW Signals in Narrow-band Gaussian Noise," *Bell Syst. Tech. J.*, Vol. 46, No. 4, 1967, pp. 755–774.

[9] Levanon, N., *Radar Principles*, New York: John Wiley & Sons, 1988.

[10] McDonough, R.N., and Whalen, A.D., *Detection of Signals in Noise*, 2nd ed., San Diego, CA: Academic Press, 1995.

Chapter 7

Detection of Signals in Colored Gaussian Noise

7.1 INTRODUCTION

The study of detection and estimation can be enhanced by using an approach based on a series representation. The general concept of series decomposition and series representation is addressed in Section 7.2. Section 7.3 examines the problem to detect known signals embedded in white Gaussian noise using a general complete ortho-normal basis set. The Gram-Schmidt procedure is introduced in Section 7.4, while in Section 7.5 the Gram-Schmidt technique is applied to detect a known signal embedded in white Gaussian noise. Section 7.6 deals with the series expansion in a colored noise background. In this context, integral equations, Mercer's theorem and the Karhunen-Loève expansion are introduced. In Section 7.7, the problem to detect a known signal embedded in colored Gaussian noise is addressed. The bilateral Laplace transform is introduced to allow the conversion of the Fredholm integral equation into a differential equation. A simple colored noise detection problem serves to illustrate the concepts. The performance of the correlator operating in a colored noise environment and the analog implementation of the whitening filter is briefly discussed. Section 7.8 examines approaches to discrete-time colored noise problems. We note that Chapter 7 deals almost exclusively with Gaussian type noise and we only deviate from the Gaussianity in the discrete-time matched filtering in Section 7.8.5.

7.2 SERIES REPRESENTATIONS

There are many ways to describe a given function, one particularly useful one is the series representation. A series representation (or expansion) refers to the use of a set of basis functions and a set of corresponding expansion coefficients to represent a particular function. The number of basis functions may be finite or countable. We are familiar with this type of representation since we often consider functional expressions by representing them as a weighted sum of basis functions. A very well known example, for periodic continuous time functions with periodicity T_0, is the synthesis equation (i.e., Fourier series representation)

$$x(t) = 1/T_0 \sum_{k=-\infty}^{\infty} X(k) \exp\left(j\omega_0 kt\right) \tag{7.1}$$

where $\omega_0 = 2\pi/T_0$ and

$$X(k) = \int_0^T x(t) \exp\left(-\, j\omega_0 kt\right) dt$$

serves as the analysis equation.

Another example is the discrete-time Fourier transform (DFT). For the DFT, we have

$$\begin{aligned} x(n) &= \frac{1}{N} \sum_{k=0}^{N-1} C_k \; e^{j\omega_0 kn} \\ &= \frac{1}{N} \sum_{k=0}^{N-1} X(k) \; e^{j(2\pi/N)kn} \end{aligned} \tag{7.2}$$

where $\omega_0 = 2\pi/N$ and

$$C_k = X(k) = \sum_{n=0}^{N-1} x(n) \; e^{-j(2\pi/N)kn}$$

is the kth Fourier coefficient. This representation is frequently used in the signal processing area, where we interpret the basis functions as a set of orthogonal basis functions (i.e., see Chapter 3, [12,19]). As discussed in Chapter 3, the DFT and FFT are the transformation that dominate the majority of signal processing applications. For the Fourier based representation, we use a set of oscillatory type basis functions where the basis functions are weighted and summed as given by Equation 7.2.

In general, an expansion (representation) in which the basis functions are orthogonal is chosen. This makes the expansion coefficients of data (i.e., signals, observation processes) that have a Gaussian type PDF, statistically independent. Hence, the algorithms (i.e., the simple likelihood ratio test) developed for independent samples apply directly, with high order joint PDFs becoming the product of one-dimensional PDFs. The expansions for finite duration data can be based on the signal set or on the noise space.

If we use an expansion based on the signal set, we can utilize one of two types of representations. Should we use an expansion based on the noise characteristics then we depend on an eigenvalue type decomposition.

(a) The basis function set spans the signal space:

Typical candidates are Fourier kernel [1,12] or any other orthogonal polynomial function [2] (i.e., Hermite, Laguerre, Chebbycheff, etc.). Usually a complete ortho-normal (C.O.N.) set is prefered for the analysis. For signal space dependent expansions, we can use an expansion along a preferred signal (coordinate) axis. We select as the first basis function one particular signal (i.e., the signal under the H_1 condition) and choose all other basis functions, in a sequential fashion, orthogonal to the first and all previously selected basis functions. All basis functions are normalized providing a set of ortho-normal basis functions (i.e., Gram-Schmidt procedure [3,13] and Chapter 7.4).

(b) The eigenfunctions (or eigenvectors) set spans the noise space:

We use the correlation function of the noise process to generate the basis functions (i.e., the Karhunen-Loève expansion). In the continuous time case (i.e., see Section 7.6), the eigenfunctions of the auto-correlation function of the noise serve as basis functions [4,5]. In the discrete-time case, we use the auto-correlation matrix of the noise and the corresponding eigenvectors [3].

Independent of which basis function set is chosen, it is essential (critical) that this set is a complete one. That is, no matter what the signal form is, it must be expressible as a countable sum of weighted basis functions as given by

$$y(t) = \sum_{k=1}^{\infty} y_k \, g_k(t) \qquad 0 \leq t \leq T \tag{7.3}$$

where $\{g_k(t)\}$ is a set of countable basis functions and y_k is the expansion coefficient that is to be paired with the kth basis function. For notational convenience, the interval is chosen to be $[0, T]$, but in general, the method can be used for any arbitrary interval $[t_0, t_0 + T]$. Note, that k can range from one to infinity or from one to K (a finite number). In some problems, the expansion count k can start at zero (i.e., see the DFT type representation in Chapter 3). In some expansions, the count variable k can also take on negative values, such as in the complex Fourier series representation. We note that in most discussions we assume that the basis functions are real valued. If complex valued basis functions are to be considered, the definition of the inner product must be generalized by conjugating one of the functions involved. This indeed guarantees that the projection of one basis function onto itself (i.e., its energy) will always be real valued.

The kth expansion coefficient, y_k, is obtained by projecting $y(t)$ onto the kth basis function

$$y_k = \langle y(t), g_k(t) \rangle = \int_0^T y(t)\ g_k(t)\ dt \tag{7.4}$$

where the angular brackets denote the inner product. Note, the usage of the angular brackets to denote the inner product. In Chapter 9, the angular brackets will be used to indicate a time averaging operation. The basis set $\{g_k(t)\}$ consists of a C.O.N. set, that is

$$\langle g_i(t), g_k(t) \rangle = \int_0^T g_i(t)\ g_k(t)\ dt = 1\delta_{ik} \tag{7.5}$$

where

$$\delta_{ik} = \begin{cases} 1 & i = k \\ 0 & \text{else} \end{cases}, \qquad \text{(i.e., the Kronecker delta)}$$

and $\{g_k(t)\}$ will span the signal space. If necessary, expressions (7.3)–(7.5) can easily be modified to allow the representation of complex valued signals or to represent real valued signals using complex valued basis functions. As the reader will recall, the basis functions of the DFT and FFT are complex valued exponentials. For some complex basis function examples see Chapter 3. In most work discussed in Chapter 7, as earlier stated, we shall deal with real valued processes.

We denote the mth approximation to the function $f(t)$ (i.e., the known signal), by $f_m(t)$, that is

$$f_m(t) = \sum_{k=1}^{m} f_k\ g_k(t) \tag{7.6}$$

such that

$$\lim_{m\to\infty} f_m(t) = \sum_{k=1}^{\infty} f_k\ g_k(t) = f(t)$$

We will re-examine the problem of a known signal embedded in white Gaussian noise to illustrate the use of series representations and the manipulations of the resulting terms.

7.3 DERIVATION OF THE CORRELATOR STRUCTURE USING AN ARBITRARY COMPLETE ORTHO-NORMAL (C.O.N.) SET

This section employs the series expansion to derive the standard correlator result. It primarily serves the purpose of exposing the reader to the application of series expressions by allowing the derivation of a known result using an alternative (the series based) approach. We shall investigate the detection of a known signal in an additive white Gaussian scenario as is addressed in Chapter 6, Section 6.2. Under H_i, we have

$$H_i \ : \quad y(t) = s_i(t) + n(t), \qquad \begin{array}{c} i = 0, 1 \\ 0 \le t \le T \end{array} \tag{7.7}$$

where $n(t)$ is white Gaussian noise. The kth projection (component along the kth basis function) is given by

$$y_k = s_{ik} + n_k \qquad i = 0, 1 \tag{7.8}$$

where

$$\begin{aligned} y_k &= \langle y(t), g_k(t) \rangle = \int_0^T y(t) \ g_k(t) \ dt \\ s_{1k} &= \langle s_1(t), g_k(t) \rangle = \int_0^T s_1(t) \ g_k(t) \ dt \\ s_{0k} &= \langle s_0(t), g_k(t) \rangle = \int_0^T s_0(t) \ g_k(t) \ dt \\ n_k &= \langle n(t), g_k(t) \rangle = \int_0^T n(t) \ g_k(t) \ dt \end{aligned}$$

and $\{g_k(t)\}$ forms a C.O.N. arbitrary basis set.

Example 7.1 *An example of a second order approximation for the received signal, when the H_1 hypothesis is true (i.e., $s_1(t)$ is present), is illustrated in Figure 7.1. Here under H_1, we have: $y(t) = s_1(t) + n(t)$ while the second order approximation is given by*

$$\begin{aligned} y_2(t) &= s_{12}(t) + n_2(t) \\ &= \sum_{k=1}^{2} s_{1k} \ g_k(t) + \sum_{k=1}^{2} n_k \ g_k(t) \end{aligned}$$

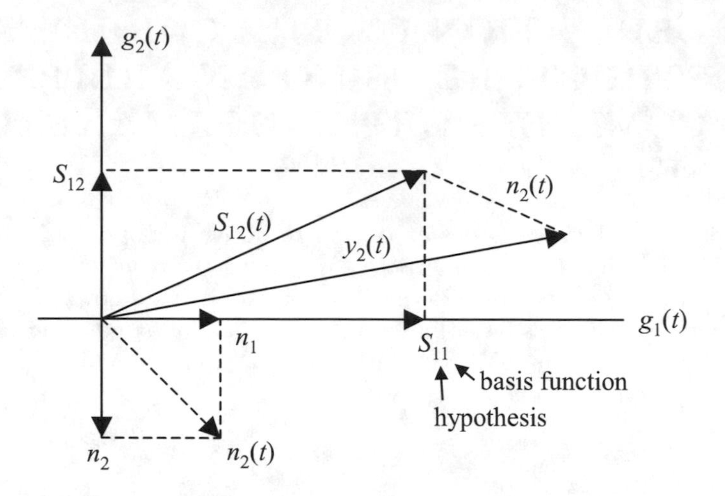

Figure 7.1: Second order approximation of $y(t)$ under the H_1 condition.

Note: y_2 is the projection of $y(t)$ onto the second basis function while $y_2(t)$ is the second order approximation of $y(t)$. Note also that we use Figure 7.1 to represent both the projections of the components onto the basis axes, as well as the resultant time dependent components. For example, s_{11} and s_{12} represent the projection of $s_{12}(t)$ onto $g_1(t)$ and $g_2(t)$, respectively. The expression $s_{12}(t)$ represents the sum of $s_{11}\ g_1(t)\ +\ s_{12}\ g_2(t)$.

In general, the expansion coefficients $y_1, y_2, \cdots$, are random variables which are defined by

$$\text{under} \quad H_0: \quad y_k = \langle y(t), g_k(t)\rangle = \langle s_0(t) + n(t), g_k(t)\rangle$$
$$= s_{0k} + n_k \tag{7.9}$$

$$H_1: \quad y_k = \langle s_1(t) + n(t), g_k(t)\rangle$$
$$= s_{1k} + n_k \tag{7.10}$$

We note that $n_k = \langle n(t), g_k(t)\rangle$ is a linear transformation of $n(t)$ and since $n(t)$ is Gaussian, (under each hypothesis) n_k, and hence, y_k will be Gaussian. All that is required is knowledge of the means and the variance. The means are given by

$$E\{y_k|H_1\} = E\{s_{1k} + n_k\} = s_{1k} \tag{7.11}$$

$$E\{y_k|H_0\} = E\{s_{0k} + n_k\} = s_{0k} \tag{7.12}$$

while the variance is described by

$$\begin{aligned}
\text{var}\ \{y_k|H_0\} &= \text{var}\ \{s_{0k} + n_k\} \\
&= E\{s_{0k} + n_k\}^2 - E^2\{s_{0k}\} \\
&= E\{s_{0k}\}^2 + E\{n_k\}^2 + 2E\{s_{0k}\ n_k\} - E\{s_{0k}\}^2
\end{aligned}$$

and since $E\{s_{0k}\ n_k\} = 0$, the variance becomes

$$\begin{aligned}
\text{var}\ \{y_k|H_0\} &= \\
&= E\{n_k\}^2 \\
&= E\{s_{1k} + n_k\}^2 - E^2\{s_{1k}\} \\
&= \text{var}\ \{y_k|H_1\}
\end{aligned} \tag{7.13}$$

In the white noise case, we would have

$$\begin{aligned}
E\{n_k\}^2 &= E\int_0^T \int_0^T n(t)\ n(s)\ g_k(t)\ g_k(s)\ dt\ ds \\
&= \int_0^T \int_0^T \frac{N_0}{2}\ \delta(t-s)\ g_k(t)\ g_k(s)\ dt\ ds \\
&= \frac{N_0}{2}\int_0^T g_k^2(t)\ dt = \frac{N_0}{2}\ \delta_{kk} = \frac{N_0}{2}
\end{aligned} \tag{7.14}$$

We can form the likelihood ratio test using the orthogonal components $(y_1 g_1(t), y_2 g_2(t), \cdots, y_k g_k(t))$ as given by

$$\begin{aligned}
\Lambda\left(y(t)\right) &= \lim_{k\to\infty} \Lambda\left(y_k(t)\right) \\
&= \lim_{k\to\infty} \frac{f_1\left(y_1 g_1(t), y_2 g_2(t), \cdots, y_k g_k(t)\right)}{f_0\left(y_1 g_1(t), y_2 g_2(t), \cdots, y_k g_k(t)\right)}
\end{aligned} \tag{7.15}$$

We recall that if a Gaussian has parameters $x \sim N(m_x,\ \sigma_x^2)$, then

$$y = \alpha x \sim N(\alpha m_x,\ \alpha^2\ \sigma_x^2)$$

therefore, the nth term for $n = 1, 2, \cdots$, is given by

$$\frac{f_1(y_n g_n(t))}{f_0(y_n g_n(t))} = \frac{f_1(y_n)}{f_0(y_n)}$$

so the LRT becomes

$$\Lambda\left(y(t)\right) = \lim_{k\to\infty} \frac{f_1\left(y_1, y_2, \cdots, y_k\right)}{f_0\left(y_1, y_2, \cdots, y_k\right)}$$

$$= \lim_{k\to\infty} \prod_{n=1}^{k} \frac{f_1(y_n)}{f_0(y_n)}$$

$$= \lim_{k\to\infty} \frac{\prod_{n=1}^{k} e^{-(1/(2N_0/2))(y_n - s_{1n})^2}}{\prod_{n=1}^{k} e^{-(1/(2N_0/2))(y_n - s_{0n})^2}} \underset{H_0}{\overset{H_1}{\gtrless}} \lambda_0 \tag{7.16}$$

Taking the natural logarithm of the kth approximation of the LRT leads to

$$\ln \Lambda\left(y_k(t)\right) = \frac{2}{N_0} \sum_{n=1}^{k} y_n \left(s_{1n} - s_{0n}\right) + \frac{1}{N_0} \sum_{n=1}^{k} \left(s_{0n}^2 - s_{1n}^2\right) \underset{H_0}{\overset{H_1}{\gtrless}} \ln \lambda_0 \tag{7.17}$$

We want to obtain an equivalent expression for the first and second summations in (7.17). To do this, we will start with the following integral

$$\int_0^T y_k(t)\ s_{ik}(t)\ dt = \int_0^T \left[\sum_{n=1}^{k} y_n\ g_n(t)\right] \left[\sum_{r=1}^{k} s_{ir}\ g_r(t)\right] dt \tag{7.18}$$

where $y_k(t)$ is the kth approximation of $y(t)$ and $s_{ik}(t)$ is the kth approximation of $s_i(t)$ for $i = 0, 1$, and $g_n(t)$ is the nth basis function for $n = 1, \cdots, k$. By rearranging the order of the linear operators of the right-hand side of (7.18), we obtain

$$\sum_{n=1}^{k} \sum_{r=1}^{k} y_n\ s_{ir} \underbrace{\int_0^T g_n(t)\ g_r(t)\ dt}_{=\left\{\begin{smallmatrix} 1 & n=r \\ 0 & \text{else} \end{smallmatrix}\right. = \delta_{nr}} = \sum_{n=1}^{k} \sum_{r=1}^{k} y_n\ s_{ir}\ \delta_{nr}$$

$$= \sum_{n=1}^{k} y_n\ s_{in}\ ; \quad \text{for } i = 0, 1 \tag{7.19}$$

Reversing the order of the equation, we obtain

$$\sum_{n=1}^{k} y_n\ s_{in} = \int_0^T y_k(t)\ s_{ik}(t)\ dt \tag{7.20}$$

By analogy it is easily shown that

$$\sum_{n=1}^{k} s_{in}^2 = \int_0^T s_{ik}^2(t)\ dt \tag{7.21}$$

Therefore, the LRT can be written as

$$\ln \Lambda\left(y_k(t)\right) = \frac{2}{N_0}\int_0^T y_k(t)\left[s_{1k}(t) - s_{0k}(t)\right] dt + \frac{1}{N_0}\int_0^T \left[s_{0k}^2(t) - s_{1k}^2(t)\right] dt \underset{H_0}{\overset{H_1}{\gtrless}} \ln\ \lambda_0 \qquad (7.22)$$

Taking the limit as $k \to \infty$, leads to

$$\begin{aligned}\lim_{k\to\infty} \ln \Lambda\left(y_k(t)\right) &= \ln \Lambda\left(y(t)\right) \\ &= \frac{2}{N_0}\int_0^T y(t)\left[s_1(t) - s_0(t)\right] dt \\ &\quad + \frac{1}{N_0}\int_0^T \left[s_0^2(t) - s_1^2(t)\right] dt \underset{H_0}{\overset{H_1}{\gtrless}} \ln\ \lambda_0 \end{aligned} \qquad (7.23)$$

where λ_0 is a function of detector philosophy (i.e., ML implies λ_0 is unity). For Bayes' formulation, we have

$$\lambda_0 = \frac{P_0\left(C_{10} - C_{00}\right)}{P_1\left(C_{01} - C_{11}\right)}$$

In general, the detector becomes

$$\int_0^T y(t)\ s_1(t)\ dt - \int_0^T y(t)\ s_0(t)\ dt \underset{H_0}{\overset{H_1}{\gtrless}} \gamma \qquad (7.24)$$

where

$$\gamma = \frac{N_0}{2}\ln \lambda_0 + \frac{1}{2}\int_0^T \left(s_1^2(t) - s_0^2(t)\right) dt$$

If a Neyman-Pearson criterion is used, then γ is selected to meet the desired P_{FA}.

Notes:

(1) Again we derived the correlation structure, the same one that was first developed in Chapter 6, Section 6.2. This time, rather than using a data sampling approach, we used a decomposition based on an arbitrary C.O.N. basis set.

(2) If only white noise is present, any C.O.N. set will be appropriate to be used as basis function set $\{g_n(t)\}$, independent of whether or not the observation interval T is finite or not. Due to its natural interpretation and the ease of its implementation [6], the Fourier basis set is usually preferred.

(3) If the signal duration T becomes very large (i.e., ranging from $-\infty$ to $+\infty$), the coefficients of the Fourier expansion of any w.s.s. random process (of any spectral shape) tend to be uncorrelated [7].

(4) A problem occurs when dealing with a colored random process if the signal is of finite duration (i.e., $T < \infty$). We cannot use the Fourier expansion since the coefficients become uncorrelated only as $T \to \infty$ [7]. The Karhunen-Loève expansion [4], to be discussed in Section 7.6.3, allows decomposition of the signals of interest such that the (colored) noise expansion coefficients are uncorrelated.

7.4 GRAM-SCHMIDT PROCEDURE

This section serves as an introduction to the Gram-Schmidt procedure, a technique frequently used in statistical digital signal processing. Suppose that we have a Hilbert space (i.e., a vector space with an inner product and a norm derived from the inner product), and allowing for complex valued functions (or vectors) we can define the inner product depending on the scenario as

$$\langle \mathbf{x}, \mathbf{y} \rangle = \begin{cases} \sum_i x_i \, y_i^* & ; \text{ for discrete time variables} \\ \int x(t) \, y^*(t) \, dt & ; \text{ for continuous time variables} \\ E \, x \, y^* & ; \text{ for random variables} \end{cases} \tag{7.25}$$

where * denotes conjugation. The norm is defined as

$$\text{norm} = \langle \mathbf{x}, \mathbf{x} \rangle^{\mathbf{1/2}} = ||\mathbf{x}|| \tag{7.26}$$

Given a vector space is spanned by the vectors, $\{\boldsymbol{\alpha}_1, \boldsymbol{\alpha}_2, \cdots\}$, we want to span the same vector space with a set of complete ortho-normal (C.O.N.) vectors $\{\mathbf{q}_1, \mathbf{q}_2, \cdots,\}$. We do this by sequentially forming orthogonal vectors $\{\mathbf{v}_1, \mathbf{v}_2, \cdots,\}$ and performing a normalization to obtain $\{\mathbf{q}_1, \mathbf{q}_2, \cdots,\}$

$$\underbrace{\{\boldsymbol{\alpha}\}}_{\substack{\text{arbitr.}\\ \text{vectors}}} \longrightarrow \underbrace{\{\mathbf{v}\}}_{\substack{\text{orthogonal}\\ \text{vectors}}} \longrightarrow \underbrace{\{\mathbf{q}\}}_{\substack{\text{ortho-}\\ \text{normal}\\ \text{vectors}}}$$

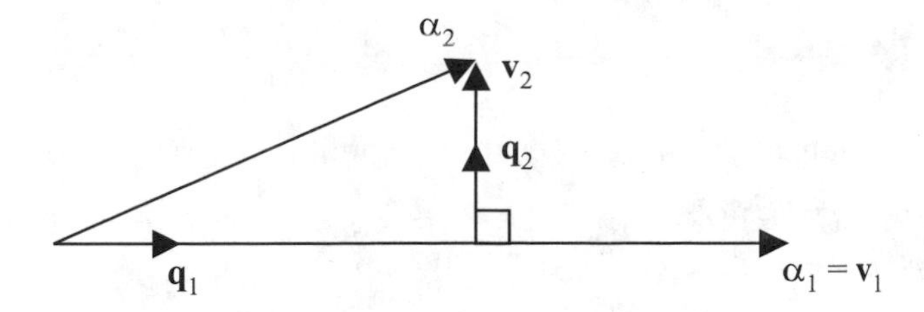

Figure 7.2: First two vector relationships when using the Gram-Schmidt procedure.

First we select a starting vector. Without loss of generality, we select α_1 as the starting vector and then sequentially determine the follow-on vector $\mathbf{v}_2$

$$\mathbf{v}_1 = \boldsymbol{\alpha}_1 \qquad \mathbf{q}_1 = \frac{\mathbf{v}_1}{||\mathbf{v}_1||} = \frac{\mathbf{v}_1}{\langle \mathbf{v}_1, \mathbf{v}_1 \rangle^{1/2}} \tag{7.27}$$

$$\mathbf{v}_2 = \boldsymbol{\alpha}_2 - k_1 \mathbf{q}_1$$

where k_1 is selected such that $\mathbf{v}_2$ and $\mathbf{q}_1$ are orthogonal ($\mathbf{v}_2$ and $\mathbf{v}_1$ are orthogonal). When vector $\mathbf{v_2} \perp \mathbf{q}_1$ (read $\mathbf{v}_2$ orthogonal to $\mathbf{q}_1$), then we have

$$\begin{aligned} \langle \mathbf{q}_1, \mathbf{v}_2 \rangle &= 0 \\ &= \langle \mathbf{q}_1, \boldsymbol{\alpha}_2 - k_1 \mathbf{q}_1 \rangle \\ &= \langle \mathbf{q}_1, \boldsymbol{\alpha}_2 \rangle - k_1 \langle \mathbf{q}_1, \mathbf{q}_1 \rangle \\ &= \langle \mathbf{q}_1, \boldsymbol{\alpha}_2 \rangle - k_1 \end{aligned} \tag{7.28}$$

We realize that $\langle \mathbf{q}_1, \mathbf{q}_1 \rangle = 1$, hence the coefficient k_1 is obtained as

$$k_1 = \langle \mathbf{q}_1, \boldsymbol{\alpha}_2 \rangle = \langle \boldsymbol{\alpha}_2, \mathbf{q}_1 \rangle \tag{7.29}$$

while the orthogonal vector $\mathbf{v}_2$ is given by

$$\begin{aligned} \mathbf{v}_2 &= \boldsymbol{\alpha}_2 - \langle \boldsymbol{\alpha}_2, \mathbf{q}_1 \rangle \, \mathbf{q}_1 \\ &= \boldsymbol{\alpha}_2 - \frac{\langle \boldsymbol{\alpha}_2, \mathbf{v}_1 \rangle}{\langle \mathbf{v}_1, \mathbf{v}_1 \rangle^{1/2}} \frac{\mathbf{v}_1}{\langle \mathbf{v}_1, \mathbf{v}_1 \rangle^{1/2}} \end{aligned} \tag{7.30}$$

and ortho-normal vector $\mathbf{q}_2$ is obtained by

$$\mathbf{q}_2 = \frac{\mathbf{v}_2}{||\mathbf{v}_2||} \tag{7.31}$$

The first two relationships are illustrated in Figure 7.2. In general, the ortho-normalization procedure is obtained from a sequential determination of orthogonal vectors for $k = 1, 2, \ldots$, as defined by

$$\mathbf{v}_k = \boldsymbol{\alpha}_k - \sum_{i=1}^{k-1} \frac{\langle \boldsymbol{\alpha}_k, \mathbf{v}_i \rangle}{\langle \mathbf{v}_i, \mathbf{v}_i \rangle^{1/2}} \frac{\mathbf{v}_i}{\langle \mathbf{v}_i, \mathbf{v}_i \rangle^{1/2}} \qquad \left(\begin{array}{c} \text{orthogonal} \\ \text{set} \end{array} \right) \tag{7.32}$$

$$\mathbf{q}_k = \frac{\mathbf{v}_k}{||\mathbf{v}_k||} = \frac{\mathbf{v}_k}{\langle \mathbf{v}_k, \mathbf{v}_k \rangle^{1/2}} \quad \left(\begin{array}{c} \text{ortho-normal} \\ \text{set} \end{array} \right) \tag{7.33}$$

where the normalization was obtained by dividing each vector by its norm.

Example 7.2 *Given* $\{\alpha_1, \alpha_2, \cdots,\} = \{t^0, t^1, t^2, \cdots,\}$ *where* $-1 \leq t \leq 1$. *Find an ortho-normal set over* $[-1, 1]$. *We note that*

$$\langle \alpha_1, \alpha_1 \rangle = \int_{-1}^{1} 1 \cdot 1 \, dt = 2 \neq 1$$

$$\langle \alpha_1, \alpha_3 \rangle = \int_{-1}^{1} 1 \cdot t^2 \, dt = \left. \frac{t^3}{3} \right|_{-1}^{1} = \frac{2}{3} \neq 0$$

Hence, the set $\{\alpha_1, \alpha_2, \cdots,\}$ *does not form an orthogonal basis set for* $|t| \leq 1$. *Let* $v_1 = \alpha_1$

$$q_1 = \frac{v_1}{||v_1||} = \frac{1}{\sqrt{\int_{-1}^{1} 1 \cdot 1 dt}} = \frac{1}{\sqrt{2}}$$

$$\begin{aligned}
v_2 &= \alpha_2 - \langle \alpha_2, q_1 \rangle \, q_1 \\
&= t - \left\langle t, \frac{1}{\sqrt{2}} \right\rangle \frac{1}{\sqrt{2}} \\
&= t - \int_{-1}^{1} t \, \frac{1}{\sqrt{2}} \, dt \frac{1}{\sqrt{2}} \; = \; t - \left. \frac{t^2}{4} \right|_{-1}^{1} = t
\end{aligned}$$

$$||v_2|| = \sqrt{\int_{-1}^{1} t \cdot t \, dt} = \sqrt{\left. \frac{t^3}{3} \right|_{-1}^{1}} = \sqrt{\frac{2}{3}}$$

$$q_2 = \frac{v_2}{||v_2||} = \sqrt{\frac{3}{2}} \, t$$

$$\begin{aligned}
v_3 &= \alpha_3 - \langle \alpha_3, q_2 \rangle \, q_2 - \langle \alpha_3, q_1 \rangle \, q_1 \\
&= t^2 - \left\langle t^2, \sqrt{\frac{3}{2}} t \right\rangle \sqrt{\frac{3}{2}} t - \left\langle t^2, \frac{1}{\sqrt{2}} \right\rangle \frac{1}{\sqrt{2}} \\
&= t^2 - \frac{3}{2} \, t \, \int_{-1}^{1} t^3 \, dt - \frac{1}{2} \int_{-1}^{1} t^2 \, dt \\
&= t^2 - \frac{1}{3}
\end{aligned}$$

$$
\begin{aligned}
||v_3|| &= \left|\left|t^2 - \frac{1}{3}\right|\right| = \sqrt{\int_{-1}^{1}\left(t^2 - \frac{1}{3}\right)^2 dt} \\
&= \sqrt{\left(\frac{t^5}{5} - \frac{t^3}{3 \cdot 3} \cdot 2 + \frac{t}{9}\right)\Big|_{-1}^{1}} = \frac{4}{3}\,\frac{1}{\sqrt{10}} \\
q_3 &= \frac{v_3}{||v_3||} = \frac{3}{4}\sqrt{10}\left(t^2 - \frac{1}{3}\right) \\
&\vdots
\end{aligned}
$$

The ortho-normal set $\{q_1, q_2, q_3, \cdots\}$ *is now of the form*

$$
\{q_1, q_2, q_3, \cdots,\} = \left\{\frac{1}{\sqrt{2}},\ \sqrt{\frac{3}{2}}\ t,\ \frac{3}{4}\sqrt{10}\left(t^2 - \frac{1}{3}\right), \cdots\right\}
$$

which is the set of ortho-normal Legendre polynomials over the interval $(-1, 1)$. *We leave it for the reader to verify that indeed*

$$
\langle q_1, q_1\rangle = \langle q_2, q_2\rangle = \langle q_3, q_3\rangle = 1
$$

and that

$$
\langle q_1, q_2\rangle = \langle q_1, q_3\rangle = \langle q_2, q_3\rangle = 0
$$

7.5 DETECTION OF A KNOWN SIGNAL IN ADDITIVE WHITE GAUSSIAN NOISE USING THE GRAM-SCHMIDT PROCEDURE

In this section, we want to apply the Gram-Schmidt ortho-normalization procedure to the problem of detecting a known signal embedded in white Gaussian noise. This enhances the understanding of series representations and leads to the general correlator result, using coordinates axes that are slightly different compared to results derived earlier.

Suppose we want to determine whether or not the signal $s_0(t)$ or $s_1(t)$ is embedded in white Gaussian noise, and that we have

$$
\text{under} \quad \begin{array}{lcl} H_0 : & y(t) & = s_0(t) + n(t) \\ H_1 : & y(t) & = s_1(t) + n(t) \end{array} \quad ; \qquad 0 \leq t \leq T \tag{7.34}
$$

We want to find the optimal detector using the Gram-Schmidt procedure. Let

$$
\varepsilon_1 = \langle s_1(t), s_1(t)\rangle = \int_0^T s_1^2(t)\ dt
$$

be the energy of $s_1(t)$

$$\varepsilon_0 = \langle s_0(t), s_0(t) \rangle = \int_0^T s_0^2(t)\, dt$$

be the energy of $s_0(t)$, and

$$r = \langle s_0(t), s_1(t) \rangle = \int_0^T s_0(t)\, s_1(t)\, dt$$

be the un-normalized cross-correlation coefficient.

We arbitrarily have chosen $s_1(t)$ to be the first basis function (note that we could have used $s_0(t)$ just as well); hence,

$$\begin{aligned} v_1(t) &= s_1(t) \\ q_1(t) &= \frac{s_1(t)}{\langle s_1(t), s_1(t) \rangle^{1/2}} = \frac{s_1(t)}{\sqrt{\varepsilon_1}} \end{aligned}$$

$$\begin{aligned} v_2(t) &= s_0(t) - \frac{\langle s_0(t), s_1(t) \rangle}{\langle s_1(t), s_1(t) \rangle^{1/2}} \frac{s_1(t)}{\langle s_1(t), s_1(t) \rangle^{1/2}} \\ &= s_0(t) - \frac{r}{\sqrt{\varepsilon_1}} \frac{s_1(t)}{\sqrt{\varepsilon_1}} = s_0(t) - \frac{r}{\varepsilon_1} s_1(t) \\ q_2(t) &= \frac{v_2(t)}{||v_2(t)||} \end{aligned}$$

The norm of $v_2(t)$ is given by

$$\begin{aligned} ||v_2(t)|| &= \sqrt{\left\langle \left(s_0(t) - \frac{r}{\varepsilon_1} s_1(t) \right), \left(s_0(t) - \frac{r}{\varepsilon_1} s_1(t) \right) \right\rangle} \\ &= \sqrt{\varepsilon_0 - \frac{2r^2}{\varepsilon_1} + \frac{r^2}{\varepsilon_1^2} \varepsilon_1} \\ &= \sqrt{\frac{\varepsilon_1 \varepsilon_0 - r^2}{\varepsilon_1}} \end{aligned}$$

hence

$$q_2(t) = \frac{s_0(t) - \dfrac{r}{\varepsilon_1} s_1(t)}{\sqrt{\varepsilon_1 \varepsilon_0 - r^2}} \sqrt{\varepsilon_1}$$

For the binary case, no additional basis functions are needed since we already have a complete ortho-normal set to describe the signals $s_1(t)$ and $s_0(t)$. If any were to be chosen, they would be orthogonal to $q_1(t)$ and $q_2(t)$. In general,

$$y(t) = \sum_{\ell=1}^{\infty} y_\ell \; q_\ell(t) \; ; \qquad 0 \le t \le T \tag{7.35}$$

where $y_\ell = \langle y(t), q_l(t) \rangle$. In our case

$$y(t) = \sum_{\ell=1}^{2} y_\ell \; q_l(t) \qquad 0 \le t \le T.$$

We realize,

(a) that y_ℓ is the projection of $y(t)$ onto the basis functions $q_\ell(t)$ for $\ell = 1, 2$, and

(b) due to the linearity of the inner product, the variables y_ℓ, for $\ell = 1, 2$, will be independent Gaussian random variables. We need the means and variances of these Gaussian random variables. Because the noise has zero mean, we have

$$\begin{aligned} E\{y_\ell | H_0\} &= \int_0^T y(t|H_0) \; q_\ell(t) \; dt \\ &= \int_0^T s_0(t) \; q_\ell(t) \; dt = s_{0\ell} \; ; \qquad \ell = 1, 2 \end{aligned} \tag{7.36}$$

Similarly, the mean under the H_1 hypothesis and the variance (independent of the hypothesis) are given by

$$E\{y_\ell | H_1\} = \int_0^T s_1(t) \; q_\ell(t) \; dt = s_{1\ell} \qquad \ell = 1, 2 \tag{7.37}$$

$$\begin{aligned} \text{var } \{y_\ell | H_i\} &= \int_0^T \int_0^T E\{n(t) \; n(s)\} \; q_\ell(t) \; q_\ell(s) \; dt \; ds \\ &= \frac{N_0}{2} \int_0^T q_\ell^2(t) \; dt = \frac{N_0}{2} \; ; \qquad \begin{matrix} \ell = 1, 2 \\ i = 0, 1 \end{matrix} \end{aligned} \tag{7.38}$$

respectively, where ℓ is basis function and i corresponds to the hypothesis.

We ortho-normalized the two original functions and need only the two ortho-normal functions and their expansion coefficients (i.e., $\Lambda(y(t)) = \Lambda(y_2(t))$.

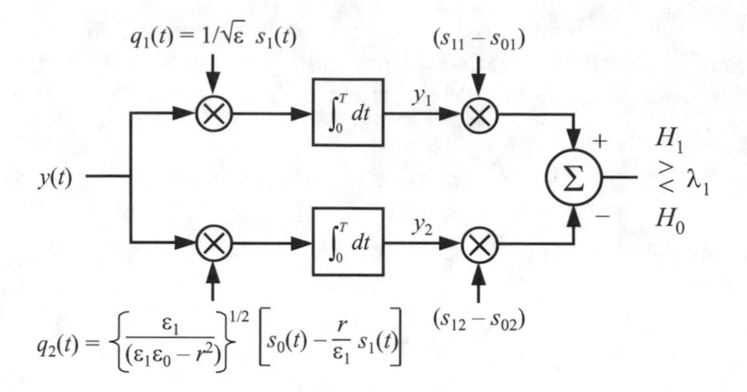

Figure 7.3: Optimal detector using Gram-Schmidt procedure ($s_{i\ell}$, for $i = 0, 1$ and $\ell = 1, 2$, denotes the ℓth coefficient of the ith signal).

If more basis functions, say M, are required such as in an M-ary problem, then $\Lambda(y(t)) = \Lambda(y_M(t))$. For our binary problem, the LRT becomes

$$\Lambda(y_2(t)) = \frac{\displaystyle\prod_{k=1}^{2} \frac{1}{\sqrt{\pi N_0}} e^{(-1/N_0)(y_k - s_{1k})^2}}{\displaystyle\prod_{k=1}^{2} \frac{1}{\sqrt{\pi N_0}} e^{(-1/N_0)(y_k - s_{0k})^2}} \underset{H_0}{\overset{H_1}{\gtrless}} \lambda_0 \tag{7.39}$$

or

$$\ln \Lambda(y_2(t)) = \frac{2}{N_0} \sum_{k=1}^{2} y_k (s_{1k} - s_{0k}) + \frac{1}{N_0} \sum_{k=1}^{2} (s_{0k}^2 - s_{1k}^2) \underset{H_0}{\overset{H_1}{\gtrless}} \ln \lambda_0 \tag{7.40}$$

or

$$\sum_{k=1}^{2} y_k \, (s_{1k} - s_{0k}) \underset{H_0}{\overset{H_1}{\gtrless}} \frac{N_0}{2} \ln \lambda_0 - \frac{1}{2} \sum_{k=1}^{2} (s_{0k}^2 - s_{1k}^2) = \lambda_1 \tag{7.41}$$

This detector is easily implemented as shown in Figure 7.3. As we see, the structure is the same as the correlator structure advocated in Chapter 6, Section 6.2 and Section 7.3. We can interpret the detector operation as a projection of the data onto basis functions (vectors) to determine which projection coefficient dominates. Of course, we account for prior information, cost, and for differential energy.

In summary, we have addressed three ways for detecting known signals in white Gaussian noise. Each approach led to the correlator structure which is a projection of the data onto a set of basis functions for continuous time signals (or vectors for discrete-time signals) in a Hilbert space setting.

7.6 SERIES EXPANSION FOR CONTINUOUS TIME DETECTION FOR COLORED GAUSSIAN NOISE

7.6.1 Introduction

This section introduces some concepts that are essential in solving the problem of detecting signals embedded in colored Gaussian noise. We will first review Mercer's theorem and the Karhunen-Loève expansion, which are needed in the derivation of the detector.

If the additive Gaussian noise process is colored and the signal duration is finite, then the use of traditional orthogonal (or ortho-normal) basis functions leads to correlated coefficients. This does not lend itself to an approach based on a simple LRT. A workable solution is available if we resort to the Karhunen-Loève expansion [4].

The Karhunen-Loève expansion is based on the correlation properties of the random process. It provides basis functions that guarantee the projections of the noise onto the basis functions to be uncorrelated. The Karhunen-Loève expansion is based on the eigenfunction of the noise auto-correlation function, where the auto-correlation function is symmetric in the arguments, non-negative, and continuous. Note, we will use the label $q(t)$ for basis functions derived via the noise space characteristics.

7.6.2 Mercer's Theorem

Any real valued function $R(t,s)$, which is symmetric, positive semi-definite, and continuous on the square $0 \leq t, s \leq T$ may be expanded into an absolutely and uniformly convergent series [8]

$$R(t,s) = \sum_{k=1}^{\infty} \lambda_k \; q_k(t) \; q_k(s) \; ; \qquad 0 \leq t, s \leq T \tag{7.42}$$

The set $\{\lambda_k \;, q_k(t)\}$ contains the pairs of eigenvalues and eigenfunctions obtained from the integral equation

$$\lambda_k \; q_k(t) = \int_0^T R(t,s) \; q_k(s) \; ds \tag{7.43}$$

where $\{q_k(t)\}$ forms a C.O.N. set over the interval $0 \leq t \leq T$. Equation (7.43) is also called the homogeneous integral equation [9].

7.6.3 Karhunen-Loève Expansion

The expansion based on the eigenfunctions of the auto-correlation function is called the Karhunen-Loève expansion. The eigenfunctions $\{q_k(t)\}$ form a complete ortho-normal set, that is

$$\int_0^T q_n(t)\, q_m(t)\, dt = \delta_{nm} = \begin{cases} 1 & n = m \\ 0 & \text{else} \end{cases} \tag{7.44}$$

The projection of the noise onto the kth basis function produces

$$n_k = \langle n(t), q_k(t) \rangle = \int_0^T n(t)\, q_k(t)\, dt \tag{7.45}$$

The noise function is given by

$$n(t) = \text{l.i.m.}_{N\to\infty} \sum_{k=1}^{N} n_k\, q_k(t)\ ; \qquad 0 \le t \le T \tag{7.46}$$

where $\text{l.i.m.}_{N\to\infty}$ denotes the limit in the mean square sense, that is

$$\lim_{N\to\infty} E\left\{ n(t) - \sum_{k=1}^{N} n_k\, q_k(t) \right\}^2 = 0$$

If $E\{n(t)\} = 0$, then

$$E\{n_k\} = \int_0^T E\, n(t)\, g_k(t)\, dt = 0$$

The second order moment is given by

$$\begin{aligned} E\{n_k\, n_m\} &= E \int_0^T \int_0^T n(t)\, n(\sigma)\, q_k(t)\, q_m(\sigma)\, dt\, d\sigma \\ &= \int_0^T \left(\underbrace{\int_0^T R_N(t,\sigma)\, q_m(\sigma)\, d\sigma}_{=\lambda_m\, q_m(t)\quad (7.43)} \right) q_k(t)\, dt \end{aligned} \tag{7.47}$$

hence

$$E\{n_k\, n_m\} = \int_0^T \lambda_m\, q_m(t)\, q_k(t)\, dt = \lambda_m\, \delta_{mk} \tag{7.48}$$

hence, if $k \neq m$

$$E\{n_k\, n_m\} = 0 \tag{7.49}$$

If $k = m$, then

$$E\{n_k\ n_m\} = \text{var}\ n_k = \lambda_k \tag{7.50}$$

We note, that the variance of the kth term of the noise expansion can also be obtained as

$$\int_0^T \int_0^T q_k(t)\ R_N(t,s)\ q_k(s)\ dt\ ds = \int_0^T q_k(t)\ \lambda_k\ q_k(t)\ dt = \lambda_k \tag{7.51}$$

and that the sum of the eigenvalues corresponds to the integral of the correlation function (potentially the time-varying instantaneous power) as shown by

$$\begin{aligned} E \int_0^T n^2(t)\ dt &= \int_0^T R_N(t,t)\ dt \\ &= \int_0^T \sum_{k=1}^{\infty} \lambda_k\ g_k(t)\ g_k(t)\ dt \\ &= \sum_{k=1}^{\infty} \lambda_k \end{aligned} \tag{7.52}$$

7.7 DETECTION OF KNOWN SIGNALS IN ADDITIVE COLORED GAUSSIAN NOISE

This section focuses on deriving the correlator structure for known signals embedded in colored Gaussian noise. The bilateral Laplace transform is reviewed and the integral equation for the leucogenic noise case is converted into a differential equation, which then can be solved. We also look at the performance assessment (Section 7.7.3) of the detector and discuss an approach based on whitening the additive noise (Section 7.7.4).

Suppose that the signal $s_i(t)$ is received in the presence of additive zero mean colored Gaussian noise. Under H_i, we have

$$y(t) = s_i(t) + n(t)\ ; \qquad \begin{matrix} i = 0,1 \\ 0 \leq t \leq T \end{matrix} \tag{7.53}$$

Looking at the projection of the noise onto the jth basis function, we have a sequence of Gaussian random variables

$$n_j = \int_0^T n(t)\ q_j(t)\ dt\ ; \qquad j = 1, 2, \cdots \tag{7.54}$$

where $E\ n_j = 0$ since $E\ n(t) = 0$. The statistical dependency of the ith and k^{th} noise component is obtained as shown in (7.48)

$$E\{n_j\ n_k\} = \int_0^T \lambda_k\ q_k(t)\ q_j(t)\ dt = \lambda_k \delta_{kj} = \begin{cases} \lambda_k\ , & k = j \\ 0\ , & \text{else} \end{cases} \tag{7.55}$$

Since the coefficients are Gaussian and uncorrelated, it follows that they are independent random variables. We see that

$$\begin{aligned} \text{under } H_0: \quad y_j &= \int_0^T y\,(t|H_0)\ q_j(t)\ dt \\ &= \int_0^T (s_0(t) + n(t))\ q_j(t)\ dt = s_{0j} + n_j\ ; \qquad j = 1, 2, \cdots \end{aligned} \tag{7.56}$$

similarly, under

$$H_1: \quad y_j = \int_0^T (s_1(t) + n(t))\ q_j(t)\ dt = s_{1j} + n_j\ ; \qquad j = 1, 2, \cdots \tag{7.57}$$

Under each hypothesis, the inner product given by y_j is Gaussian, hence, we can set up the LRT as in the previous sections, once we obtain means and variances. The means of the random variables y_j are given by

$$E\{y_j|H_0\} = E\,(s_{oj} + n_j) = s_{0j} \tag{7.58}$$

and

$$E\{y_j|H_1\} = s_{1j} \tag{7.59}$$

The variance is independent of the mean, hence

$$\text{var}\ (y_j|H_i) = \text{var}\ (y_j) = \text{var}\ (n_j)\ , \qquad i = 0, 1 \tag{7.60}$$

Using (7.60) and (7.55)

$$\text{var}\ (y_j) = \lambda_j \tag{7.61}$$

At this point, we can formulate the kth approximation of the LRT

$$\Lambda\,(y_k(t)) = \frac{\displaystyle\prod_{j=1}^{k} \frac{1}{\sqrt{2\pi\lambda_j}}\ e^{(-1/2)(y_j - s_{1j})^2/\lambda_j}}{\displaystyle\prod_{j=1}^{k} \frac{1}{\sqrt{2\pi\lambda_j}}\ e^{(-1/2)(y_j - s_{0j})^2/\lambda_j}} \begin{array}{c} H_1 \\ > \\ < \\ H_0 \end{array} \lambda_0 \tag{7.62}$$

Canceling common terms, taking the natural logarithm, and letting $k \to \infty$ results in

$$\ln \Lambda\,(y(t)) = \sum_{j=1}^{\infty} \frac{y_j}{\lambda_j}(s_{1j} - s_{0j}) + \frac{1}{2} \sum_{j=1}^{\infty} \frac{\left(s_{0j}^2 - s_{1j}^2\right)}{\lambda_j} \begin{array}{c} H_1 \\ > \\ < \\ H_0 \end{array} \ln\ \lambda_0 \tag{7.63}$$

By replacing the Karhunen-Loève coefficients with their corresponding inner products, this can be written as

$$y_j = \int_0^T y(t)\, q_j(t)\, dt \quad \text{and} \quad n_j = \int_0^T n(t)\, q_j(t)\, dt, \text{ etc.},$$

hence, (7.63) becomes

$$\sum_{j=1}^{\infty} \frac{1}{\lambda_j} \int_0^T y(t_1) q_j(t_1) dt_1 \int_0^T (s_1(t_2) - s_0(t_2))\; q_j(t_2) dt_2$$
$$+\frac{1}{2} \sum_{j=1}^{\infty} \frac{1}{\lambda_j} \left\{ \int_0^T \int_0^T s_0(t')\, s_0(\sigma)\, q_j(t')\, q_j(\sigma)\, dt'\, d\sigma \right. \tag{7.64}$$
$$\left. - \int_0^T \int_0^T s_1(t')\, s_1(\sigma)\, q_j(t')\, q_j(\sigma)\, dt'\, d\sigma \right\} \underset{H_0}{\overset{H_1}{\gtrless}} \ln \lambda_0$$

Reordering, (7.64) leads to

$$\int_0^T \int_0^T y(t_1)\, (s_1(t_2) - s_0(t_2)) \sum_{j=1}^{\infty} \frac{1}{\lambda_j}\, q_j(t_1)\, q_j(t_2)\, dt_1\, dt_2$$
$$\underbrace{+\frac{1}{2} \int_0^T \int_0^T [s_0(t')\, s_0(\sigma) - s_1(t')\, s_1(\sigma)] \sum_{j=1}^{\infty} \frac{1}{\lambda_j} q_j(t') q_j(\sigma) dt' d\sigma}_{\substack{\text{This is independent of the received data,} \\ \text{hence, it is equal to a constant } K.}} \gtrless \ln \lambda_0 \tag{7.65}$$

hence, we get

$$\int_0^T y(t_1) \left[\int_0^T (s_1(t_2) - s_0(t_2)) \sum_{j=1}^{\infty} \frac{q_j(t_1)}{\lambda_j} q_j(t_2) dt_2 \right] dt_1 \underset{H_0}{\overset{H_1}{\gtrless}} \ln \lambda_0 - K = T_1 \tag{7.66}$$

or

$$\int_0^T y(t_1) \sum_{j=1}^{\infty} \frac{1}{\lambda_j} \int_0^T [s_1(t_2) - s_0(t_2)]\, q_j(t_2)\, dt_2\, q_j(t_1)\, dt_1 \underset{H_0}{\overset{H_1}{\gtrless}} T_1 \tag{7.67}$$

which becomes

$$\int_0^T y(t_1) \underbrace{\sum_{j=1}^{\infty} \frac{1}{\lambda_j}(s_{1j} - s_{0j})\, q_j(t_1)}_{h(t_1)}\, dt_1 \quad \underset{H_0}{\overset{H_1}{\gtrless}} \quad T_1 \tag{7.68}$$

If we label the terms in the integrand of (7.68), excluding $y(t_1)$, as $h(t_1)$ and use t as the dummy variable of integration, we can write

$$\int_0^T y(t)\, h(t)\, dt \quad \underset{H_0}{\overset{H_1}{\gtrless}} \quad T_1 \tag{7.69}$$

We have a correlation receiver where the standard replica is replaced by $h(t)$ which is given by

$$\begin{aligned} h(t) &= \sum_{j=1}^{\infty} \frac{s_{1j} - s_{0j}}{\lambda_j} q_j(t) \\ &= \sum_{j=1}^{\infty} \frac{s_{1j}}{\lambda_j}\, q_j(t) - \sum_{j=1}^{\infty} \frac{s_{0j}}{\lambda_j}\, q_j(t) \\ &= h_1(t) - h_0(t) \end{aligned} \tag{7.70}$$

We note, if the λ_j's equal a constant, for all j, such as it is the case with white noise, then (7.70) would become the scaled difference between $s_1(t)$ and $s_0(t)$. If desired, we can split $h(t)$ into $h_1(t)$ and $h_0(t)$, as indicated in (7.70). This will provide the two legs of the correlator structure, one for each signal as shown in Figure 7.4.

To use (7.70), we would have to obtain $\{\lambda_j\}$ and $\{q_j(t)\}$ for all j. This approach is very tedious. Of course, one could try to approximate $h(t)$ of (7.70) with a finite number of terms. But then the question remains: which terms should be used and how many? Rather than solving for $h(t)$ in this fashion, we propose the following [4,18,21]:

To solve for $h(t)$, we take $h(t)$ and its defining equation, multiply by the noise kernel and integrate over the signal duration. Assuming that the noise is w.s.s., we have

$$\int_0^T R_N(t-\tau)\, h(\tau)\, d\tau = \sum_{j=1}^{\infty} \left(\frac{s_{1j} - s_{0j}}{\lambda_j} \right) \underbrace{\int_0^T R_N(t-\tau)\, q_j(\tau) d\tau}_{\text{via (7.43) } \lambda_j\, q_j(t)}$$

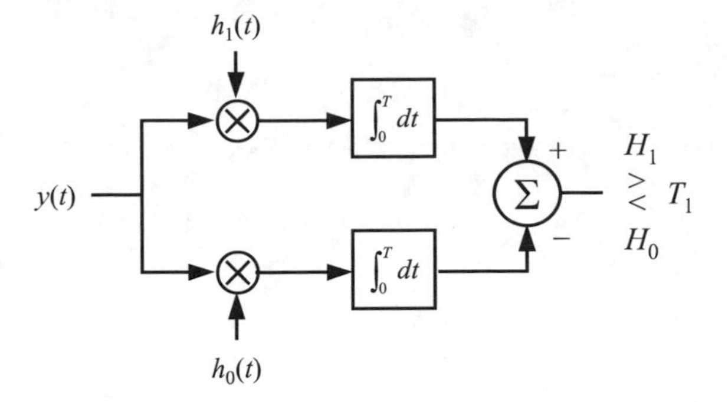

Figure 7.4: Correlator structure (colored Gaussian noise).

$$= \sum_{j=1}^{\infty} \frac{s_{1j} - s_{0j}}{\lambda_j} \lambda_j \; q_j(t) \tag{7.71}$$

$$= \sum_{j=1}^{\infty} (s_{1j} - s_{0j}) \; q_j(t) = s_1(t) - s_0(t) \mathrel{\hat{=}} s(t)$$

Hence, we obtain a Fredholm integral equation of the first kind [10]

$$\int_0^T R_N(t-\tau) \; h(\tau) \; d\tau = s(t) \; ; \qquad 0 \leq t \leq T \tag{7.72}$$

Note, if $R_N(\tau) = N_0/2 \; \delta(\tau)$ (i.e., the white noise case) then

$$\int_0^T \frac{N_0}{2} \; \delta(t-\tau) \; h(\tau) \; dt = s(t) \; ; \qquad 0 \leq t \leq T$$

becomes

$$\frac{N_0}{2} \; h(t) = s(t)$$

or

$$h(t) = \frac{2}{N_0} \; s(t) \qquad \text{(a scaled version of } s(t))$$

Therefore, in the white noise case, we would obtain our original result, that is we correlate the received data with a replica of the signal that we are testing for. This paraphrases the note that follows (7.70).

We can split the $h(t)$ and $s(t)$ functions into their "0" and "1" components, as given by

$$\int_0^T R_N(t-\tau) \; h_1(\tau) \; d\tau = s_1(t) \qquad \text{for } 0 \leq t \leq T \tag{7.73}$$

and

$$\int_0^T R_N(t-\tau)\; h_0(t)\; dt = s_0(t) \qquad \text{for } 0 \le t \le T \tag{7.74}$$

We see that the optimum receiver (detector) is the correlator, where the functions $h_i(t)$, for $i = 0, 1$, have to be computed by solving a Fredholm integral equation of the first kind [10]. In general, we need to solve the following equation for the quantity $h(t)$

$$\int_0^T R_N(t, u)\; h(u)\; du = s(t)\;; \qquad 0 \le t \le T \tag{7.75}$$

where $R_N(t, u)$ is the correlation function of the noise and $s(t) = s_1(t) - s_0(t)$. In the stationary case $R_N(t, u)$ becomes $R_N(t-u)$. One can also solve (7.75) for $h_1(t)$ and $h_0(t)$ by using $s_1(t)$ and $s_0(t)$ as drivers (also called excitation or input), respectively.

To be able to solve the integral equations we will first review some of the properties of the bilateral Laplace transform.

7.7.1 Bilateral Laplace Transform

This section serves as a quick review on a topic that is used to convert the integral to an equivalent differential equation. It may also serve as an introduction to the general Laplace transformation, allowing analysis of two-sided functions, such as correlation, power spectral density or probability density functions. The bilateral Laplace transform of the function $f(t)$ is given by [11]

$$\mathcal{L}_B\left\{f(t)\right\} \hat{=} F(s) = \int_{-\infty}^{\infty} f(t)\; e^{-st}\; dt \tag{7.76}$$

where $\sigma_0 < \text{Re } s < \sigma_1$ lies in the region of convergence (R.O.C.) (see Figure 7.5). The region of convergence guarantees the existence of $F(s)$. The transform variable $F(s)$ can be decomposed into two parts:

$$F(s) = F_+(s) + F_-(s) \tag{7.77}$$

where

$$F_+(s) = \int_0^{\infty} f(t)\; e^{-st}\; dt \tag{7.78}$$

and

$$F_-(s) = \int_{-\infty}^{0} f(t)\; e^{-st}\; dt \tag{7.79}$$

Let $\mathcal{L}_+\{\quad\}$ denote a conventional one-sided Laplace transform, then

$$F_+(s) \hat{=} \mathcal{L}_+\left\{f(t)\right\} = \int_0^{\infty} f(t)\; e^{-st}\; dt \tag{7.80}$$

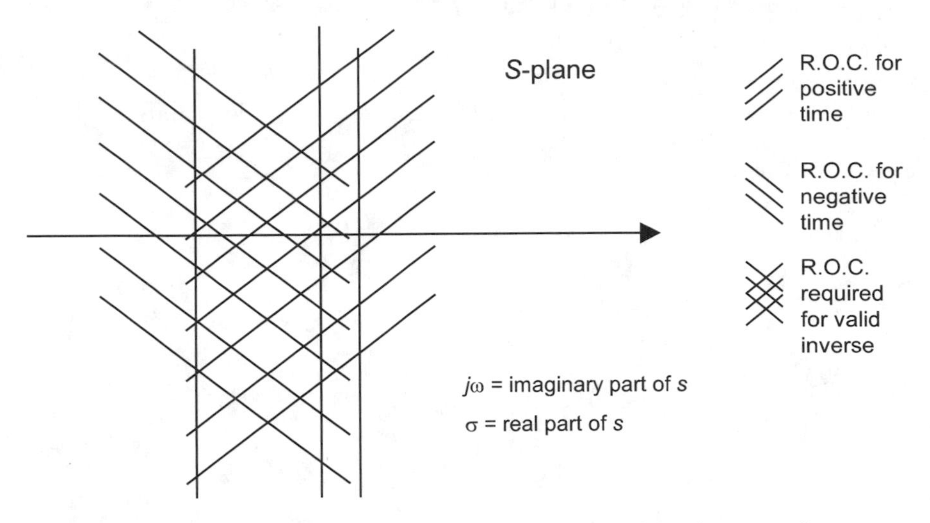

Figure 7.5: Region of convergence for the bilateral Laplace transform.

and

$$F_-(s) = \mathcal{L}_+ \{f_-(-t)\}|_{s \to -s} \tag{7.81}$$

where $f_-(t)$ is $f(t)$ for negative time arguments, and $f_-(-t)$ is $f(t)$ for negative time arguments with t being replaced by $t \to -t$.

To verify that the Laplace transform of the anti-causal part (negative time function) is done properly, we take a look at the expression

$$\begin{aligned}
\mathcal{L}_+ \{f_-(-t)\}|_{s \to -s} &= \int_0^\infty f_-(-t)\ e^{-st}\ dt\Big|_{s \to -s} \\
&= \int_0^\infty f_-(-t) e^{-s(-t)} dt \\
&= \int_{-\infty}^0 f(t)\ e^{-st}\ dt
\end{aligned}$$

which is the desired result.

To obtain the inverse transform in a bilateral sense, an overlapping region of convergence for both positive and negative time segments must exist (see the cross-hatched strip in Figure 7.5). We perform the inverse in four steps:

(1) For Re $s > \sigma_0$, take the one-sided conventional inverse and call it $f_+(t)\ u(t)$.

(2) For Re $s < \sigma_1$, replace $s \to -s$, take the conventional one-sided inverse, and call the result $f_-(-t)$.

(3) Replace $t \to -t$ in 2) and call it $f_-(t)$.

(4) $f(t) = f_+(t)\, u(t) + f_-(t)\, u(-t)$.

To illustrate the procedure we will look at a few simple examples.

Example 7.3 *Forward transform:*

$$f(t) = e^{-\gamma|t|} \qquad \gamma > 0$$

or

$$\begin{aligned} f(t) &= e^{-\gamma t}\, u(t) + e^{\gamma t}\, u(-t) \\ &= f_+(t)\, u(t) + f_-(t)\, u(-t) \end{aligned}$$

$$\begin{aligned} F_+(s) &= \mathcal{L}_+\{f_+(t)\} \\ &= \mathcal{L}_+\{e^{-\gamma t}\, u(t)\} = \int_0^\infty e^{-\gamma t}\, e^{-st}\, dt \\ &= \frac{1}{s+\gamma} \qquad ;\text{for Re (s)} > -\gamma \qquad \text{(R.O.C.)} \end{aligned}$$

$$\begin{aligned} F_-(s) &= \mathcal{L}_+\{f_-(-t)\}\big|_{s\to -s} \\ &= \left\{\int_0^\infty e^{\gamma t} u(-t)\Big|_{t\to -t} e^{-st} dt\right\}\Bigg|_{s\to -s} \\ &= \left\{\int_0^\infty e^{-\gamma t} e^{-st} dt\right\}\Bigg|_{s\to -s} = \frac{1}{s+\gamma}\Bigg|_{s\to -s} \\ &= \frac{1}{-s+\gamma} \end{aligned}$$

for Re $(s) < \gamma$ *(R.O.C.) so*

$$F(s) = \frac{1}{s+\gamma} + \frac{1}{-s+\gamma} = \frac{2\gamma}{-s^2+\gamma^2}$$

for Re $(s) > -\gamma$ *for* Re $(s) < \gamma$. *R.O.C.:* $|\text{Re }(s)| > \gamma$.

Inverse transform: Given

$$F(s) = \frac{2\gamma}{-s^2+\gamma^2}$$

R.O.C.: $|\text{Re }(s)| < \gamma$ *where* $\gamma > 0$. *By partial fraction expansion, we have*

$$F(s) = \frac{1}{s+\gamma} + \frac{1}{-s+\gamma}$$

Re $(s) > -\gamma$ Re $(s) < \gamma$

(1) $f_+(t) = \mathcal{L}_+^{-1}\left\{\frac{1}{s+\gamma}\right\} = e^{-\gamma t}u(t)$

(2) $f_-(-t) = \mathcal{L}_+^{-1}\left\{\left.\frac{1}{-s+\gamma}\right|_{s\to -s}\right\} = \mathcal{L}_+^{-1}\left\{\frac{1}{s+\gamma}\right\} = e^{-\gamma t}$

(3) $f_-(t) = e^{\gamma t}$

(4) $f(t) = e^{-\gamma t}\ u(t) + e^{\gamma t}\ u(-t) = e^{-|\gamma| t}$; *for all* t.

Example 7.4

$$F(s) = \frac{5}{-s^2+s+6} \qquad \text{R.O.C.}: \ -2 < \text{ Re }(s) < 3$$

Note: there is only for a valid pole assignment.

$\Longrightarrow$ the causal pole responsible for the positive time function is at $s = -2$

$\Longrightarrow$ the anti-causal pole responsible for the negative time function is at $s = 3$

Using partial fraction expansion, we have

$$F(s) = \frac{1}{s+2} + \frac{1}{-s+3} = F_+(s) + F_-(s)$$

R.O.C. Re $(s) > -2$ R.O.C. Re $(s) < 3$

(1) $f_+(t) = \mathcal{L}_+^{-1}\left\{\frac{1}{s+2}\right\} = e^{-2t}u(t)$

(2) $f(-t) = \mathcal{L}_+^{-1}\left\{\left.\frac{1}{-s+3}\right|_{s\to -s}\right\} = \mathcal{L}_+^{-1}\left\{\frac{1}{s+3}\right\} = e^{-3t}$

(3) $f_-(t) = e^{3t}$

(4) $f(t) = e^{-2t}\ u(t) + e^{3t}\ u(-t)$.

7.7.2 Integral Equations

When the noise is a second order random process and can be thought of as being obtained as the output of a physically realizable filter that is driven by white noise, then we can convert the integral (7.72) into a differential equation. This section deals with the Fredholm integral equation of the first and second kind [10]. We changed the notation for the right-hand side of the integral equation by letting $x(t)$ replace $s(t)$. This is to avoid confusion, since the capital letter S stands for power spectral density when we evaluate the expressions in the s-domain. The equation of the form

$$\int_{t_0}^{t_f} R_N(t,u)\; h(u)\; du = x(t) \qquad t_0 \le t \le t_f \tag{7.82}$$

is the Fredholm equation of the first kind, where $R_N(t,u)$ and $x(t)$ are known functions and $h(t)$ is the unknown quantity. For Section 7.7.2, we prefer the more general limits t_0 and t_f and evaluate them at $t_0 = 0$ and $t_f = T$, when convenient.

If the colored noise has an underlying stationary white noise component, then the noise auto-correlation function has an impulse-like component. The noise correlation function can then be written as

$$R_N(t,u) = \frac{N_0}{2}\; \delta(t-u) + R_{N_c}(t,u) \tag{7.83}$$

where $R_N(t,u)$ is the correlation function of the noise process, $N_0/2\; \delta(t-u)$ is the correlation function of the white noise component, and $R_{N_c}(t,u)$ is the correlation function due to the colored noise component. Then the integral equation becomes

$$\frac{N_0}{2}\; h(t) + \int_{t_0}^{t_f} R_N(t,u)\; h(u)\; du = x(t)\;; \qquad t_0 \le t \le t_f \tag{7.84}$$

This is called the Fredholm integral equation of the second kind. There is also the homogeneous integral (7.43) which was discussed earlier.

Usually, solutions to integral equations are tedious undertakings. However, under some reasonable conditions, we can obtain solutions by translating to and solving an equivalent differential equation. This approach is applicable when one can interpret the colored noise as being the output of a linear time invariant system that is driven by stationary white noise (i.e., leucogenic noise as coined in [4]). In this case, the colored noise is stationary and its power spectral density (in the s-domain) is the ratio of two even ordered polynomials. The degree of the denominator is greater than the degree of the numerator, guaranteeing that the power of the noise process is finite. We can express the Laplace transform of the correlation function as

$$S_N(s) \quad = \quad \int_{-\infty}^{\infty} R_N(\tau)\; e^{-s\tau}\; d\tau$$

$$= \frac{N(s^2)}{D(s^2)} \tag{7.85}$$

We recall that

$$\mathcal{L}\{\delta(t)\} = \int_{-\infty}^{\infty} \delta(t)\, e^{-st}\, dt = 1 \tag{7.86}$$

hence,

$$\mathcal{L}\{\delta(t-\tau)\} = \int_{-\infty}^{\infty} \delta(t-\tau)\, e^{-st}\, dt = e^{-s\tau} \tag{7.87}$$

with an inverse given by

$$\mathcal{L}^{-1}\left\{e^{-s\tau}\right\} = \int_{-\infty}^{\infty} e^{-s\tau} e^{st}\, ds$$

and

$$\delta(t-\tau) = \int_{-\infty}^{\infty} e^{s(t-\tau)}\, ds$$

Now

$$\begin{aligned} \frac{d}{dt}\,\delta(t-\tau) &= \int_{-\infty}^{\infty} \frac{d}{dt}\, e^{s(t-\tau)}\, ds \\ &= \int_{-\infty}^{\infty} s\, e^{s(t-\tau)}\, ds \end{aligned} \tag{7.88}$$

Hence, in general

$$\frac{d^{(k)}}{dt^{(k)}}\,\delta(t-\tau) = \int_{-\infty}^{\infty} s^k\, e^{s(t-\tau)}\, ds \tag{7.89}$$

Using the notation $p \mathrel{\hat{=}} d/dt$

$$N(p^2)\,\delta(t-\tau) = \int_{-\infty}^{\infty} N(s^2)\, e^{s(t-\tau)}\, ds \tag{7.90}$$

where $N(p^2)$ is an even ordered polynomial of differential operations and $N(s^2)$ is a corresponding polynomial in even powers of s (i.e., if $N(p^2) = a_0p^0 + a_2p^2$ then $N(s^2) = a_0 + a_2\, s^2$). We also recall

$$R_N(u) = \int_{-\infty}^{\infty} S_N(s)\, e^{su}\, ds \tag{7.91}$$

where $R_N(u)$ is the correlation function and $S_N(s)$ is the spectral density in the s-domain. A shifted version then looks like (i.e., let $u = t - \tau$)

$$R_N(t-\tau) = \int_{-\infty}^{\infty} S_N(s)\, e^{s(t-\tau)}\, ds \tag{7.92}$$

Application of $D(p^2)$ to both sides of (7.92) leads to

$$D(p^2)\ R_N(t-\tau) = \int_{-\infty}^{\infty} D(s^2)\ S_N(s)\ e^{s(t-\tau)}\ ds \tag{7.93}$$

Realizing that

$$S_N(s) = \frac{N(s^2)}{D(s^2)}$$

or

$$N(s^2) \ = \ D(s^2)\ S_N(s) \tag{7.94}$$

we obtain

$$D(p^2)\ R_N(t-\tau) = \int_{-\infty}^{\infty} N(s^2)\ e^{s(t-\tau)}\ ds \tag{7.95}$$

This expression becomes, using the results from (7.90) on the right-hand side

$$D(p^2)\ R_N(t-\tau) \ = N(p^2)\ \delta(t-\tau) \tag{7.96}$$

where $p = d/dt$. We can use this result to solve (7.82) for stationary noise kernels, which is reproduced here for convenience.

$$\int_{t_0}^{t_f} R_N(t-\tau)\ h(\tau)\ d\tau = x(t) \qquad t_0 \le t \le t_f \tag{7.97}$$

Hence, when we operate on both sides of (7.97) with $D(p^2)$, we obtain

$$\int_{t_0}^{t_f} D(p^2)\ R_N(t-\tau)\ h(\tau)\ d\tau = D(p^2)\ x(t)\ ; \qquad t_0 \le t \le t_f \tag{7.98}$$

Using (7.96) this becomes

$$\int_{t_0}^{t_f} N(p^2)\ \delta(t-\tau)\ h(\tau)\ d\tau = D(p^2)\ x(t)\ ; \qquad t_0 \le t \le t_f \tag{7.99}$$

or

$$N(p^2)\ h(t) = D(p^2)\ x(t)\ ; \qquad t_0 \le t \le t_f \tag{7.100}$$

Equation (7.100) is a differential equation. Essentially, we converted an integral equation problem into a differential equation problem, which can be solved using standard techniques.

The solution of the differential equation (7.100) consists of three parts $h(t) = h_h(t) + h_p(t) + h_e(t)$, where $h_h(t)$ is the homogeneous solution (i.e., no excitation), $h_p(t)$ is the particular solution (i.e., due to the input $x(t)$), and $h_e(t)$ is the end point solution [4,10,18,21].

The solution to (7.97) is obtained by solving (7.100). We will sketch out the steps and provide an example. Given that the differential equation is of the form

$$N(p^2)\ h(t) = D(p^2)\ x(t)\ ; \qquad t_0 \le t \le t_f$$

we solve

(a) The homogeneous part

$$N(p^2)\ h(t) = 0 \tag{7.101}$$

$h_h(t) =$ solution of (7.101)

(b) $N(p^2)\ h(t) = D(p^2)\ x(t)$ can be solved using Laplace transform techniques

$$\begin{aligned} N(s^2)\ H(s) &= D(s^2)\ X(s) \\ H(s) &= \frac{D(s^2)}{N(s^2)} X(s) \\ &= S_N^{-1}(s)\ X(s) \\ h_p(t) &= \mathcal{L}^{-1}\left\{S_N^{-1}(s)\ X(s)\right\} \end{aligned} \tag{7.102}$$

(c) The end point solution is given by

$$h_e(t) = \sum_{i=0}^{2(p-q-1)} \left[a_i\ \delta^{(i)}(t - t_0) + b_i\ \delta^{(i)}(t - t_f)\right] \tag{7.103}$$

where $\delta^{(i)}(t)$ is the ith derivative of the delta function, p is the order of $D(s^2)$, and q is the order of $N(s^2)$.

Numerical values of the coefficients contained in the set $\{a_i, b_i\}$ are obtained by evaluating (7.97) and using

$$h(t) = h_h(t) + h_p(t) + h_e(t)$$

Example 7.5 *A correlation based receiver for a known signal in additive colored Gaussian noise ($0 \le t \le T$) is desired. We are given*

$$y(t) = x(t) + n(t)$$

where $x(t) = x_1(t) - x_0(t)$ is the difference signal

$$R_N(\tau) = Ae^{-\gamma\ |\tau|} \longrightarrow S_N(s) = \frac{2A\ \gamma}{-s^2 + \gamma^2}$$

$$\int_0^T R_N(t - \tau)\ h(\tau)\ d\tau = x(t)\ ; \qquad 0 \le t \le T$$

(This is a Fredholm integral equation of the first kind.)

$$\begin{aligned} D(s^2) &= -s^2 + \gamma^2 s^0 \implies p = 1 \qquad \text{(order of } s^2\text{)} \\ N(s^2) &= 2A\gamma s^0 \implies q = 0 \qquad \text{(order of } s^2\text{)} \end{aligned}$$

Solution:

$$N(p^2)\ h(t) = D(p^2)\ x(t)\ ; \qquad 0 \le t \le T$$

we have

(a) $N(p^2)\ h(t) = 0$ *(homogeneous part), or* $2\ A\ \gamma\ h(t) = 0$. *This is not a differential equation, just an algebraic one. Hence, (a),* $h_h(t)$ *is of the form*

$$h_h(t) = 0$$

(b)

$$\underbrace{-\frac{d^2}{dt^2}x(t) + \gamma^2 x(t)}_{D(p^2)x(t)} = \overbrace{2A\gamma}^{N(p^2)}\ h(t) \qquad \text{(particular solution)}$$

is an algebraic equation in $h(t)$ *(a differential equation in* $x(t)$*). Hence, (b),* $h_p(t)$ *is of the form*

$$h_p(t) = \frac{1}{2A\gamma}\left[\gamma^2 x(t) - \ddot{x}(t)\right]$$

(c) *End points*

$$i \text{ has a range from } 0 \text{ to } 2(p-q-1) = 0 \quad \Longrightarrow \quad \delta^{(i)}(t) = \delta^{(0)}(t) = \delta(t)$$

$$\begin{aligned} h_e(t) &= \sum_{i=0}^{0} a_i\ \delta^{(i)}(t) + b_i\ \delta^{(i)}(t-T) \\ &= a_0\ \delta(t) + b_0\ \delta(t-T) \end{aligned}$$

The total solution is given by

$$\begin{aligned} h(t) &= h_p(t) + h_h(t) + h_e(t) \\ h(t) &= \frac{1}{2A\gamma}\left(\gamma^2 x(t) - \ddot{x}(t)\right) + a_0\ \delta(t) + b_0\ \delta(t-T) \end{aligned}$$

We still need to solve for a_0 *and* b_0*. Here* $x(t)$ *is the known signal,* $y(t)$ *is the received signal, and* $n(t)$ *is the additive colored Gaussian noise. From (7.97), we have*

$$\begin{aligned} x(t) &= \int_{0_-}^{T^+} R_N(t-\tau)h(\tau)d\tau \qquad \text{(includes end points)} \\ &= \int_{0_-}^{T^+} Ae^{-\gamma|t-\tau|}\left\{\frac{1}{2A\gamma}\left(\gamma^2 x(\tau) - \ddot{x}(\tau)\right) + a_0\delta(\tau) + b_0\delta(\tau-T)\right\} d\tau \end{aligned}$$

The integral is split into two regions to eliminate the absolute value symbol.

$$x(t) = \int_{0_-}^{t} Ae^{-\gamma(t-\tau)} \left\{ \frac{1}{2A\gamma} \left(\gamma^2 x(\tau) \quad - \ddot{x}(\tau) \right) + a_0 \delta(\tau) \right\} d\tau$$

$$+ \int_{t}^{T^+} Ae^{+\gamma(t-\tau)} \left\{ \frac{1}{2A\gamma} \left(\gamma^2 x(\tau) - \ddot{x}(\tau) \right) + b_0 \delta(\tau - T) \right\} d\tau$$

$$= a_0 \; Ae^{-\gamma t} + b_0 \; A \; e^{\gamma(t-T)} + \int_{0_-}^{t} Ae^{-\gamma(t-\tau)} \frac{1}{2A\gamma} \left(\gamma^2 x(\tau) - \ddot{x}(\tau) \right) d\tau$$

$$+ \int_{t}^{T^+} Ae^{+\gamma(t-\tau)} \frac{1}{2A\gamma} \left(\gamma^2 \; x(\tau) - \ddot{x}(\tau) \right) d\tau$$

Now

$$\int_{0_-}^{t} e^{\gamma\tau} \; \ddot{x}(\tau) d\tau = e^{\gamma t} \; \dot{x}(t) \; - \dot{x}(0) - \gamma \; e^{\gamma t} \; x(t) - \gamma \; x(0)$$

$$+ \int_{0}^{t} \gamma^2 e^{\gamma \; \tau} \; x(\tau) \; d\tau$$

and

$$\int_{t}^{T^+} e^{-\gamma\tau} \ddot{x}(\tau) d\tau = e^{-\gamma T} \dot{x}(T) - e^{-\gamma t} \dot{x}(t) + \gamma e^{-\gamma T} x(T) - \gamma e^{-\gamma t} x(t)$$

$$+ \int_{t}^{T^+} \gamma^2 e^{-\gamma\tau} x(\tau) d\tau$$

so we obtain

$$x(t) = x(t) + \left\{ e^{\gamma t} \left[a_0 A + \frac{\dot{x}(0)}{2\gamma} - \frac{x(0)}{2} \right] \right.$$

$$\left. e^{+\gamma t} e^{-\gamma T} \left[b_0 A - \frac{\dot{x}(T)}{2\gamma} - \frac{x(T)}{T} \right] \right\}$$

$$\text{for } 0 \leq t \leq T$$

But $x(t) \equiv x(t) \implies$ *curly bracket terms* <u>*must*</u> *be zero for all* $0 \leq t \leq T$.

$$\left. \begin{aligned} \implies a_0 A + \frac{\dot{x}(0)}{2\gamma} - \frac{x(0)}{2} = 0 \\ b_0 A - \frac{\dot{x}(T)}{2\gamma} - \frac{x(T)}{2} = 0 \end{aligned} \right\}$$

can split and set each term to zero since second term in the last expression has a time varying term (i.e., $e^{\gamma t}$ *and* $e^{-\gamma t}$)

or

$$\begin{aligned} a_0 &= \frac{1}{2A\gamma}\left(\gamma x(0) - \dot{x}(0)\right) \\ b_0 &= \frac{1}{2A\gamma}\left(\gamma x(T) + \dot{x}(T)\right) \end{aligned}$$

Hence, the desired reference correlation (i.e., replica) function is given by

$$h(t) = \frac{1}{2A\gamma}\left\{\left(\gamma x(0) - \dot{x}(0)\right)\delta(t) + \left(\gamma x(T) + \dot{x}(T)\right)\delta(t-T) + \gamma^2 x(t) - \ddot{x}(t)\right\}$$

for $0 \leq t \leq T$.

7.7.3 Performance of the Optimum Receiver for Known Signals in Colored Gaussian Noise

This section shows how one can solve in principle for P_{FA} and P_D to test for the presence of a known signal in additive colored Gaussian noise. From Figure 7.4, we can ascertain that the output of the optimum detector will be a Gaussian random variable. This is true under each hypothesis, since a linear operation on a Gaussian random variable results in a Gaussian random variable. Replicating (7.69), for convenience and denoting the output Gaussian random variable as z, we have

$$z = \int_0^T y(t)\ h(t)\ dt \ \underset{H_0}{\overset{H_1}{\gtrless}}\ T_1 \tag{7.104}$$

The mean of z under the H_0 hypothesis is given by

$$\begin{aligned} E_0\ z &= \int_0^T (s_0(t) + n(t))\ h(t)\ dt \\ &= \int_0^T s_0(t)\ h(t)\ dt \end{aligned} \tag{7.105}$$

while the mean of z the H_1 hypothesis is given by

$$E_1\ z = \int_0^T s_1(t)\ h(t)\ dt \tag{7.106}$$

The variance, invariant under each hypothesis is given by

$$\sigma_z^2 \ = \ \sigma_z^2|H_0 \ = \ \sigma_z^2|H_1$$

$$
\begin{aligned}
&= E_i\{[\int_0^T (s_i(t) + n(t))\ h(t)\ dt - \int_0^T s_i(t)\ h(t)\ dt]^2\} \\
&= E_i\{[\int_0^T n(t)\ h(t)\ dt]^2\} \\
&= \int_0^T \int_0^T R_N(t,s)\ h(t)\ h(s) dt\ ds \\
&= \int_0^T \int_0^T R_N(t-s) \sum_i 1/\lambda_i\ (s_{1i} - s_{0i}) g_i(t) \\
&\quad \sum_j 1/\lambda_j\ (s_{1j} - s_{0j}) g_j(s)\ dt\ ds \\
&= \sum_i \sum_j 1/(\lambda_i \lambda_j)(s_{1i} - s_{0i})(s_{1j} - s_{0j}) \int_0^T \lambda_j g_j(t) g_j(t) dt \\
&= \sum_i 1/\lambda_i\ (s_{1i} - s_{0i}) \qquad (7.107)
\end{aligned}
$$

So the false alarm rate can be expressed as

$$P_{FA} = Q((T_1 - E_0\ z)/\sigma_z) \qquad (7.108)$$

while the probability of detection is given by

$$P_D = Q((T_1 - E_1\ z)/\sigma_z) \qquad (7.109)$$

where the means and the variance are defined above. To obtain actual values one has to solve the homogeneous integral equation, which at best is a tedious undertaking.

7.7.4 Whitening Filter

Whitening, which is also sometimes called spectral shaping or pre-whitening, is a simple technique which, as the name implies, converts the colored noise to white noise. If possible, the whitener will correct the spectral shape of the noise and convert the problem into one that can then be solved with the standard matched filter or correlator. One can solve for a whitening filter by assuming that the impulse response of the whitener is of the form [18]

$$h_W(t,u) = \sum_j h_j\ q_j(t)\ q_j(u) \qquad (7.110)$$

where $\{q_j(t)\}$ is a C.O.N. set based on the noise kernel eigendecomposition. Note that this impulse response admits a non-stationary behavior. With some work [18] one can show that $\{h_j\} = \{1/\lambda_j\}$. Hence,

$$h_W(t,u) = \sum_j 1/\lambda_j \;\; q_j(t) \; q_j(u) \tag{7.111}$$

To obtain the eigenvalues and eigenfunctions, the homogeneous integral equation has to be solved.

A practical solution exists when the noise is stationary and the signal duration T much larger than the correlation width of the noise, or the observation interval $>> T$, and the noise spectrum is of a rational (leucogenic) form. If we let the start time approach $-\infty$, then the whitening filter of (7.110) can be made time-invariant, that is

$$h_W(\tau) = h_W(t-u) \tag{7.112}$$

where $\tau = t - u$. Hence, given the noise correlation function $R_N(\tau)$, the spectral density is given by $S_N(s) = S_N^+(s) \; S_N^-(s)$, where $S_N^+(s)$ has all its poles and zeros in the left-hand s-plane (i.e., the causal contribution) and $S_N^-(s)$ has all its poles and zeros in the right-hand s-plane (i.e., the anti-causal contribution), then the transfer function of the whitening filter is given by

$$H_W(s) = 1/S_N^+(s) \tag{7.113}$$

and the corresponding impulse response is defined by

$$h_W(t) = \mathcal{L}^{-1} \, \{1/S_N^+(s)\} \tag{7.114}$$

If this filter is used to whiten the received data, it must also be applied to the replica signals (see Figure 7.6) to allow the signals to correlate when they are present.

7.8 DISCRETE-TIME DETECTION — KNOWN SIGNALS EMBEDDED IN COLORED GAUSSIAN NOISE

7.8.1 Introduction

This section will focus on the detection of known signals in colored Gaussian noise, where the observations are processed in discrete-time (i.e., at sample instances). We assume that the noise process has a spectral density function that is a rational polynomial, hence the process will always satisfy the Paley-Wiener condition. This implies that the power spectral density has no extended regions in which it takes on a zero value [3]. Depending on the spectral region of interest, and relative to the total spectral region available, one can partition the detection problem into two classes:

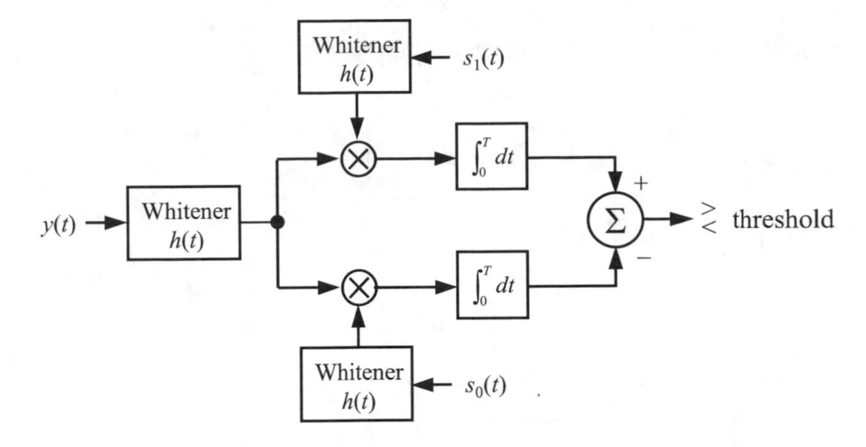

Figure 7.6: Two channel realization of the analog whitener.

(a) The spectral region of the signal(s) of interest, i.e., the bandwidth, is narrow relative to the center frequency, and the PSD of the noise is almost constant over the band. In this particular case, as far as the receiver is concerned, the noise (after proper filtering) can be treated as a bandlimited white noise process, where the PSD outside the band takes on very small values. Hence, the noise correlation function is approximately a sinc-function, whose first zero crossing occurs at lag one. This says that $R_N(m) \cong 0$, for $|m| \geq 1$ since all other correlation lags will have low correlation values. Hence, due to their Gaussian nature, the uncorrelated samples can be assumed to be statistically independent. We can solve this class of problems, as was already done in Chapter 6 (see the development of the discrete-time correlator).

(b) The PSD of the noise exhibits variations in magnitude over the band of interest. Then one can whiten (also called pre-whiten) the data so that at the output of the whitener the noise will have a constant spectral height. In some literature the whitener is also referred to as a spectral equalizer. At this point in the processing chain, we have converted the original problem of detecting a known signal embedded in colored Gaussian noise into the problem of detecting the known filtered signal embedded in white Gaussian noise. Hence, the standard approach of replica correlation applies, as was derived in Chapter 6, Section 6.2. The major difference is that now the replica function will be replaced with a filtered version of the original signal. This technique is illustrated in Figure 7.7.

We note that one can combine the whitening and filtering operations into a one-step operation. For clarity, we typically do not implement this type of

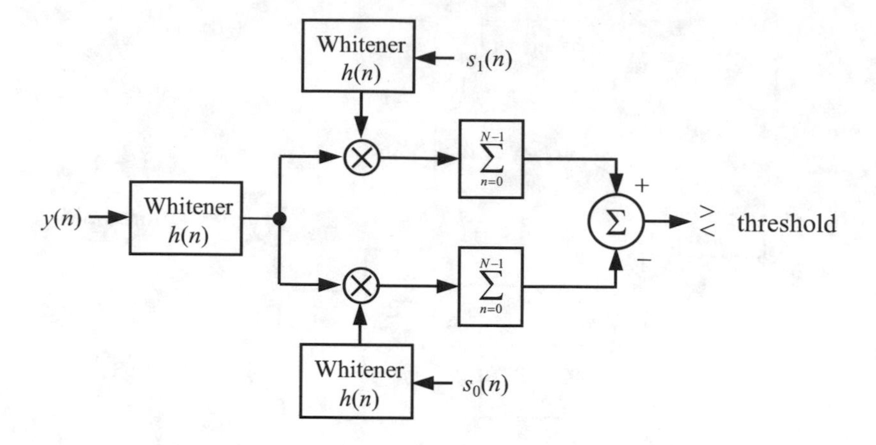

Figure 7.7: Discrete-time colored noise detector.

streamlined operation. Rather than that, we usually prefer to keep the building blocks (an explicit whitener and the correlator) separate. We also want to point out that the wavelet transformation tends to whiten the noise [20], but the different filters will have a different bandwidth and output sampling rate (i.e., see Chapter 3 and Appendix E).

7.8.2 Whitening via Spectral Factorization

There are several ways to accomplish the whitening of the noise. Many of them are championed in the digital signal processing literature [3,13–16]. Knowledge of the noise auto-correlation function is equivalent to knowing the power spectral density of the noise. Most of the time, the PSD can be factored into its pole and zero constituents. This allows for the separation of the poles and zeros that lie inside the unit circle from those that do not. We assume that the PSD of the colored noise does not take on a zero value; hence, there are no zeros on the unit circle. There cannot be any poles on the unit circle since that would result in spectral resonance (i.e., a response due to a sinusoid). So in this discussion only poles and zeros inside and outside the unit circle will be considered. We further assume that the observation (or processing) time is much larger than the signal duration, unless it is called for otherwise. The PSD can be written as

$$S_N(z) = \mathcal{Z}\{R_N(m)\} \tag{7.115}$$

which can be factored as

$$S_N(z) = K \;\; S_N^+(z) \; S_N^-(z) \tag{7.116}$$

where $S_N^+(z)$ contains all the poles and zeros that reside inside the unit circle, $S_N^-(z)$ contains all poles and zeros outside the unit circle, and K is a positive

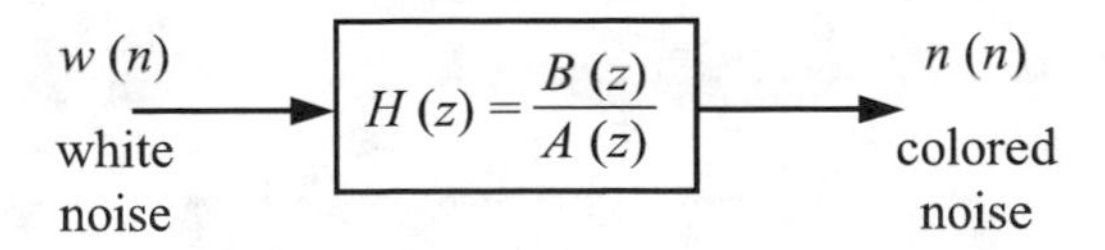

Figure 7.8: Generation of the colored noise sequence $n(n)$ from the white noise sequence $w(n)$.

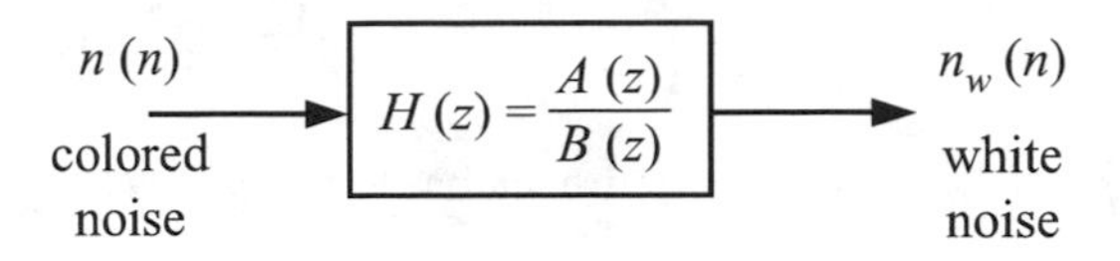

Figure 7.9: Whitener (spectral equalization).

valued constant. If we let $H(z) = S_N^+(z)$, then we can filter the noise process using the inverse of the filter transfer function (i.e., $H^{-1}(z)$) to obtain a constant power spectral density.

This is equivalent to the leucogenic process in analog systems [4] and its generating mechanism is shown in Figure 7.8.. One assumes that the colored noise is generated by processing stationary white noise with a system that has a transfer function with a finite number of poles and zeros. All poles and zeros must lie within the unit circle (i.e., the stable region). The requirement on the order of the poles relative to the zeros is relaxed. It is permissable to have more zeros than poles (i.e., FIR filter transfer functions have a non-trivial numerator polynomial) since the region of support extends only from $-\pi$ to π radians. The wording non-trivial polynomial refers to having roots not located at the origin. A valid model for the noise process is given by Figure 7.8. The zeros of the numerator $A(z)$ and denominator $B(z)$ polynomials are all inside the unit circle.

The whitening is accomplished as the data sequence $n(n)$ is equalized (see Figure 7.9), that is, the output becomes spectrally flat.

The filter response is obtained by using the factored power spectral density:

$$H_1(z) = S_N^+(z)^{-1} = \frac{A(z)}{B(z)} \tag{7.117}$$

Now the whitened noise is given by

$$n_W(n) = n(n) \; * \; h_1(n) \tag{7.118}$$

hence,

$$\begin{aligned} S_{N_W}(z) &= S_N(z)\,|H_1(z)|^2 = |B(z)/A(z)|^2 \; |A(z)/B(z)|^2 \\ &= \text{constant} \end{aligned} \tag{7.119}$$

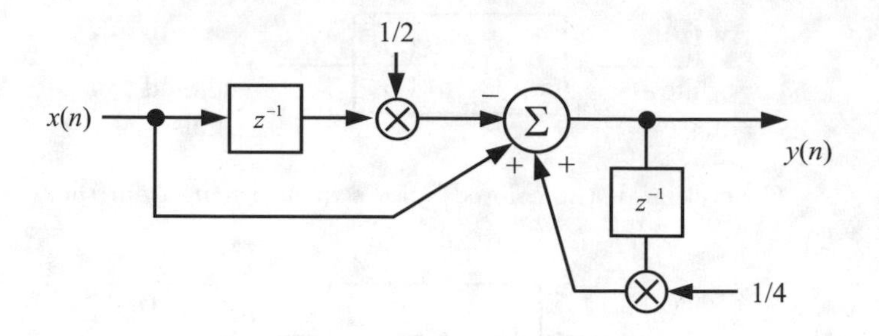

Figure 7.10: Example of a whitener based on spectral factorization.

Therefore, the noise has been whitened. Of course, the signal will also have been filtered, that is, the output signal is the prevailing input signal convolved with the impulse response $h_1(n)$. This approach leads to a whitener that operates in real time (i.e., on line) in a contiguous fashion (i.e., sample by sample). Now that the additive noise component is white, the standard approach (see Chapter 6.2) applies to the filtered signals embedded in white Gaussian noise.

Example 7.6

$$S_N(z) = \frac{z - (1/4 + 1/16) + z^{-1}}{z - (1/2 + 1/4) + z^{-1}}$$

or

$$S_N(z) = \frac{1 - 1/4\ z^{-1}}{1 - 1/2\ z^{-1}} \frac{1 - 1/4\ z}{1 - 1/2\ z}$$

hence

$$S_N^+ = \frac{1 - 1/4\ z^{-1}}{1 - 1/2\ z^{-1}} = \frac{z - 1/4}{z - 1/2}$$

The transfer function of the whitener is

$$H(z) = \frac{1}{S_N^+(z)} = \frac{z - 1/2}{z - 1/4}$$

The difference equation is easily derived via

$$Y(z)\ (1 - 0.25\ z^{-1}) = X(z)\ (1 - 0.5\ z^{-1})$$

resulting in

$$y(n) = 1/4\ y(n-1) + x(n) - 1/2\ x(n-1)$$

The filter that uses this transfer function is shown in detail in Figure 7.10. Since this filter shapes the embedded signal, the filter must also be applied to the stored replica to allow effective correlation.

7.8.3 Whitening Using Correlation Domain Information

If the observation time is relatively short (i.e., $T_{\text{observ}} \approx T_{\text{signal}} = T$), then an approach based on the decomposition of the noise auto-correlation function can be used. In this case we can factorize the correlation matrix and use the resultant matrix to filter a finite data segment. A good candidate is the $\mathbf{L}\ \mathbf{D}_L\ \mathbf{L}^H$ decomposition. The $\mathbf{L}\ \mathbf{D}_L\ \mathbf{L}^H$ decomposition is the special case of the $\mathbf{L}\ \mathbf{D}_L\ \mathbf{U}^H$ decomposition, when the matrix to be factored is symmetric. Using the matrix obtained via the factorization, one can transform (filter) the data of length L (i.e., $L \times L$ is the size of the correlation matrix) into L uncorrelated samples [3,16,17]. The transformation is obtained by factoring the correlation matrix as given by

$$\mathbf{R}_N = \mathbf{L}\ \mathbf{D}_L\ \mathbf{L}^H \tag{7.120}$$

where H denotes the Hermitian transpose (i.e., complex conjugate transpose). The matrix $\mathbf{L}$ is a unit lower triangular matrix, with unit elements on the diagonal. Pre-multiplying by $\mathbf{L}^{-1}$ and post-multiplying by $(\mathbf{L}^H)^{-1}$, we obtain

$$\mathbf{L}^{-1}\ \mathbf{R}_N\ (\mathbf{L}^H)^{-1} = \mathbf{L}^{-1}\ \mathbf{L}\ \mathbf{D}_L\ \mathbf{L}^H\ (\mathbf{L}^H)^{-1} = \mathbf{D}_L \tag{7.121}$$

Hence, $\mathbf{D}_L$ is a correlation matrix of the random vector given by

$$\mathbf{w} = \mathbf{L}^{-1}\ \mathbf{n} \tag{7.122}$$

Since $\mathbf{D}_L$ is a diagonal matrix, the components of the vector $\mathbf{w}$ are orthogonal (ortho-normal, if $\mathbf{D}_L$ is the identity matrix). Since $\mathbf{L}$ is a lower triangular matrix, $\mathbf{L}^{-1}$ is also a lower triangular matrix, which will perform the transformation shown in (7.122). As far as the noise is concerned, the end goal is to obtain independent random variables (i.e., uncorrelated and Gaussian). Any choice of transformation of the form

$$\mathbf{w} = \mathbf{A}\mathbf{n} \tag{7.123}$$

will do, as long as the matrix $\mathbf{A}$ makes the random variables contained in $\mathbf{w}$ orthogonal (or ortho-normal). The decomposition, $\mathbf{L}\ \mathbf{D}_L\ \mathbf{L}^T$ is just one of the transformations that can be used. For example, the Cholesky decomposition [17] can also be used to obtain the whitening effect.

Example 7.7 *Suppose the noise correlation function is given by the exponential expression*

$$R_N(i) = e^{-|i|} \qquad \text{for } i = 0, 1, 2, 3$$

The one sided correlation vector is given by

$$\mathbf{R}_N = [1.0000, 0.3679, 0.1353, 0.0498]^T$$

The $\mathbf{L}\ \mathbf{D}_L\ \mathbf{L}^H$ *decomposition provides an* $\mathbf{L}$ *matrix of the form*

$$\mathbf{L} = \begin{bmatrix} 1.0000 & 0.0000 & 0.0000 & 0.0000 \\ 0.3679 & 1.0000 & 0.0000 & 0.0000 \\ 0.1353 & 0.3679 & 1.0000 & 0.0000 \\ 0.0498 & 0.1353 & 0.3679 & 1.0000 \end{bmatrix}$$

So the desired transformation matrix is given by

$$\mathbf{L}^{-1} = \begin{bmatrix} 1.0000 & 0.0000 & 0.0000 & 0.0000 \\ -.3679 & 1.0000 & 0.0000 & 0.0000 \\ 0.0000 & -.3679 & 1.0000 & 0.0000 \\ 0.000 & 0.0000 & -.3679 & 1.0000 \end{bmatrix}$$

which can be used in (7.123). If we were to use the Cholesky decomposition, that is $\mathbf{R}_N = \mathbf{C}\ \mathbf{C}^T$*, then* $\mathbf{C}$ *is given by*

$$\mathbf{C} = \begin{bmatrix} 1.0000 & 0.0000 & 0.0000 & 0.0000 \\ 0.3679 & 0.9299 & 0.0000 & 0.0000 \\ 0.1353 & 0.3421 & 0.9299 & 0.0000 \\ 0.0498 & 0.1258 & 0.3421 & 0.9299 \end{bmatrix}$$

Then the desired transformation matrix is given by

$$\mathbf{C}^{-1} = \begin{bmatrix} 1.0000 & 0.0000 & 0.0000 & 0.0000 \\ -.3956 & 1.0754 & 0.0000 & 0.0000 \\ 0.0000 & -.3956 & 1.0754 & 0.0000 \\ 0.000 & 0.0000 & -.3956 & 1.0754 \end{bmatrix}$$

which can be used in (7.123).

Example 7.8 *Suppose the noise correlation function is given by the expression*

$$R_N(i) = 0.5^{|i|} \qquad \text{for } i = 0, 1, 2, 3$$

The one-sided correlation vector is given by

$$\mathbf{R}_N = [1.0000, 0.500, 0.2500, 0.1250]^T$$

The $\mathbf{L}\ \mathbf{D}_L\ \mathbf{L}^H$ *decomposition provides an* $\mathbf{L}$ *matrix of the form*

$$\mathbf{L} = \begin{bmatrix} 1.0000 & 0.0000 & 0.0000 & 0.0000 \\ 0.5000 & 1.0000 & 0.0000 & 0.0000 \\ 0.2500 & 0.5000 & 1.0000 & 0.0000 \\ 0.1250 & 0.2500 & 0.5000 & 1.0000 \end{bmatrix}$$

So the desired transformation matrix is given by

$$\mathbf{L}^{-1} = \begin{bmatrix} 1.0000 & 0.0000 & 0.0000 & 0.0000 \\ -.5000 & 1.0000 & 0.0000 & 0.0000 \\ 0.0000 & -.5000 & 1.0000 & 0.0000 \\ 0.000 & 0.0000 & -.5000 & 1.0000 \end{bmatrix}$$

which can be used in (7.123). If we were to use the Cholesky decomposition, that is $\mathbf{R}_N = \mathbf{C}\ \mathbf{C}^T$, *then* $\mathbf{C}$ *is given by*

$$\mathbf{C} = \begin{bmatrix} 1.0000 & 0.0000 & 0.0000 & 0.0000 \\ 0.5000 & 0.8660 & 0.0000 & 0.0000 \\ 0.2500 & 0.4330 & 0.8660 & 0.0000 \\ 0.1250 & 0.2165 & 0.4330 & 0.8660 \end{bmatrix}$$

Then the desired transformation matrix is given by

$$\mathbf{C}^{-1} = \begin{bmatrix} 1.0000 & 0.0000 & 0.0000 & 0.0000 \\ -.5774 & 1.1547 & 0.0000 & 0.0000 \\ 0.0000 & -.5774 & 1.1547 & 0.0000 \\ 0.000 & 0.0000 & -.5774 & 1.1547 \end{bmatrix}$$

which can be used in (7.123).

Example 7.9 *Suppose the noise correlation function is given as the linear combination of two exponential type terms*

$$R_N(i) = (-0.5)^{|i|} + 0.1^{|i|}$$

for $i = 0, 1, 2, 3$. *The one-sided correlation vector is given by*

$$\mathbf{R}_N = [2.0000, -0.4000, 0.2600, -0.1240]^T$$

The $\mathbf{L}\ \mathbf{D}_L\ \mathbf{L}^H$ *decomposition provides an* $\mathbf{L}$ *matrix of the form*

$$\mathbf{L} = \begin{bmatrix} 1.0000 & 0.0000 & 0.0000 & 0.0000 \\ -.2000 & 1.0000 & 0.0000 & 0.0000 \\ 0.1300 & -.1813 & 1.0000 & 0.0000 \\ -.0620 & 0.1225 & -.1793 & 1.0000 \end{bmatrix}$$

So the desired transformation matrix is given by

$$\mathbf{L}^{-1} = \begin{bmatrix} 1.0000 & 0.0000 & 0.0000 & 0.0000 \\ 0.2000 & 1.0000 & 0.0000 & 0.0000 \\ -.0938 & 0.1813 & 1.0000 & 0.0000 \\ 0.0207 & -.0900 & 0.1793 & 1.0000 \end{bmatrix}$$

which can be used in (7.123). If we were to use the Cholesky decomposition, that is $\mathbf{R}_N = \mathbf{C}\,\mathbf{C}^T$*, then* $\mathbf{C}$ *is given by*

$$\mathbf{C} = \begin{bmatrix} 1.4142 & 0.0000 & 0.0000 & 0.0000 \\ -.2828 & 1.3856 & 0.0000 & 0.0000 \\ 0.1838 & -.2511 & 1.3795 & 0.0000 \\ -.0877 & 0.1697 & -.2474 & 1.3792 \end{bmatrix}$$

Then the desired transformation matrix is given by

$$\mathbf{C}^{-1} = \begin{bmatrix} 0.7071 & 0.0000 & 0.0000 & 0.0000 \\ 0.1443 & 0.7217 & 0.0000 & 0.0000 \\ -.0680 & 0.1314 & 0.7249 & 0.0000 \\ 0.0150 & -.0653 & 0.1300 & 0.7250 \end{bmatrix}$$

which can be used in (7.123).

The matrix decomposition method will work for segments of length L (i.e., the dimension of the noise correlation matrix which equals the signal duration). This implies that the start and stop times of the signal should line up with the segment under consideration (i.e., known signal timing parameters). If we segment the data and the expected signal duration is of length L, then it is possible to be misaligned as much as by 1/2 of the length of the data points. If the signal energy is homogeneous in time, potentially half of the energy could be lost (i.e., a 3 dB loss). This loss can be reduced by performing overlap processing (choosing overlapping segments of data for processing). Of course, the price paid is the additional processing cost, i.e., for a 50 percent overlap factor, the processing cost doubles relative to an approach in which contiguous non-overlapping segments are used. The non-overlap segment processing lends itself to scenarios in which signal synchronization is ensured (i.e., binary communication signals, fixed pulse repetition frequency (PRF) radar signals, sonar signals, etc.). Noise samples from segment to segment will be correlated but within one segment the L samples will be independent. We note, that if one were to let this filter operate in a contiguous mode (i.e., shift one data point to define a new segment), output samples separated by a number of samples smaller than the filter length will be correlated, hence be colored.

7.8.4 Whitening via Auto-Regressive Modeling

An important aspect of digital signal processing is auto-regressive modeling. Here one assumes that the data is generated using an auto-regressive model (i.e., an IIR filter driven by a white noise source). As indicated in Figure 7.11, we see that the observation noise is obtained by processing white noise

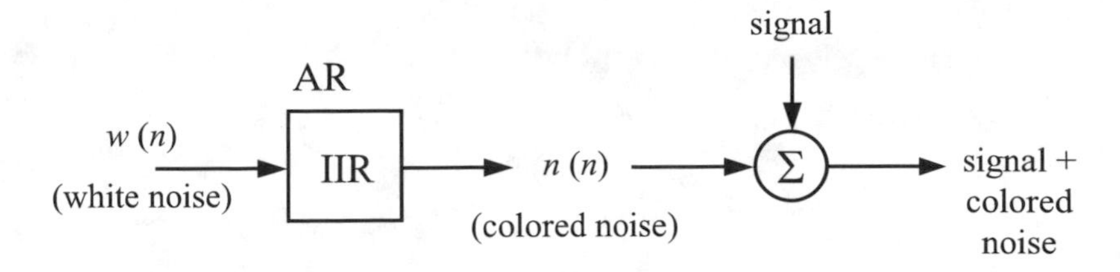

Figure 7.11: AR model.

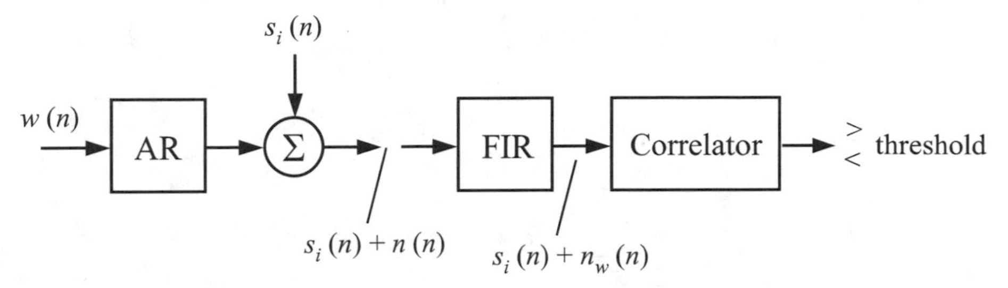

Figure 7.12: AR generation, whitener, and optimal detector.

with an IIR filter. This noise is also called an auto-regressive process and the shaping filter is called an all pole model (disregarding zeros due to the trivial numerator polynomial with roots only at the origin).

We already know from the earlier discussion in this section, that proper filtering of the colored noise sequence produces a white noise sequence. Figure 7.12 shows the processing stream to perform detection of the signal components embedded in colored Gaussian noise.

Here $n(n)$ is a colored noise sequence, and $s_i'(n)$ is the filtered signal component. To obtain the optimal filter weights one has to solve the Wiener (Yule-Walker, Normal) equations. These equations are intimately related, see for example [3]. We can solve for the optimal filter by studying the model in Figure 7.8. We have

$$n(n) = h(n) * w(n)$$

where $h(n)$ is the general impulse response, $w(n)$ is the white noise input sequence and $n(n)$ is the colored noise output sequence. We can express the system difference equation as

$$n(n) - \sum_{i=1}^{p} a_i \; w(n-i) = w(n) \qquad (7.124)$$

where a_i denotes the variable $h(i)$, for all required values of i. Post-multiplying by $n(n-l)$ and taking expectation leads to

$$R_N(l) - \sum_{i=1}^{p} a_i \; R_N(l-i) = R_{WN}(l) \qquad (7.125)$$

Now

$$\begin{aligned} R_{WN}(l) &= E\{w(n)\, n(n-l)\} \\ &= E\{w(n) \sum_{i=1}^{p} a_i\, w(i) h(n-l-i)\} \\ &= \sum_{i=1}^{p} \sigma_W^2\, \delta(n-i)\, h(n-l-i) \\ &= \sigma_W^2\, h(-l) \end{aligned}$$

So (7.125) becomes

$$\begin{array}{lll} R_N(0) & -\sum_{i=1}^{p} a_i\, R_N(-i) = \sigma_W^2\,, & \text{for } l = 0 \\ R_N(-1) & -\sum_{i=1}^{p} a_i\, R_N(-i+1) = 0\,, & \text{for } l = 1 \\ R_N(-2) & -\sum_{i=1}^{p} a_i\, R_N(-i+2) = 0\,, & \text{for } l = 2 \\ \vdots & \vdots & \vdots \\ R_N(-p) & -\sum_{i=1}^{p} a_i\, R_N(-i+p) = 0\,, & \text{for } l = p \end{array}$$

In matrix form, we can write

$$\mathbf{R}_N\, \mathbf{a} = \mathbf{r} \tag{7.126}$$

$$\mathbf{R}_N = \begin{bmatrix} R_N(0) & R_N(-1) & R_N(-2) & \cdots & R_N(-P+1) \\ R_N(1) & R_N(0) & R_N(-1) & \cdots & R_N(-P+2) \\ R_N(2) & R_N(1) & R_N(0) & \cdots & R_N(-P+3) \\ \vdots & \vdots & \vdots & & \vdots \\ R_N(P-1) & R_N(P-2) & R_N(P-3) & \cdots & R_n(0) \end{bmatrix}$$

with

$$\mathbf{a}^T = [a(1), a(2), \cdots, a(p)]$$

and

$$\mathbf{r}^T = [R_N(1)\; R_N(2) \cdots R_N(P)]$$

This approach works with sequential data samples (i.e., real time), hence there is no need to segment the data. All noise output samples of the FIR whitener will be statistically independent once the FIR filter has processed L data samples (i.e., has processed sample $(L-1)$). The whitened output can then be correlated with the filtered $s_i'(n)$ components, (i.e., $s_i'(n) = s_i(n) * h(n)$, the FIR filter weights). We also note that the auto-regressive modeling can be interpreted as a spectral factorization, where only the denominator $A(z)$ is a non-trivial polynomial.

Example 7.10 *Suppose we have the one-sided correlation vector as given by*

$$\mathbf{R}_N = [1.0000, 0.800, 0.0640]^T$$

We can easily verify that the corresponding correlation function given by

$$R_N(i) = 0.8^{|i|} \qquad \text{for } i = 0, 1, 2$$

Suppose we let the model order be $P = 1$, then we have

$$R_N(0)\ a_1 = R_N(1)$$

hence

$$a_1 \;=\; R_N(1)/R_N(0) \;=\; 0.8/1 \;=\; 0.8$$

Suppose we let the model order be $P = 2$, then we have

$$\begin{aligned} R_N(0)\ a_1 \;+\; R_N(1)\ a_2 &= R_N(1) \\ R_N(1)\ a_1 \;+\; R_N(0)\ a_2 &= R_N(2) \end{aligned}$$

hence

$$\begin{aligned} 1.0\ a_1 \;+\; 0.8\ a_2 &= 0.8 \\ 0.8\ a_1 \;+\; 1.0\ a_2 &= 0.64 \end{aligned}$$

Solving these two simultaneous equations leads to

$$a_1 = 0.8 \quad \text{and} \quad a_2 = 0$$

If we would increase the model order P, we will see that all coefficients past the first one (i.e., a_1 will be zero). So we see we can easily extract the proper pole location, even when the assumed model order exceeds the true model order.

7.8.5 Whitening by Discretizing the Continuous Time Karhunen-Loève Equations

We can also approach the colored noise problem using the results obtained in Section 7.7. Suppose that we do have continuous time processes so that

(7.72) is applicable. Further, assume that we sample the data in the Nyquist sense, then $s(n)$, any signal component, can be accurately represented by the DFT representation (i.e., the discrete Fourier synthesis equation)

$$s(n) = \frac{1}{N} \sum_{k=0}^{N-1} S(k)\ e^{j(2\pi/N)kn} \tag{7.127}$$

where $S(k)$ is the complex amplitude spectrum and $2\pi k/N = \omega_k$. The highest frequency component (i.e., $\omega_{\max} = \omega_0$) has a correlation function component of the form

$$R(m) = A \cos(\omega_0\ m)$$

where ω_0 corresponds to the highest frequency present. One can sample the correlation function at the same rate as would be appropriate for the analog (continuous time) data. Hence, (7.72) can be represented by

$$\sum_{m=0}^{L-1} R_N(l-m)\ h(m) = s(l) = s_1(l) - s_0(l) \qquad 0 \leq l \leq L-1 \tag{7.128}$$

In matrix form we have, using $\mathbf{s} = \mathbf{s}_1 - \mathbf{s}_0$ as the signal difference sequence

$$\mathbf{R}_N\ \mathbf{h} = \mathbf{s} \tag{7.129}$$

where we replaced the continuous version of $s(t)$ in (7.72) with the appropriate sampled version $s(l) = s_1(l) - s_0(l)$. The matrix equation is similar to the matrix equation of the auto-regressive approach (i.e., see (7.126)), but with $\mathbf{h}$ and $\mathbf{r}$ represented by

$$\mathbf{a}^T = \mathbf{h}^T = [h_1(0) - h_0(0), h_1(1) - h_0(1), \cdots, h_1(L-1) - h_0(L-1)]$$

and

$$\mathbf{r}^T = \mathbf{s}^T = [s_1(0) - s_0(0), s_1(1) - s_0(1), \cdots, s_1(L-1) - s_0(L-1)]$$

One can also start directly with the discrete-time sequences, and produce the discrete-time equivalent of the Karhunen-Loève expansion. With some work one can show the result is equivalent to the last matrix equation. We note, if the noise process is white, then (7.128) can be solved as given by

$$h(i) = \frac{s_1(i) - s_0(i)}{R_N(0)}$$

resulting in the correlator solution.

We also note that many times the analytic form of the noise correlation function, or the noise PSD are not available. One can estimate the correlation function, noting that the signal may be present and that its effects must be nullified. Some application of adaptive filtering techniques might provide ways to get a useful approximation of the correlation function. This approach tries to take advantage of the temporal and spectral differences between signal and noise. At high SNR levels, one can use time-frequency distributions (i.e., Wigner-Ville distributions [22]) to mask out temporal and spectral regions that may have signal components.

If one has noise only data cuts, one can obtain numerical estimates of the correlation or spectral density function. These estimates are not the same as the analytic expressions. To solve the Yule Walker, Normal or Wiener equations, or in general use correlation based approaches, numerical estimates of the correlation function will do. It is not as simple as when one wants to extract pole and zero locations to design a whitener based on (7.117).

7.8.6 Matched Filtering

For the known signal case (i.e., deterministic signal) in colored noise reference [3] derives the optimal FIR filter to maximize the output SNR. In this particular case the whitening and correlation are combined to occur simultaneously while performing the FIR filter operation. For real valued signals, the optimal filter weights are given by

$$\mathbf{h} = (\mathbf{s}^T \mathbf{R}_N^{-1}\ \mathbf{s})^{-1/2}\ \mathbf{R}_N^{-1}\ \tilde{\mathbf{s}} \tag{7.130}$$

where $\tilde{\mathbf{s}}$ is the time reversed vector $\mathbf{s}$. For white noise, the FIR filter weights become

$$\mathbf{h} = (\mathbf{s}^T \mathbf{s})^{-1/2}\ \tilde{\mathbf{s}} \tag{7.131}$$

which is the scaled time reversed version of the signal under consideration. This particular approach works also if the noise is not Gaussian, but note the optimum receiver is based on maximizing the output SNR and not on minimizing the Bayes' cost. In the non-Gaussian case, these two approaches do not provide the same solution.

7.9 SUMMARY

Chapter 7 discusses the detection of signals when the noise is not white but still Gaussian. At first, detection in a white noise background is re-examined in the context of C.O.N. basis sets. Then the basis sets are changed using the Gram-Schmidt procedure. In Section 7.4, the Gram-Schmidt procedure is applied to the white Gaussian noise case. All of this is used to familiarize the reader with series representations. Section 7.5 addresses the series expansion for colored Gaussian noise. In particular, the Karhunen-Loève expansion and Mercer's theorem are introduced. Section 7.6 applies the concepts introduced in Section 7.5 to the colored Gaussian noise case, leading to the integral equation (7.78). In this context, the bilateral Laplace transformation is introduced and the integral equations, in particular Fredholm integral equations, are discussed. The Laplace transform is used to convert the Fredholm integral equation of the first kind into a differential equation. One standard problem is worked out to illustrate the steps of the technique. Section 7.8 addresses discrete-time problems and introduces some of the algorithms used in the statistical digital signal processing area. This allows whitening of the discrete-time colored noise sequence. Consequently, the colored noise becomes white noise, the signals are spectrally shaped, and the problem becomes a standard problem that is solved earlier in Chapter 6.

7.10 PROBLEMS

1. Given the power series: $1/0!, t^1/1!, t^2/2!, \cdots, t^i/i!, \cdots$ for $0 \leq t \leq 1$
   (a) Obtain the Gram-Schmidt expansion for the first four terms.
   (b) Plot the basis functions for the original and the modified functions (using MATLAB or any other software).
2. Assume λ_k are the eigenvalues associated with the kernel $R(t_1, t_2)$.
   (a) Show
   $$\int_0^T R(t_1, t_1)dt_1 = \sum_k \lambda_k$$
   (b) Show
   $$\int_0^T \int_0^T R^2(t_1, t_2)dt_1 dt_2 = \sum_k \lambda_k^2$$
   (c) Show, if
   $$R_N^{-1}(t, \tau) = \sum_i \frac{1}{\lambda_i} g_i(t) g_i(\tau)$$
   then the integral
   $$\int_0^T R_N^{-1}(t, \tau) R_N(\tau, \sigma) d\tau = ?$$
   Show all work.
3. Given that we have a discrete-time, real valued, and proper autocorrelation function (i.e., non-negative, symmetric, $R(0) \geq R(m)$), are the eigenvectors:

(a) real valued

(b) orthogonal

(c) normal

Convince the reader.

References

[1] Papoulis, A., *The Fourier Integral and its Applications*, New York: McGraw-Hill, 1962.

[2] Johnson, D.E., and Johnson, J.R., *Mathematical Methods in Engineering and Physics — Special Functions and Boundary Value Problems*, New York: The Rondale Press Co., 1965.

[3] Therrien, C.W., *Discrete Random Signals and Statistical Signal Processing*, Englewood Cliffs, NJ: Prentice-Hall, 1992.

[4] Helstrom, C.W., *Statistical Theory of Signal Detection*, 2nd Ed., Headington Hill Hall, Oxford, London: Pergamon Press Ltd., 1968.

[5] Karhunen, K., "Ueber lineare Methoden in der Wahrscheinlichkeitsrechnung," Annales Academiae Scientiarum Fennicae, Series A1, *Mathematica-Physica*, 37, 1947. English translation by I. Selin, "On Linear Methods in Probability Theory," The Rand Corp., Doc. T-131, August, 11, 1960.

[6] Papoulis, A., *Probability, Random Variables, and Stochastic Processes*, New York: McGraw-Hill, 1965.

[7] Davenport, W.B., and Root, W.L., *An Introduction to the Theory of Random Signals and Noise*, New York: McGraw-Hill, 1958.

[8] Riesz, F., and Sz.-Nagy, B., *Functional Analysis*, New York: Fredrick Ungar Publishing Company, 1955, also reprinted by Dover Publications Inc., New York, 1960.

[9] Lovitt, W.V., *Linear Integral Equations*, New York: McGraw-Hill Book Company, 1924, also reprinted by Dover Publications Inc., New York, 1950.

[10] Roach, G.F., *Green's Functions: An Introductory Theory with Applications*, New York: Van Nostrand Reinhold Company, 1970.

[11] LePage, W.R., *Complex Variables and the Laplace Transform for Engineers*, New York: McGraw-Hill, 1961.

[12] Oppenheim, A.V., Willsky, A.S., and Nawab, S.H., *Signals an Systems*, Englewood Cliffs, NJ: Prentice-Hall, 1997.

[13] Orfanidis, S.J., *Optimum Signal Processing: An Introduction*, 2nd Ed., New York: MacMillan, 1988.

[14] Hayes, M.H., *Statistical Digital Signal Processing and Modeling*, New York: John Wiley & Sons, 1996.

[15] Moon, T.K., and Stirling, W.C., *Mathematical Methods and Algorithms for Signal Processing*, Englewood Cliffs, NJ: Prentice-Hall, 2000.

[16] Manolakis, D.G., Ingle, V.K., and Kogon, S.M., *Statistical and Adaptive Signal Processing: Spectral Estimation, Signal Modeling, Adaptive Filtering and Array Processing*, New York: McGraw-Hill, 2000.

[17] Golub, G.H., and VanLoan, C.F., *Matrix Computations*, 2nd Ed., Baltimore and London: The Johns Hopkins University Press, 1989.

[18] Srinath, M.D., Rajasekaran, P.K., and Viswanathan, R., *Introduction to Statistical Signal Processing with Applications*, Englewood Cliffs, NJ: Prentice-Hall, 1996.

[19] Proakis, J.G., and Manolakis, D.G., *Digital Signal Processing: Principles, Algorithms, and Applications*, 3rd Ed., Upper Saddle River, NJ: Prentice-Hall, 1996.

[20] Borda, M., and Isar, D., "Whitening with Wavelets," *Proc. ECCTD'97 Conf.*," Budapest, August 1997.

[21] Barkat, M., *Signal Detection and Estimation*, London: Artech House, 1991.

[22] Cohen, L., *Time-Frequency Analysis*, Englewood Cliffs, NJ: Prentice-Hall, 1995.

Chapter 8

Estimation

8.1 INTRODUCTION

In previous chapters we addressed the detection of a particular signal (or component), now we want to obtain an estimate of its value. This chapter attempts to offer a rudimentary introduction into the topic of estimation theory. For additional details, we refer the reader to [1–5,19]. As shown in Figure 8.1, we can look at the estimation problem and its solution in two steps. The first step is the mapping of the parameter (the ones to be estimated) to the observation space, while the second step is the mapping from the observation space to the estimation space. The estimation of the parameter provides a rule (algorithm) to obtain an optimal value of the parameter of interest. Since the observations, by nature, are corrupted by noise, the observations are random variables. Any operation on these random variables will result in a new random variable; hence, part of this chapter will examine ways to address the goodness of the particular estimator in question. We distinguish between the estimate and the estimator in the sense that the estimate is the result of operating on data with an estimator. We also note that we do attach the properties of the estimate to that of the estimator (i.e., an unbiased estimate equates to an unbiased estimator).

8.2 BASIC ESTIMATION SCHEMES

This section takes a cursory look at three types of estimators: the MAP, ML, and Bayes' estimator [1–8]. Similarly to the detection approaches, we can choose to deal with the *a priori* probability density (i.e, MAP) or a cost function (i.e., Bayes'), or neither one (i.e., ML). We shall deal only with scalar

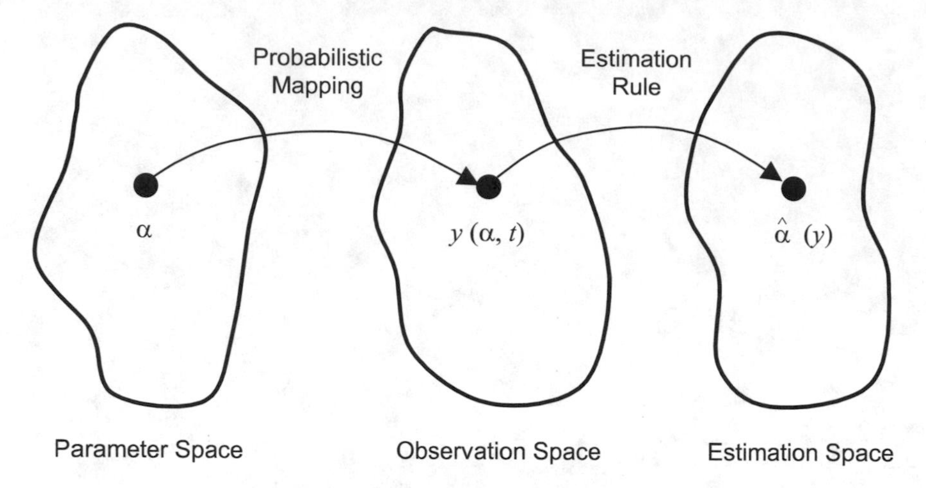

Figure 8.1: General estimation.

parameters. Vector extensions, in principle, are straightforward, but may be difficult to use in illustrations and as examples.

8.2.1 MAP Estimation

MAP estimation is an optimization technique that maximizes the *a posteriori* probability. That is, we find the most likely value of the parameter α given the observation. This is expressed as

$$\max_{\{\alpha\}} f(\alpha|\mathbf{y}) = \max_{\{\alpha\}} \left\{ \frac{f(\mathbf{y}|\alpha) f(\alpha)}{f(\mathbf{y})} \right\} \tag{8.1}$$

It is obvious that we need the PDF of the parameter α (i.e., $f(\alpha)$) to maximize expression (8.1). A typical conditional PDF scenario is shown in Figure 8.2, where the location of the maximum provides the MAP estimate of the parameter α. No cost function is required.

Example 8.1 *Given that we have the observations*

$$\begin{aligned} y_i &= a + n_i \qquad\qquad i = 0, \cdots, N-1 \\ n_i &\sim N(0, \sigma_n^2), \quad \text{(i.i.d.)} \end{aligned}$$

a is a Gaussian random variable, independent of n_i having a PDF given by $a \sim N(0, \sigma_a^2)$.

Find the MAP estimate of a

$$f(\mathbf{y}|a) = \prod_{i=0}^{N-1} \frac{1}{\sqrt{2\pi\sigma_n^2}} \, e^{-(y_i-\alpha)^2/2\sigma_n^2}$$

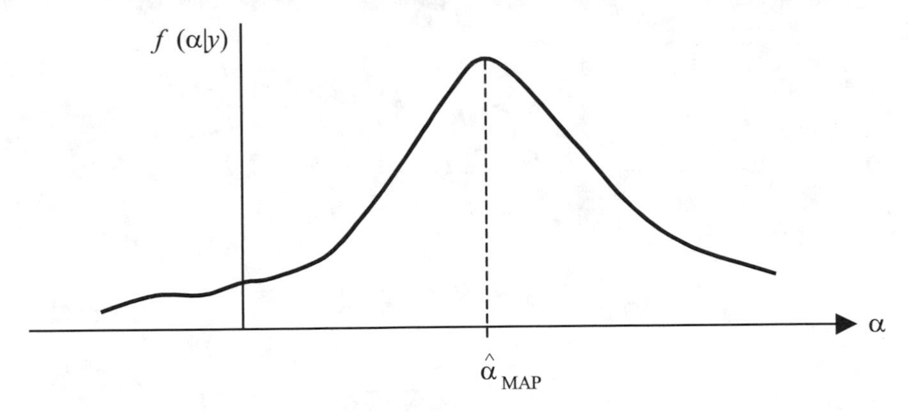

Figure 8.2: MAP estimation.

$$
\begin{aligned}
f(a) &= \frac{1}{\sqrt{2\pi\sigma_\alpha^2}}\, e^{-a^2/2\sigma_a^2} \\
f(a|\mathbf{y}) &= \frac{f(\mathbf{y}|a)f(a)}{f(\mathbf{y})} \\
&= \left(\prod_{i=0}^{N-1} \frac{1}{\sqrt{2\pi}\,\sigma_n}\right) e^{-\sum_{i=0}^{N-1}(y_i-a)^2/2\sigma_n^2} \frac{e^{-a^2/2\sigma_a^2}}{f(\mathbf{y})\sqrt{2\pi}\,\sigma_a} \\
f(a|\mathbf{y}) &= k(\mathbf{y})\, e^{-(1/2\sigma^2)(a-\sigma^2/\sigma_n^2\sum_{i=0}^{N-1} y_i)^2}
\end{aligned}
$$

where

$$\sigma^2 = \frac{\sigma_a^2\,\sigma_n^2}{N\sigma_a^2 + \sigma_n^2}$$

and $k(y)$ is a function of y only.

The best estimate (i.e., MAP) for the parameter a is the value where $f(a|y)$ peaks, which happens when

$$a = \frac{\sigma^2}{\sigma_n^2}\sum_{i=0}^{N-1} y_i$$

hence

$$
\begin{aligned}
\hat{a}_{\mathrm{MAP}} &= \frac{\sigma^2}{\sigma_n^2}\sum_{i=0}^{N-1} y_i \\
&= \frac{\sigma_a^2\sigma_n^2}{(N\sigma_a^2+\sigma_n^2)\sigma_n^2}\sum_{i=0}^{N-1} y_i
\end{aligned}
$$

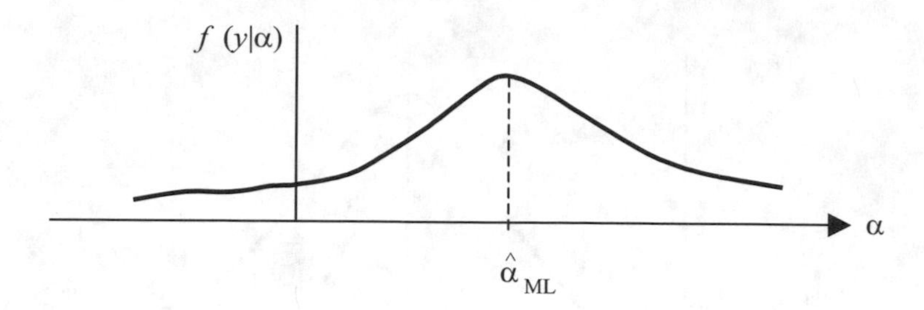

Figure 8.3: ML estimation.

$$= \frac{\sigma_a^2}{\sigma_a^2 + \sigma_n^2/N} \frac{1}{N} \sum_{i=0}^{N-1} y_i$$

We note, if $\sigma_a^2 \gg \sigma_n^2/N$, *then*

$$\hat{a}_{\text{MAP}} \approx \frac{1}{N} \sum_{i=0}^{N-1} y_i \qquad \text{(the sample mean)}$$

8.2.2 ML Estimation

ML estimation is an optimization technique that maximizes the likelihood function $f(\mathbf{y}|\alpha)$, i.e., finds the peak of the likelihood function, as shown in Figure 8.3.

This is expressed as

$$\max_{\{\alpha\}} f(\mathbf{y}|\alpha) \tag{8.2}$$

As is obvious in the MAP estimator definition, one can look at the ML estimator as a version of the MAP estimator in which the PDF of α (i.e., $f(\alpha)$) has no effect on the actual outcome. The ML estimator does not require a cost function nor the PDF of $f(\alpha)$.

Example 8.2 *Given*

$$\begin{aligned} y_i &= a + n_i \qquad\qquad i = 0, 1, \cdots, N-1 \\ n_i &\sim N(0, \sigma_n^2), \quad \text{(i.i.d.)} \end{aligned}$$

Estimate the mean of sequence of data points y_i, *where* $i = 0, 1, \cdots, N-1$. *Since we do not have a PDF description for* a, *we assume that the conditional PDF* $f(a|\mathbf{y})$, *as far as any functional interactions are concerned, is essentially independent of* $f(a)$. *Hence, an ML estimate is appropriate, i.e., maximize* $f(\mathbf{y}|a)$

$$\max f(\mathbf{y}|a) = \max \prod_{i=0}^{N-1} \frac{1}{\sqrt{2\pi}\,\sigma_n} e^{-(y_i - a)^2/2\sigma_n^2}$$

or, equivalently minimize the logarithm of $f(\mathbf{y}|a)$ *(i.e.,* $\min \ln f(\mathbf{y}|a)$*). The natural logarithm of* $f(\mathbf{y}|a)$ *can be written as*

$$\ln f(\mathbf{y}|a) = k - \sum_{i=0}^{N-1} \frac{(y_i - a)^2}{2\sigma_n^2}$$

where k *is independent of* α*, and hence*

$$\frac{\partial}{\partial a} \ln f(\mathbf{y}|a) = 2 \sum_{i=0}^{N-1} \frac{(y_i - a)}{2\sigma_n^2} = 0$$

or

$$\sum_{i=0}^{N-1} (y_i - a) = 0$$

or finally

$$\hat{a}_{\text{ML}} = \frac{1}{N} \sum_{i=0}^{N-1} y_i \qquad \text{(the sample mean)}$$

8.2.3 Bayes' Estimator

The Bayes' estimator uses an average cost function which will be minimized. The cost of an error is defined as $C(\alpha_e)$, which is a non-negative term that is a function of the estimate and its true value (i.e., $C(\alpha_e) = C(\alpha, \hat{\alpha})$. The average cost is defined as

$$C(\hat{\alpha}) = \iint C(\alpha, \hat{\alpha})\ f(\alpha|\mathbf{y})\ f(\mathbf{y})\ d\mathbf{y}\ d\alpha \tag{8.3}$$

where we assume that all variables are continuous in nature. We see it is similar in form to the average cost

$$C = \sum_{i=0}^{M-1} \sum_{j=0}^{M-1} C_{ij}\ Pr(D_i|H_j)\ P_j \tag{8.4}$$

as advocated in (4.26). It can be interpreted as the generalization of the M-ary hypothesis test to a test where a continuum of outcomes is possible. Usually, three typical cost terms are advocated: linear (absolute error) cost, constant (uniform) error cost, and quadratic error cost. Figure 8.4 shows the cost as a function of error for these three cases.

Figure 8.4a shows that the penalty (i.e., cost) for the absolute error criterion will increase linearly with the magnitude of the error. The uniform error cost, Figure 8.4b will tolerate an error in magnitude less than $\Delta/2$, that is, a zero cost (or penalty) is applied as long as the error is smaller than the

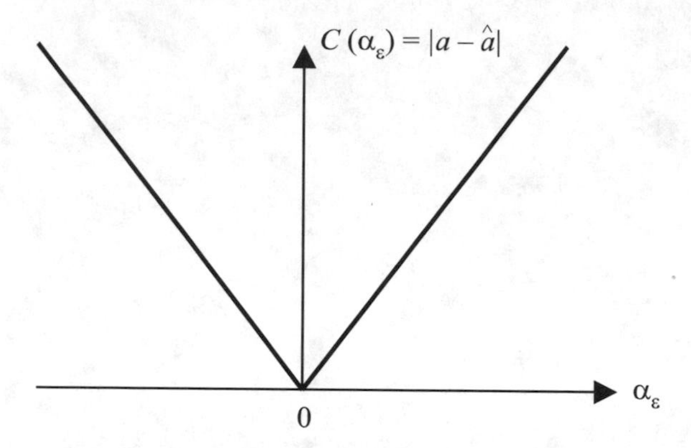

(a) Absolute (linear) error cost function.

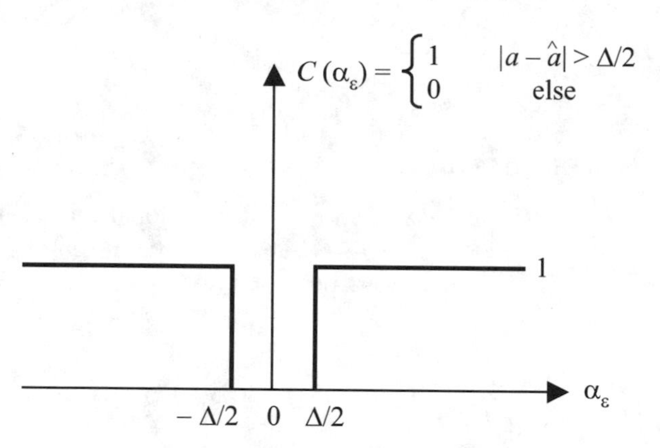

(b) Uniform (constant) error cost function.

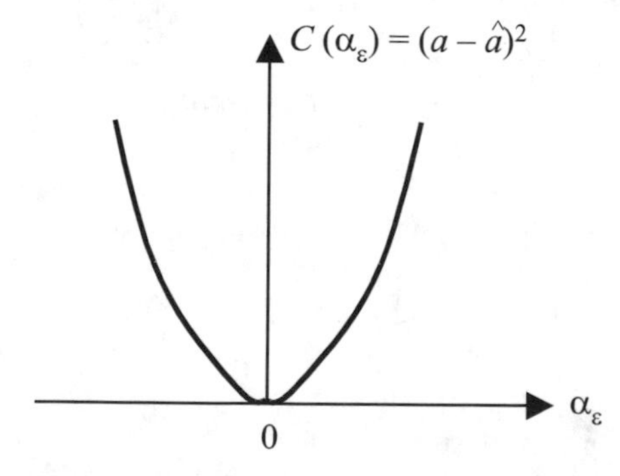

(c) Quadratic error cost function.

Figure 8.4: Cost functions.

guard band $\Delta/2$. As the name implies, the quadratic cost (Figure 8.4c) will penalize any error proportional to the square of the error. Of the three cost functions, the quadratic cost usually is the one that will be used. The main reasons are

(a) The expressions are analytically tractable.

(b) The error that is produced is the minimum mean squared error (MMSE).

We can write (8.3) as follows

$$C(\hat{\alpha}) = \int C(\hat{\alpha}|\mathbf{y})\; f(\mathbf{y})\; d\mathbf{y} \tag{8.5}$$

where $C(\hat{\alpha}|\mathbf{y})$ is the so-called conditional cost given by

$$C(\hat{\alpha}|\mathbf{y}) = \int C(\alpha, \hat{\alpha})\; f(\alpha|\mathbf{y})\; d\alpha \tag{8.6}$$

The estimator $\hat{\alpha}$ that minimizes the total cost $C(\hat{\alpha})$ is called the Bayes' estimator. The total cost is minimized by minimizing $C(\hat{\alpha}|\mathbf{y})$. Using the preferred cost function (i.e., the quadratic cost), the conditional cost (8.6) becomes

$$C(\hat{\alpha}|\mathbf{y}) = \int (\alpha - \hat{\alpha})^2\; f(\alpha|\mathbf{y})\; d\alpha \tag{8.7}$$

This expression is minimized with respect to α by taking a partial derivative with respect to $\hat{\alpha}$ and setting the result equal to zero, as given by

$$\frac{\partial}{\partial \hat{\alpha}} C(\hat{\alpha}|\mathbf{y}) = \int \frac{\partial}{\partial \hat{\alpha}} (\alpha - \hat{\alpha})^2\; f(\alpha|\mathbf{y})\; d\alpha = 0 \tag{8.8}$$

or

$$\frac{\partial}{\partial \hat{\alpha}} C(\hat{\alpha}|\mathbf{y}) = 2 \int (\alpha - \hat{\alpha})\; f(\alpha|\mathbf{y})\; d\alpha \tag{8.9}$$

This can be solved to show that the best estimator is the conditional mean, as shown by

$$\hat{\alpha} \int f(\alpha|\mathbf{y})\; d\alpha = \int \alpha\; f(\alpha|\mathbf{y})\; d\alpha \tag{8.10}$$

or

$$\hat{\alpha} = \int \alpha\; f(\alpha|\mathbf{y})\; d\alpha = E\left\{\alpha|\mathbf{y}\right\} \tag{8.11}$$

The controls literature is also advocating estimation based on the so-called H_∞ norm. For some detail we refer the interested reader to the book by Burl [20]. This technique has also been successfully applied to estimate the frequency of a complex sinusoid in white noise [21].

8.3 PROPERTIES OF ESTIMATORS

The estimates are always functions of the data, where the data consists at least in part of a random component. This makes the estimate, no matter which estimation technique is used, a random variable. At this point, it is of interest to investigate the quality of the estimate by evaluating its mean and its standard deviation. The mean conveys information about the expected value of the estimate. The standard deviation indicates the degree of scattering of the estimate's value about its mean (i.e., the less scattering the better). If no error would occur then the mean of the estimate would be the true value of the quantity to be estimated and the variance (or standard deviation) would be close to zero. We recall that random quantities have a PDF where the behavior of the random quantities, in part, is determined by the mean and standard deviation. We will introduce some of the definitions that are customarily used.

Definition 1: Unbiased Estimate

If the mean of the estimate equals the true value, then the estimate is unbiased. The difference between the true value and the expected value is called the bias.

Definition 2: Asymptotically Unbiased Estimate

If the estimate possesses the property that

$$\lim_{N\to\infty} E\{\hat{\alpha}_N|\alpha\} = \alpha$$

then the estimate is called asymptotically unbiased. The subscript of the estimate indicates the functional dependency on the sample size N.

Definition 3: Consistent Estimate

If, as the number of observations increases, the PDF of an unbiased estimate becomes more and more peaked (i.e., the variance keeps decreasing with an increase in the number of observations) then the estimate is called a consistent estimate. More precisely, $\hat{\alpha}$ is called a consistent estimate if

$$\lim_{N\to\infty} Pr\{(\hat{a} - a) > \epsilon\} = 0 \tag{8.12}$$

Definition 4: Minimum Variance Estimate

The unbiased estimate $\hat{\alpha}$ is the minimum variance estimate $\hat{\alpha}_{\text{MV}}$ if

$$\text{var } \hat{\alpha}_{\text{MV}} \leq \text{var } \hat{\alpha}_{(\text{any other estimation scheme})} \tag{8.13}$$

Definition 5: Cramer-Rao (CR) Bound

The Cramer-Rao bound is a bound on the variance of the estimate given by

$$\sigma_{\hat{\alpha}}^2 \geq \frac{1}{E\left\{\left(\frac{\partial \ln f(\mathbf{y}|\alpha)}{\partial \alpha}\right)^2\right\}} = \frac{-1}{E\left\{\frac{\partial^2 \ln f(\mathbf{y}|\alpha)}{\partial^2 \alpha}\right\}} \tag{8.14}$$

Definition 6: Efficient Estimate

An unbiased estimate that meets the CR bound, is called an efficient estimate.

There is a host of other definitions that apply to the area of estimation [2,4], which we leave for the interested reader to pursue. The body of information on estimators and their properties is large. We are primarily interested in this topic since we do use parameter estimates in nuisance type likelihood ratio problems. A second reason comes about in that some typical problems, detection and estimation, are closely coupled. See for example, the periodogram (power spectral density estimator) in Chapter 9. It is the ideal detector for low SNR narrow-band signals which are embedded in white Gaussian noise, where the phase and frequency are assumed to be uniformly distributed. That is, the phase and frequency have the worst possible uncertainty. We shall illustrate some of the definitions using simple examples.

Example 8.3 *Let us re-examine Example 8.2, the maximum likelihood estimator of the mean and evaluate its expected value and its variance.*

$$\begin{aligned}
\hat{a}_{\mathrm{ML}} &= \frac{1}{N}\sum_{i=0}^{N-1} y_i \\
E\{\hat{a}_{\mathrm{ML}}\} &= \frac{1}{N}\sum_{i=0}^{N-1} E\{y_i\} \qquad \text{where we use } E\ y_i \hat{=} E\ y \\
&= \frac{1}{N}\sum_{i=0}^{N-1} a = a \\
\Rightarrow \quad \hat{a}_{\mathrm{ML}} &= \frac{1}{N}\sum_{i=0}^{N-1} y_i \qquad \text{is an unbiased estimate}
\end{aligned}$$

$$\mathrm{var}\ \hat{a}_{\mathrm{ML}} = E\left\{(\hat{a}_{\mathrm{ML}})^2\right\} - (E\{\hat{a}_{\mathrm{ML}}\})^2$$

$$= \frac{1}{N^2}\left\{\sum_i \sum_j E\{y_i\ y_j\}\right\} - \left(\frac{1}{N}\sum_i E\{y_i\}\right)^2$$

$$= \frac{1}{N^2}\left\{\underbrace{(N-1)N\left(E\{y\}\right)^2}_{\text{off-diagonal terms}} + \underbrace{NE\{y^2\}}_{\text{diagonal term}} - N^2\left(E\{y\}\right)^2\right\}$$

$$= \frac{1}{N^2}\left\{-N\left(E\{y\}\right)^2 + NE\{y^2\}\right\}$$

$$= \frac{1}{N}\left\{E\{y^2\} - E\{y\}\right\}$$

$$= \frac{1}{N}\,\text{var}_y = \frac{1}{N}\,\sigma_n^2$$

hence, we can say that

$$\hat{a}_{\text{ML}} = \frac{1}{N}\sum_{i=0}^{N-1} y_i \qquad \text{is a consistent estimate}$$

To see if the sample mean is also an efficient estimate, we will obtain the CR bound.

$$\frac{\partial}{\partial a}\ln f(\mathbf{y}|a) = \frac{\partial}{\partial a}\left\{k - \sum_{i=0}^{N-1}\frac{(y_i - a)^2}{2\sigma_n^2}\right\}$$

$$= -\frac{1}{2\sigma_n^2}(-2)\sum_{i=0}^{N-1}(y_i - a)$$

$$= \frac{1}{\sigma_n^2}\sum_{i=0}^{N-1}(y_i - a)$$

$$\frac{\partial^2 \ln f(\mathbf{y}|a)}{\partial^2 a} = \frac{\partial}{\partial a}\,\frac{1}{\sigma_n^2}\sum_{i=0}^{N-1}(y_i - a)$$

$$= \frac{1}{\sigma_n^2}\sum_{i=0}^{N-1}(-1) = -\frac{N}{\sigma_n^2}$$

The CR bound is given by

$$-\frac{1}{E\left\{\dfrac{\partial^2}{\partial^2\alpha}\ln f(\mathbf{y}|a)\right\}} = -\frac{1}{-\dfrac{N}{\sigma_n^2}} = \frac{\sigma_n^2}{N}$$

The variance of the estimator var $\hat{a}_{\mathrm{ML}} = \sigma_n^2/N$; *hence, the sample mean is an efficient estimate.*

Example 8.4 *Given* $y(t) = s(t,\alpha) + n(t)$, *for* $0 \leq t \leq T$, *where* $s(t,\alpha)$ *is known except for* α, *which is an unknown parameter. The additive noise is a white Gaussian random process, with zero mean and a spectral height of* $N_0/2$. *We want to find the ML estimate of the unknown parameter* α.

$$\max_{\{\alpha\}} f(y|\alpha) = \max_{\{\alpha\}} k\ e^{-1/N_0 \int_0^T (y(t)-s(t,\alpha))^2/2\sigma^2} dt$$

or

$$\min \ \ln f(y|a) = \frac{2}{N_0} \int_0^T \frac{(y(t) - s(t,\alpha))^2}{2\sigma^2} \frac{\partial}{\partial\alpha} s(t,\alpha)\ dt = 0$$

(a) Suppose $s(t,\alpha) = \alpha s(t)$, *where* $s(t)$ *is known, hence the partial derivative becomes*

$$\frac{\partial}{\partial\alpha}\ s(t,\alpha) = \frac{\partial}{\partial\alpha}\ \alpha s(t) = s(t)$$

hence, we minimize the logarithm of the conditional PDF via

$$\int_0^T (y(t) - \alpha s(t))\, s(t) dt = 0$$

or

$$\hat{\alpha}_{\mathrm{ML}} = \frac{\displaystyle\int_0^T y(t)\ s(t)\ dt}{\displaystyle\int_0^T s(t)\ s(t)\ dt}$$

Note, this is the cross-correlation of $y(t)$ *and* $s(t)$ *normalized by the auto-correlation coefficient of* $s(t)$.

(b) Suppose $s(t,\alpha) = A\cos(\omega_0 t + \alpha)$, *where* ω_0 *and* A *are known and* α, *the phase angle, is the desired unknown parameter. Then the partial derivative becomes*

$$\begin{aligned}\frac{\partial}{\partial a} s(t,\alpha) &= \frac{\partial}{\partial\alpha} A\cos(\omega_0 t + \alpha) \\ &= -A\sin(\omega_0 + \alpha)\end{aligned}$$

hence, we minimize the logarithm of the conditional PDF via

$$\int_0^T (y(t) - A\cos(\omega_0 t + \alpha))\,(-A\sin(\omega_0 t + \alpha))\, dt = 0$$

the cross-term $A^2 \int_0^T \sin(\omega_0 t + \alpha)\cos(\omega_0 t + \alpha) dt$ *equals to zero if*

$\omega_0 T = kT$ (k is integer) and is approximately equal to zero if $2\pi/\omega_0 \ll T$. We assume that this cross-term can be neglected, so we have

$$\int_0^T y(t) A \sin(\omega_0 t + \alpha) dt = 0$$

or

$$\sin\alpha \int_0^T y(t)\cos(\omega_0 t)dt = \cos\alpha \int_0^T y(t)\sin(\omega_0 t)dt$$

The ML estimate becomes

$$\hat{\alpha}_{\mathrm{ML}} = \tan^{-1}\left(\frac{\sin\alpha}{\cos\alpha}\right) = \tan^{-1}\left(\frac{\int_0^T y(t)\sin(\omega_0 t)dt}{\int_0^T y(t)\cos(\omega_0 t)dt}\right)$$

Note, if this were a discrete-time problem, then

$$\begin{aligned}
\hat{\theta}_{\mathrm{ML}} &= \tan^{-1}\left(\frac{\sum_{n=0}^{N-1} y(n)\sin\left(\frac{2\pi}{N}kn\right)}{\sum_{n=0}^{N-1} y(n)\cos\left(\frac{2\pi}{N}kn\right)}\right) \\
&= \tan^{-1}\left(\frac{\mathrm{Im}\ Y(k)}{Re Y(k)}\right)
\end{aligned}$$

where $Y(k)$ = the DFT or FFT of $y(n)$, $n = 0, 1, \cdots, N-1$, and k closely corresponds to ω_0.

8.4 CRAMER-RAO BOUND

This section will provide the derivation of the Cramer-Rao (CR) bound for a non-random unbiased parameter case. The random parameter case is treated in van Trees [6]. Assuming that we have an unbiased estimator, then

$$E\{\hat{\alpha} - \alpha)|\alpha\} = \int_{-\infty}^{\infty} (\hat{\alpha} - \alpha) f(\mathbf{y}|\alpha)\ d\mathbf{y} = 0 \tag{8.15}$$

Taking the derivative with respect to α leads to

$$\begin{aligned}
\frac{\partial}{\partial\alpha}\int_{-\infty}^{\infty} (\hat{\alpha} - \alpha) f(\mathbf{y}|\alpha)\ d\mathbf{y} &= \int_{-\infty}^{\infty} \frac{\partial}{\partial\alpha}(\hat{\alpha} - \alpha)\ f(\mathbf{y}|\alpha)\ d\mathbf{y} \\
&= -\int_{-\infty}^{\infty} f(\mathbf{y}|\alpha)\ d\mathbf{y} + \int_{-\infty}^{\infty} (\hat{\alpha} - \alpha)\frac{\partial f(\mathbf{y}|\alpha)}{\partial\alpha}\ d\mathbf{y}
\end{aligned} \tag{8.16}$$

Since $du/dx = u\ d(\ln u)/dx$, we can rewrite (8.15) as

$$\int_{-\infty}^{\infty} (\hat{\alpha} - \alpha) \left[\frac{\partial}{\partial \alpha} \ln f(\mathbf{y}|\alpha) \right] f(\mathbf{y}|\alpha)\ d\mathbf{y} = 1 \tag{8.17}$$

Rearranging (8.17) leads to

$$1 = \int_{-\infty}^{\infty} \left[\frac{\partial}{\partial \alpha} \ln f(\mathbf{y}|\alpha) \right] \sqrt{f(\mathbf{y}|\alpha)} \cdot \left(\sqrt{f(\mathbf{y}|\alpha)} (\hat{\alpha} - \alpha) \right) d\mathbf{y} \tag{8.18}$$

Recalling the Schwarz inequality,

$$\left(\int_{-\infty}^{\infty} g(x)\ h(x)\ dx \right)^2 \leq \int_{-\infty}^{\infty} g^2(x)\ dx \int h^2(x)\ dx$$

where the terms are equal when $g(x)$ is linearly related to $h(x)$. Applying the Schwarz inequality to (8.17) leads to

$$\int_{-\infty}^{\infty} \left(\frac{\partial}{\partial \alpha} \ln f(\mathbf{y}|\alpha) \right)^2 f(\mathbf{y}|\alpha) d\mathbf{y} \int_{-\infty}^{\infty} (\hat{\alpha} - \alpha)\ f(\mathbf{y}|\alpha)\ d\mathbf{y} \geq 1 \tag{8.19}$$

Using the definition for the moment of the function of a random variable, say $E\{g(x)\} = \int g(x) f(x) dx$, we obtain

$$E(\hat{\alpha} - \alpha)^2 \geq \frac{1}{E\left\{ \left(\frac{\partial}{\partial \alpha} \ln f(\mathbf{y}|\alpha) \right)^2 \right\}} \tag{8.20}$$

An equivalent definition for the CR bound is given in (8.21). The derivation is not given here, the interested reader is referred to [18]. The second form of the CR bound is given by

$$\sigma_{\hat{\alpha}}^2 \geq \frac{-1}{E\left\{ \frac{\partial^2 \ln f(\mathbf{y}|\alpha)}{\partial^2 \alpha} \right\}} \tag{8.21}$$

8.5 WAVEFORM ESTIMATION

Many times it is desirable to recover the original waveform or some function of it, where usually the waveform is embedded in noise. Typical measurements (data, observations) may be amplitude or position as a function of time. We see that one possible data estimation scheme might recover the amplitude or position, as the case may warrant. In the case of positional data, it may be desirable to estimate the velocity or acceleration directly from the data. This corresponds to the first and second differential of the true positional values,

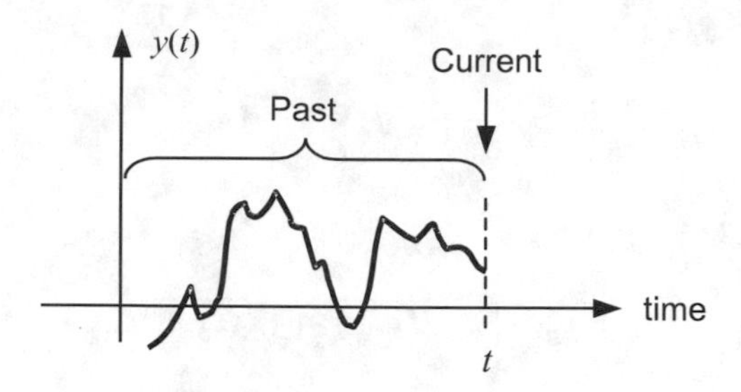

Figure 8.5: Filtering scenario.

respectively. Hence, we see that estimation of functions of the true data do naturally come up in some applications.

Estimation problems can be separated into three different basic types (configurations, solutions, setups), which are addressed in the literature. These three basic problems are filtering, predicting, and smoothing. We will briefly discuss these three processing types and refer the reader to Figures 8.5–8.7.

(a) Filtering:

This technique (see Figure 8.5) uses all past and the current data information. As the name implies, filtering manipulates current and past information (data) to obtain an optimal estimate of the true message value at time t.

(b) Predicting:

Predicting uses all information available at time t, just as the filtering operation does. To obtain an optimal predicted value of the message or signal some ϵ seconds into the future, it uses past and present information. Figure 8.6 shows the geometry of the predicting process.

(c) Smoothing:

The smoothing operation is an off-line processing technique. It is assumed that all information regarding a particular experiment is made available to the estimation procedure. Having all the information available essentially employs future data, relative to a given point in time (i.e., it is non-causal as far as the point at time t is concerned). This is why it is an off-line processing technique, since from a processing point of view all data points occur in the past. Figure 8.7 provides the geometry of this estimation problem. Clearly, at any time t, all information is made available to the processor.

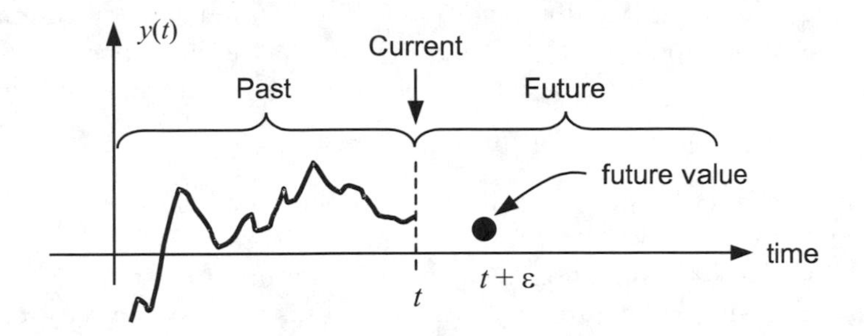

Figure 8.6: Predicting scenario.

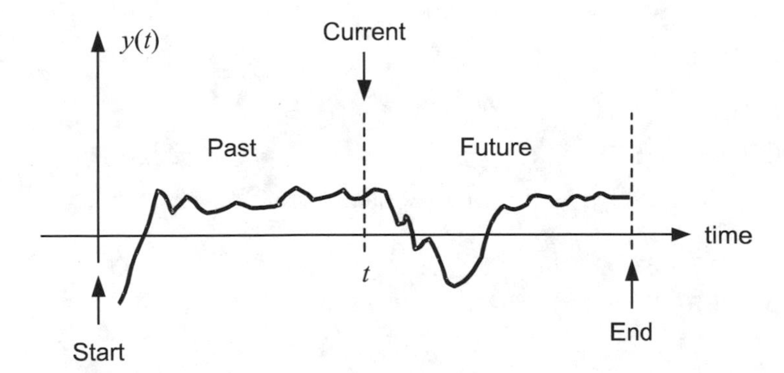

Figure 8.7: Smoothing scenario.

Historically, two different solutions have been formulated. The first one is obtained in the frequency domain, while the second one is obtained in the time domain. The frequency domain approach was formulated by Wiener [7] and leads to a description of the optimal processor in terms of second order moments (i.e., correlation or spectral properties) [8,9,19]. The estimate is non-recursive. The time domain approach was formulated by Kalman [10] and is based on an update of the estimate, a gain factor, and an error criterion. For the discrete-time case, the governing equation for filtering is of the form

$$\hat{s}(n) = \Phi\ \hat{s}(n-1) + K(n)\ [y(n) - C\ \Phi\ \hat{s}(n-1)]$$

where Φ and C are the state transition and observation matrices, respectively. The quantity $\hat{s}(n-1)$ is the estimate at the previous sample instant, the term $y(n) - C\ \Phi\ \hat{s}(n-1)$ corresponds to the error occuring at time n, and $K(n)$ is the time varying Kalman gain. The estimate is recursive in nature. Many books focus on the Kalman filtering technique, see for example [3,11,19].

8.5.1 Wiener Filtering

A fairly simple solution is obtained when the orthogonality principle (see Papoulis [12]) is invoked. Typically, the minimum mean squared error (MMSE) criterion is used, which is of the form

$$MMSE = E\left\{(g(t) - \hat{g}(t))^2\right\} \tag{8.22}$$

where *MMSE* denotes the minimum mean squared error, $g(t)$ is the desired quantity, and $\hat{g}(t)$ is its estimate. In our problems, the data is of the form

$$y(t) = s(t) + n(t) \qquad 0 \le t \le T \tag{8.23}$$

The interval is $[0, T]$, but can easily be adapted to any interval $[t_0, t_f]$ and the desired quantity is

$$\begin{aligned} g(t) &= s(t+\alpha), \quad \alpha > 0 \Rightarrow \text{a predicting problem} \\ g(t) &= s(t) \Rightarrow \text{a filtering problem} \\ g(t) &= s(t+\alpha), \quad 0 \le (\alpha +\ t) \le T \Rightarrow \text{a smoothing problem} \end{aligned}$$

Example 8.5 *Given $y(t)$, predict $y(t+\alpha)$ for $\alpha > 0$*

$$\begin{aligned} g(t) &= y(t+\alpha) \\ \hat{g}(t) &= \hat{y}(t+\alpha) = a\, y(t) \end{aligned}$$

$$\begin{aligned} \min_{\{a\}} E\left\{(g(t) - \hat{g}(t))^2\right\} &= \min_{\{a\}} E\left\{(y(t+\alpha) - a\, y(t))^2\right\} \\ &= \min_{\{a\}} J \end{aligned}$$

where $J = E\left\{(y(t+\alpha) - a\,y(t))^2\right\}$. *Setting*

$$\frac{\partial J}{\partial a} = 0$$

produces

$$2E\left\{(y(t+\alpha) - a\,y(t))\,(-y(t))\right\} = 0$$

or

$$R_Y(\alpha) - a\,R_Y(0) = 0$$

hence

$$\hat{a}_{MSSE} = \frac{R_Y(\alpha)}{R_Y(0)}$$

MMSE is given by

$$\begin{aligned}
MMSE &= E\left\{(g(t) - \hat{g}t))^2\right\} \\
&= E\left\{(y(t+\alpha) - a\,y(t))^2\right\} \\
&= E\left\{\left(y(t+\alpha) - \frac{R_y(\alpha)}{R_y(0)}y(t)\right)^2\right\} \\
&= R_Y(0) + \left(\frac{R_Y(\alpha)}{R_Y(0)}\right)^2 R_Y(0) - \frac{2R_Y(\alpha)}{R_Y(0)}R_Y(\alpha) \\
&= R_Y(0) + \frac{R_Y^2(\alpha) - 2R_Y^2(\alpha)}{R_Y(0)} \\
&= R_Y(0) - \frac{R_Y^2(\alpha)}{R_Y(0)}
\end{aligned}$$

Note, the correlation function $R_Y(\cdot)$ *can be simplified to* $R_Y(\cdot) = R_S(\cdot) + R_N(\cdot)$ *if the processes* $s(t)$ *and* $n(t)$ *are statistically independent.*

The orthogonality principle [12], a compact version of minimizing the mean squared error, says:

(a) $E\{\text{error} \cdot \text{ data}\} = 0$, hence solve for the pertinent parameters

(b) MMSE $= E\{\text{error} \cdot \text{ quantity to be estimated}\} = MMSE$.

Example 8.6 *We want to rework the Example 8.5 using the orthogonality principle. The error is given by*

$$\text{error} = y(t+\alpha) - a\,y(t)$$

hence

$$E\{\text{error} \cdot \text{ data}\} = E\{(y(t+\alpha) - a\,y(t))\,y(t)\} = 0$$

or

$$\begin{aligned} E\{y(t+\alpha)y(t)\} - a\,E\,y(t)y(t) = 0 \\ = R_Y(\alpha) - a\,R_Y(0) \end{aligned}$$

and the optimal solution is given by

$$\hat{a}_{MMSE} = \frac{R_Y(\alpha)}{R_Y(0)}$$

The MMSE is given by

$$\begin{aligned} MMSE &= E\{\text{error} \cdot \text{ quantity to be established}\} \\ &= E\left\{\left(y(t+\alpha) - \frac{R_Y(\alpha)}{R_Y(0)}y(t)\right)(y(t+\alpha))\right\} \\ &= R_Y(0) - \frac{R_Y^2(\alpha)}{R_Y(0)} \end{aligned}$$

We note that the same results are obtained as in the previous example.

Example 8.7 *Given $y(t) = s(t) + n(t)$, we want to find the best estimate of $s(t)$ using $y(t)$, i.e., $\hat{s}(t) = a\,y(t)$*

$$\begin{aligned} E\{\text{error} \cdot \text{ data}\} &= E\{(s(t) - a\,y(t))\,y(t)\} = 0 \\ &= E\,s(t)y(t) - a\,E\,y(t)y(t) \end{aligned}$$

hence

$$\begin{aligned} \hat{a}_{MMSE} = \frac{R_{SY}(0)}{R_Y(0)} &= \frac{R_S(0) + R_{SN}(0)}{R_S(0) + R_N(0) + R_{SN}(0) + R_{NS}(0)} \\ &= \frac{R_S(0)}{R_S(0) + R_N(0)} \end{aligned}$$

if signal and noise are statistically independent. Assuming that signal noise are statistically independent, the MMSE is given by

$$\begin{aligned} MMSE &= E\left\{\left(s(t) - \frac{R_{SY}(0)}{R_Y(0)}y(t)\right)s(t)\right\} \\ &= R_S(0) - \frac{R_{SY}(0)}{R_Y(0)}R_{YS}(0) \\ &= R_S(0) - \frac{R_S^2(0)}{R_S(0) + R_N(0)} = \frac{R_S(0)R_N(0)}{R_S(0) + R_N(0)} \end{aligned}$$

Example 8.8 *Given* $y(t) = s(t) + n(t)$

(a) find

$$\hat{s}(t) = \int_{-\infty}^{\infty} y(t-\tau)\ h(\tau)\ d\tau$$

Note, this implies that $y(t)$ *and* $h(t)$ *are non-causal.*

$$E\left\{\left(s(t) - \int_{-\infty}^{\infty} y(t-\tau)h(\tau)d\tau\right) y(t-\sigma)\right\} = 0$$

$$= R_{SY}(\sigma) - \int_{-\infty}^{\infty} h(\tau)R_Y(\sigma-\tau)\ d\tau$$

Taking the Fourier transform, we obtain

$$S_{SY}(f) - H(f)S_Y(f) = 0$$

where $S_{SY}(f)$ *and* $S_Y(f)$ *are the cross- and auto-power spectral densities, respectively. The optimal (non-causal) filter is given by*

$$H(f) = \frac{S_{SY}(f)}{S_Y(f)} = \frac{S_{SS}(f)}{S_S(f) + S_N(f)}$$

if signal and noise are statistically independent and

$$\hat{h}(t) = \mathcal{F}^{-1}\{H(f)\}$$

where we dropped the subscript indicating that the impulse response is due to minimizing the mean squared error. The MMSE is given, assuming signal and noise are statistically independent, by

$$\begin{aligned} MMSE &= E\left\{\left(s(t) - \int_{-\infty}^{\infty} y(t-\tau)\ \hat{h}(\tau)\ d\tau\right) s(t)\right\} \\ &= R_S(0) - \int_{-\infty}^{\infty} R_{SY}(\tau)\ \hat{h}(\tau)\ d\tau \\ &= R_S(0) - \int_{-\infty}^{\infty} R_S(\tau)\ \hat{h}(\tau)\ d\tau \end{aligned}$$

(b) Suppose

$$\hat{s}(t) = \int_0^t y(t-\tau)\ h(\tau)\ d\tau$$

$$E\left\{\left(s(t) - \int_0^t y(t-\tau)\ h(\tau)\ d\tau\right) y(t-\alpha)\right\} = 0 \quad ,0 \le \alpha \le t$$

$$R_{SY}(\alpha) - \int_0^t h(\tau)\, R_Y(\alpha - \tau)\, d\tau$$

which needs to be solved for $\hat{h}(t)$.

The MMSE is given by

$$\begin{aligned} MMSE &= E\left\{\left(s(t) - \int_0^t y(t-\tau)\, h(\tau)\, d\tau\right) s(t)\right\} \\ &= R_S(0) - \int_0^t h(\tau)\, R_Y(\tau)\, d\tau \end{aligned}$$

Example 8.9 *Given that we know $s(t)$ and its derivative $s\prime(t)$, we want to predict $s(t+\alpha)$ for $\alpha \geq 0$, in terms of $s(t)$ and $s\prime(t)$. Hence, the estimate is given by*

$$\hat{s}(t+\alpha) = a\, s(t) + b\, s\prime(t)$$

The orthogonality principle is utilized in two equations, that is

$$E\left\{(s(t+\alpha) - (a\, s(t) + b\, s\prime(t))\, s(t)\right\} = 0$$

and

$$E\left\{(s(t+\alpha) - (a\, s(t) + b\, s\prime(t)))\, s\prime(t)\right\} = 0$$

This leads to

$$R_{SS}(\alpha) - a\, R_{SS}(0) - b\, R_{S\prime S}(0) = 0$$

and

$$R_{SS\prime}(\alpha) - a\, R_{SS\prime}(0) - b\, R_{S\prime S\prime}(0) = 0$$

Since $R_{SS\prime}(0) = R_{S\prime S}(0) = 0$, due to the correlation property of a function and its derivative at the zero lag, we have

$$\begin{aligned} R_{SS}(\alpha) - a\, R_{SS}(0) &= 0 \\ R_{SS\prime}(\alpha) - b\, R_{S\prime S\prime}(0) &= 0 \end{aligned}$$

$$\begin{aligned} \hat{a}_{MMSE} &= \frac{R_{SS}(\alpha)}{R_{SS}(0)} \\ \hat{b}_{MMSE} &= \frac{R_{SS\prime}(\alpha)}{R_{S\prime S\prime}(0)} \end{aligned}$$

The MMSE is given by

$$\begin{aligned} MMSE &= E\left\{(s(t+\alpha) - a\, s(t) - b\, s\prime(t))\, s(t+\alpha)\right\} \\ &= R_{SS}(0) - a\, R_{SS}(\alpha) - b\, R_{SS\prime}(\alpha) \end{aligned}$$

where we insert the optimal values for $a = \hat{a}_{MMSE}$ and $b = \hat{b}_{MMSE}$ from the optimal solutions set.

8.5.2 Discrete-Time Waveform Estimation

The concept of continuous time Wiener filtering (i.e., filtering, predicting, smoothing) directly extends to the discrete-time case. Ample reference material can be found in the books on statistical signal processing [13–17]. The typical problem is that of a signal (message) embedded in additive noise, defined by

$$y(n) = s(n) + n(n), \qquad n = 0, \cdots, N-1 \tag{8.24}$$

Usually, one is interested in filtering out $s(n)$ or predicting $s(n+m)$, the value of $s(\cdot)$, m-steps into the future. One can also obtain a smoothed version of $s(n)$ for all $0 \le n \le N-1$, where the data time index n ranges from 0 to $N-1$. A very convenient way to obtain the necessary filter is via the orthogonality principle [12,14]. The orthogonality principle says that

(a) The expected values of the error and the data are orthogonal.

(b) The MMSE is obtained by projecting the error onto the quantity to be estimated.

In equation form, we have

$$E\{\epsilon(n)\ y(j)\} = 0 \qquad j = 0, \cdots, N-1 \tag{8.25}$$

and the MMSE is given by

$$MMSE \;=\; E\{\epsilon(n) g(n)\} \tag{8.26}$$

where $g(n)$ is the quantity to be estimated and $\epsilon(n) \;=\; g(n) \;-\; \hat{g}(n)$.

Example 8.10 *The received data is given by* $y(n) = s(n) + n(n)$, *for* $n = 0, \ldots, N-1$ *where* $s(n)$ *and* $n(n)$ *are statistically independent random sequences (i.e.,* $R_{NS}(i) = 0, \forall i$*). Estimate* $s(n)$ *using a weighted sum of the most recent* M *data points (i.e., a FIR filter of length* M*, where we need to find the optimum values of the filter weights) as given by*

$$\begin{aligned}
\hat{s}(n) &= \sum_{i=0}^{M-1} h(i)\ y(n-i) \qquad\qquad n = 0, 1, \cdots, N-1 \\
\epsilon(n) &= \hat{s}(n) - s(n) \\
E\{\epsilon(n)\ y(n-j)\} &= 0 \qquad\qquad j = 0, 1, \cdots, N-1
\end{aligned}$$

or

$$E\left\{\left(\sum_{i=0}^{M-1} h(i)\ y(n-i) - s(n)\right) y(n-j)\right\} = 0$$

hence

$$\sum_{i=0}^{M-1} h(i)\ R_Y(j-i) - R_{SY}(j) \ = \ 0$$

or

$$\sum_{i=0}^{M-1} h(i)\ R_Y(j-i) = R_{SY}(j)$$

In matrix form, we have

$$\begin{pmatrix} R_Y(0) & R_Y(-1) & \cdots & R_Y(-M+1) \\ R_Y(1) & R_Y(0) & \cdots & R_Y(-M+2) \\ \vdots & & \vdots & \ddots \\ R_Y(M-1) & R_Y(M-2) & \cdots & R_Y(0) \end{pmatrix} \begin{pmatrix} h(0) \\ h(1) \\ \vdots \\ h(M-1) \end{pmatrix}$$

$$= \begin{pmatrix} R_{SY}(0) \\ R_{SY}(1) \\ \vdots \\ R_{SY}(M-1) \end{pmatrix}$$

or

$$\mathbf{Rh} = \mathbf{r}$$

where $\mathbf{R}$ *is the auto-correlation matrix of* y*,*

$$\mathbf{h} = [h(0), h(1), \cdots, h(M-1)]^T$$

and

$$\mathbf{r} = [R_S(0), R_S(1), \cdots, R_S(M-1)]^T$$

The optimal FIR filter weights (in the MMSE sense) are given by

$$\mathbf{h} = \mathbf{R}^{-1} r$$

The MMSE is given by

$$\begin{aligned} MMSE &= E\epsilon_n s_n \\ &= E\left\{ \left(s(n) - \sum_{i=0}^{M-1} h(i) y(n-i) \right) s(n) \right\} \\ &= R_S(0) - \sum_{i=0}^{M-1} h(i)\ R_{YS}(-i) \\ &= R_S(0) - \sum_{i=0}^{M-1} h(i)\ R_{SS}(i) \end{aligned}$$

Example 8.11 *Given $y(n) = s(n) + n(n)$, for $n = 0, 1, \cdots, N-1$, where the signal and the noise are statistically independent. Use the M most recent data points to estimate $s(n+1)$ the desired quantity.*

Predict $s(n+1)$ based on $s(n-i)$, $i = 0, 1, \cdots, M-1$

$$\epsilon(n+1) = \hat{s}(n+1) - s(n+1)$$

$$E\left\{\left(\sum_{i=0}^{M-1} h(i)\ y(n-i) - s(n+1)\right) y(n-j)\right\} = 0 \qquad j = 0, 1, \cdots, M-1$$

or

$$= \sum_{i=0}^{M-1} h(i)\ R_Y(j-i) - R_{SY}(j+1) = 0$$

or

$$= \sum_{i=0}^{M-1} h(i)\ R_Y(i-j) - R_S(j+1)$$

where $R_Y(m) = R_S(m) + R_N(m)$.

In matrix form, we have

$$\mathbf{Rh} = \mathbf{r}$$

where

$$\mathbf{R} = \begin{pmatrix} R_Y(0) & \cdots & R_Y(-M+1) \\ R_Y(1) & \cdots & R_Y(-M+2) \\ \vdots & \ddots & \vdots \\ R_Y(M-1) & \cdots & R_Y(0) \end{pmatrix}$$

$$\mathbf{h} = [h(0), h(1), \cdots, h(M-1)]^T$$

$$\mathbf{r} = [R_S(1), R_S(2), \cdots, R_S(M-1)]^T$$

The MMSE is given by

$$MSSE = E\left\{\left(s(n+1) - \sum_{i=0}^{M-1} h(i)\ y(n-i)\right) (s(n+1))\right\}$$

$$= R_S(0) - \sum_{i=0}^{M-1} h(i)\ R_S(1+i)$$

Note, we could have computed a total MMS error by accounting for the MMS error at every time increment n. Note, we only examined the FIR Wiener filter solution. One can also address problems where the Wiener filter is of the IIR form. A non-causal solution is simple. The causal solution is obtained using what is called the Wiener-Hopf technique. Details can be found in [12,14,19], which require factorization of the PSD.

8.6 SUMMARY

Chapter 8 serves as an introduction to the area of estimation. We use estimation in some detection problems (i.e., use the estimate in place of the true value in the nuissance detection problems). Wiener filters are discussed in some detail while Kalman filters are mentioned for completeness sake. Both types of filters are very important processing tools in the signal processing, communications, and control areas. Section 8.2 addresses the basic classes (i.e., MAP, ML, and Bayes') of estimators, which are detailed in different subsections. In Section 8.3, we define several elementary properties that are used in identifying the characteristics of the estimate. These definitions are basic ones. In books specializing on estimation theory, additional definitions can be found. Section 8.4 derives the CR bound for the non-random unbiased parameter case. In Section 8.5, we focus on waveform estimation (i.e., predicting, filtering, and smoothing). Wiener filtering is illustrated using simple examples. Finally, the Wiener filter is examined in the discrete-time case and demonstrated with a Wiener FIR filter example.

8.7 PROBLEMS

1. We estimate the second moment of a zero mean white Gaussian sequence using:

$$Z = \frac{1}{N} \sum_{n=0}^{N-1} x^2(n)$$

which is an estimate of the second order moment.

(a) What is the bias of Z?

(b) What is the variance of Z?

2. Given the optimal detector for a pulsed sinusoid of known duration (unknown: frequency, phase, arrival time) in white Gaussian noise at low SNR is shown below:

$$y(n) \longrightarrow \text{FFT} \longrightarrow Y(k) \longrightarrow \text{magn. sq.} + \text{normaliz.} \longrightarrow 1/N\ |Y(k)|^2$$

Determine whether or not this estimator is biased (show work).

3. Given $n(i)$, for $i = 0, 1, 2, 3, \cdots, N-1$, where $n(i)$ are i.i.d. Gaussian random variables with zero mean and a variance of σ^2

(a) Obtain the ML estimate of the variance of the sequence, denoted by σ^2_{ML} (assume the true mean is zero).

(b) Compute the bias of this estimate.

(c) Compute its variance (i.e., the variance of the estimate).

4. The random processes $x(t)$ and $y(t)$ are real valued, w.s.s., and jointly Gaussian (with zero means) (i.e., $f(x(t)y(s)) = f(x(t)f(y(s))$ for all t and s). Let's define the estimate of the cross-correlation function (based on the time average) as

$$r_{XY}(\tau) = 1/T \int_0^T x(t)y(t+\tau)dt$$

where the data is available for time $(0, T+\tau)$.

(a) What is the bias of this estimate (show all work)?

(b) What is the variance of this estimator (show all work)? *Useful hint:*

$$\int_0^T \int_0^T g(t-s)dt\,ds = T \int_{-T}^{T} \left(1 - \frac{|u|}{T}\right) g(u)\,du$$

References

[1] Sage, A.P., *Estimation Theory with Applications to Communications and Control*, New York: McGraw-Hill, 1971.

[2] Kay, S.M., *Fundamentals of Statistical Signal Processing: Estimation Theory*, Englewood Cliffs, NJ: Prentice-Hall, 1993.

[3] Jazwinski, A.H., *Stochastic Processes and Filtering Theory*, New York: Academic Press, 1970.

[4] Nahi, N.E., *Estimation Theory and Applications*, New York: John Wiley & Sons, 1969.

[5] Deutsch, R., *Estimation Theory*, Englewood Cliffs, NJ: Prentice-Hall, 1965.

[6] van Trees, H.L., *Detection, Estimation, and Modulation Theory*, Part 1, New York: John Wiley & Sons, 1968.

[7] Wiener, N., *The Extrapolation, Interpolation, and Smoothing of Time Series with Engineering Applications*, New York: John Wiley & Sons, 1949.

[8] Wainstain, L.A., and Zubakov, V.D., *Extraction of Signals from Noise*, New York: Dover Publications, Inc., 1962.

[9] Davenport, W.B., and Root, W.L., *An Introduction to the Theory of Random Signals and Noise*, New York: McGraw-Hill, 1958.

[10] Kalman, R.E., "A New Approach to Linear Filtering and Prediction Problems," *Trans. ASME, J. Basic Eng.*, vol. 82D, pp. 34–45, March 1960a.

[11] Brown, R.G., and Hwang, Y.C., *Introduction to Random Signals and Applied Kalman Filtering*, 3rd Ed., New York: John Wiley & Sons, 1997.

[12] Papoulis, A., *Probability, Random Variables, and Stochastic Processes*, New York: McGraw-Hill, 1965.

[13] Orfanidis, S.J., *Optimum Signal Processing: An Introduction*, 2nd Ed., New York: MacMillan, 1988.

[14] Therrien, C.W., *Discrete Random Signals and Statistical Signal Processing*, Englewood Cliffs, NJ: Prentice-Hall, 1992.

[15] Hayes, M.H., *Statistical Digital Signal Processing and Modeling*, New York: John Wiley & Sons, 1996.

[16] Moon, T.K., and Stirling, W.C., *Mathematical Methods and Algorithms for Signal Processing*, Englewood Cliffs, NJ: Prentice-Hall, 2000.

[17] Manolakis, D.G., Ingle, V.K., and Kogon, S.M., *Statistical and Adaptive Signal Processing; Spectral Estimation, Signal Modeling, Adaptive Filtering and Array Processing*, New York: McGraw-Hill, 2000.

[18] Barkat, M., *Signal Detection and Estimation*, London: Artech House, 1991.

[19] Kamen, E.W., and Su, J.K., *Introduction to Optimal Estimation*, London: Springer-Verlag, 1999.

[20] Burl, J.B., *Linear Optimal Control: H_2 and H_∞ Methods*, Englewood Cliffs, NJ: Prentice-Hall, 1999.

[21] Nishiyama, K., "Robust Estimation of a Single Complex Sinusoid in White Noise-H_∞ Filtering Approach," *IEEE Trans. on Signal Processing*, vol. 47, pp. 2853–2856, October 1999.

Chapter 9

Applications to Detection, Parameter Estimation, and Classification

9.1 INTRODUCTION

This chapter introduces some of the modern signal processing ideas (algorithms) that can be used to perform detection and/or parameter estimation. As was discussed in the earlier chapters, the basic idea is to focus the energy (or power) of the signal of interest at one location in the decision or estimation space while spreading noise, interference, and jamming as much as possible in the adopted coordinate system. Usually, the discussions will deal with energy. Extensions to power are easily obtained by normalizing by the length, that is by the number of samples involved. Power is just normalized energy. Normalizing the energy requires one more division per output data point; hence, if we can save in the processing effort (i.e., do not normalize), we prefer to do so. If the energy at some location exceeds a given threshold, then detection has taken place and estimation can take place.

All of the applications deal with discrete-time data (i.e., discrete-time processes or sequences), since by definition an application implies using hardware and/or software to process the data. There are some analog type detectors, but the speed, size, power consumption, error correction and storage abilities, and robustness to physical perturbation (i.e., heat, shock, age deterioration) make digital implementations the preferred choice. It is difficult to switch analog devices from one processing job to another, due to an electronic charge buildup or removal. For digital implementations, a processing job change (i.e., use data from a different sensor or to obtain a different computational result) requires only the temporary storage of intermediate results in some memory device (RAM, hard drive, etc.) [1–3]. We note that in the literature, the discussion of most of the algorithms are done using continuous time arguments (i.e., continuous time, continuous frequency, etc.). Closed form expressions are much more readily obtained by using analog rather than discrete arguments. Typically, we show only discrete processing schemes, since those are the ones that do get implemented. There is sufficient information provided so that the reader will get familiar with the processing schemes. To fully understand these algorithms, we refer the reader to the references. In earlier chapters we used $\exp(j\pi kn)$ to denote an exponential terms. Now we prefer to use the more compact notation $e^{j2\pi kn}$ in this chapter. We define the correlation function based on the time average as given in Problem 8.4; that is, for a positive lag, the second variable uses a positive shift.

9.2 THE PERIODOGRAM AND THE SPECTROGRAM

We start the discussion by examining two powerful techniques used in many applications, the periodogram and the spectrogram. The periodogram and the spectrogram provide information about power (or energy) as a function of frequency and as a function of frequency and time, respectively.

9.2.1 Periodogram

We know that for sinusoids embedded in white additive Gaussian noise, the periodogram is the optimal processing algorithm (see the arguments in Chapter 6, Section 6.6). If only white Gaussian noise is present, the energy (power) obeys an exponential probability density function (PDF) at each spectral bin (see Chapter 2, Section 2.3.5). The periodogram, disregarding the normalization, is given by

$$P_X(k) = \left| \sum_{n=0}^{N-1} x(n)\; e^{\left(-j\frac{2\pi}{N}kn\right)} \right|^2 \tag{9.1}$$

where k denotes the spectral location and N is the integration or data length. To speed up the operation (compare (9.1) with (3.20)), the normalization has been left out. If the need arises, we can always normalize. The block diagram for the periodogram is given in Figure 9.1.

In a given bin, when a signal is present, the bin's output will be governed by a non-central chi-squared probability density function. We also know that for white Gaussian noise, the bin's PDF is of an exponential form, also called chi-squared with two degrees of freedom. The mean of the exponential

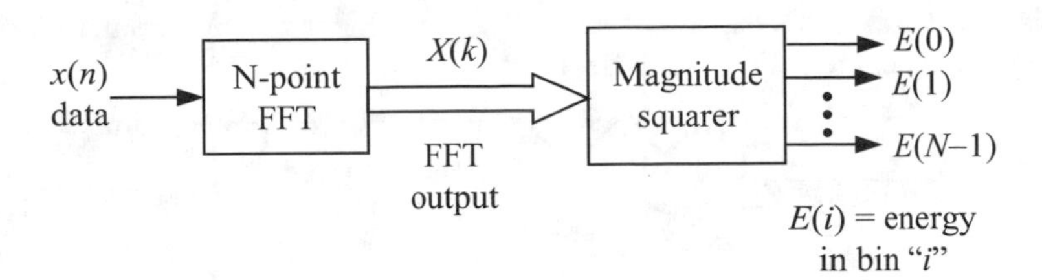

Figure 9.1: The block diagram of the periodogram.

PDF equals to one standard deviation (see the probability summary table in Appendix D). Hence, we can talk about the probability of detection and the probability of false alarm. We also know if we take the average of the noise only background across the bins, three and more standard deviations of peak to noise mean separation are required to have a reasonable false alarm rate. One could obtain an exact threshold based on the noise only PDF to guarantee a desired false alarm rate and hence, determine the probability of detection based on the prevailing SNR level.

The processing gain of the periodogram is essentially 3 dB per doubling of the FFT length, provided the sinusoid under investigation stays in one spectral bin, where the spectral bin width is halved every time the integration time is doubled (i.e., double the FFT size). All spectral components, including stable sinusoids, have an inherent frequency stability and bandwidth. If the sinusoid is time limited then the spectral peak of the sinusoid has a bandwidth inversely proportional to the duration. In a situation like this, exceeding the integration time will add additional noise to the integration process during the time epoch when the signal disappears. Even if the signal exists for the duration of the transform, the bin width may be so small that the frequency of the sinusoid wanders outside its spectral bin. When signal energy is diverted outside the principle spectral bin, then the spectral representation suffers from what is called over-resolving.

9.2.2 Spectrogram

The spectrogram allows multiple looks of the spectral content as a function of time. The algorithm, disregarding the normalization, is given by

$$SP_X(k,m) = \left| \sum_{n} x(n)\; w(n-m)\; e^{\left(-j\frac{2\pi}{N}kn\right)} \right|^2 \tag{9.2}$$

where $w(n)$ is a suitably chosen window, N is the transform length, and k and m correspond to the spectral location and the time index, respectively. The spectral output, as a function of time, is achieved by segmenting the data

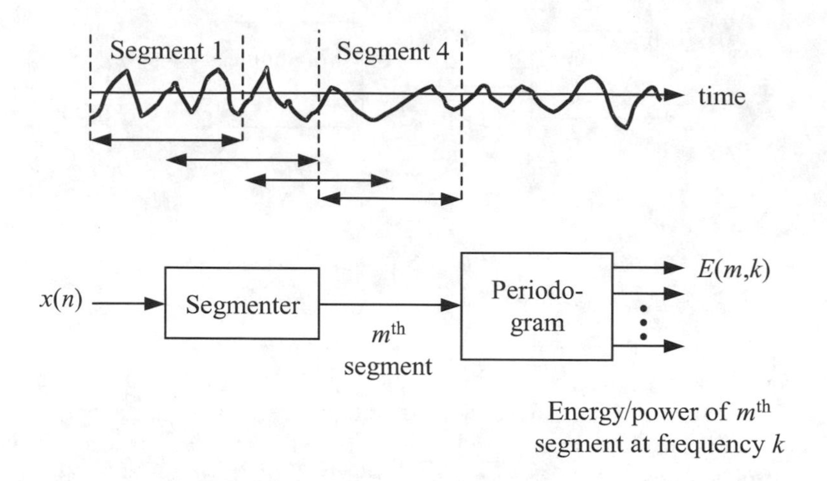

Figure 9.2: The block diagram of the spectrogram.

in an overlapping fashion and computing a periodogram for each segment; hence, it can also be interpreted as a sequence of periodograms. Figure 9.2 shows the block diagram for the spectrogram.

In a cathode ray type (CRT or scope) display, the spectrogram results in what is called a waterfall display since each new segment is plotted on top of a row of older spectra, with the oldest ones sliding down the screen and eventually dropping out of view at the bottom line. In a sonar application, this type of display is called a Lofargram, where paper is darkened according to the intensity (energy) of the spectrum. The term Lofar (Low Frequency Analysis and Recording) has its roots in military applications, that is, in the detection of submarines, torpedoes, boats, and ships which primarily was accomplished at low audio frequencies.

These displays show energy (power) versus time and frequency. Narrowband components, (i.e., sinusoids or tonals) will provide a display in which energy is high in a given spectral location for the duration of the signal. If the window, $w(n)$ in (9.2), is a rectangular window, we have

$$w(n) = \begin{cases} 1\,; & -N/2 \le n \le N/2-1 \\ 0\,; & \text{else} \end{cases} \qquad \text{(the rectangular window)}$$

The rectangular window will select data values from $N/2$ points to the left of time location m, to $N/2\ -1$ points to the right of m. If a Gaussian type window is used and the Fourier transform is not magnitude squared, then this particular version of the short time Fourier transform (STFT) is called the Gabor transform. An application of the Gabor representation to detect transient signals can be found in [48]. The choice of the Gaussian window allows

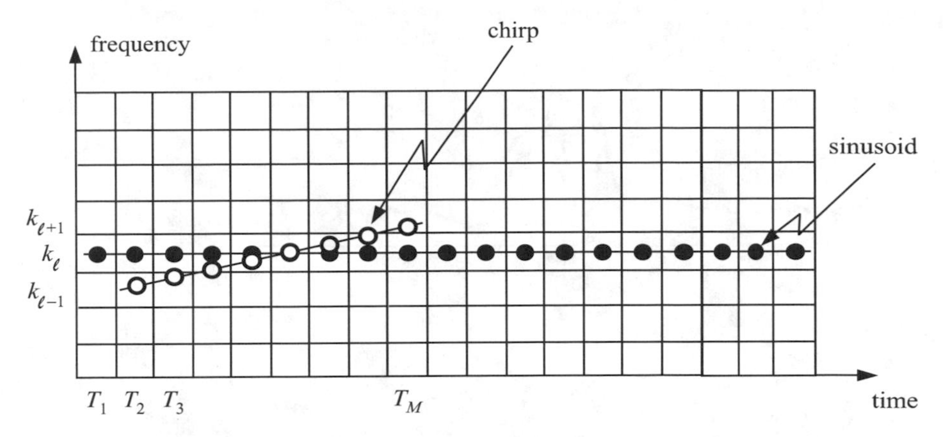

Figure 9.3: The spectrogram of a sinusoid and of a linear chirp.

a time-frequency product which achieves the lower bound of the uncertainty inequality (i.e., best time frequency localization) [7]. For discrete-time applications, the parameter m is stepped in increments of N (no overlap), $N/2$ (50 percent overlap), or $N/4$ (75 percent overlap), where N denotes the FFT size. The display may be in the form as demonstrated in Figure 9.3, which illustrates the one-sided spectral display for two signals. A sinusoid, for all time, at spectral bin location k_ℓ and a linearly chirped sinusoid ramping from spectral bin location $k_{\ell-1}$ at time T_2 to spectral bin location k_{L+1} at time T_M are present. We can further enhance the detection/estimation procedure by incoherently averaging over time or over time and frequency.

If we expect the presence of constant frequency sinusoids, we can average each spectral bin of the PSD over time. Assuming that the individual spectra are statistically independent, the output PSD will have a chi-squared PDF with two times the number of averages as the degree of freedom. Asymptotically, we expect an improvement by 1.5 dB per doubling (see discussion in Chapter 6, Section 6.8).

If we expect that a linearly chirped sinusoid is present, we can initialize an integration along slant lines in the spectrogram. This is done by integrating over the desired (expected) signal duration starting at time T_i and accumulating energy along the hypothetical path of the chirp. If the chirp rate is unknown, we can initialize an averager for start frequency and for slew rates starting at time T_i. An averager is obtained by integrating for the length of the signal. Since we do not know the start time in general, we initialize a bank of averagers for every i hence for all T_i of interest. The outcome of the integration is thresholded (CFAR fashion). When the threshold is exceeded, we establish the presence of the chirp using a pre-set false alarm rate. We can also extract the parameters of interest, such as, start and stop time, duration,

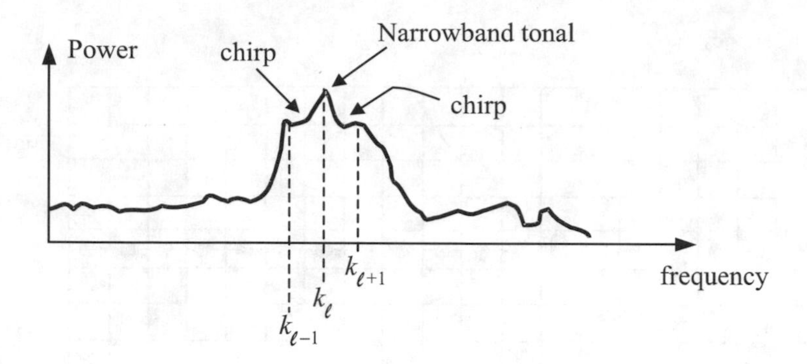

Figure 9.4: Energy of the averager.

slew rate, start and stop frequencies, and energy.

Figure 9.3 shows the so-called time frequency tiling where each tile has the same time duration and frequency bandwidth; hence, the BT (bandwidth-time) product is constant. Figure 9.4 shows a representative output of the incoherent averager (i.e., the time dependency is averaged out). If the energy stays in one spectral bin for the duration of the incoherent average, then variance reduction allows the detection of very weak (but stable) components.

We can also modify the periodogram to obtain coherent processing gain for a stable sinusoidal or a linear chirp signal. This is accomplished by lengthening the integration time of the periodogram or equivalently summing the periodogram or spectrogram outputs prior to the magnitude square operation. For a chirp waveform, we apply pre-hetrodyning to the chirp waveform (i.e., de-chirping), as shown in (9.3), assuming that start time and duration are known as given by the following equation

$$P_X(k,\alpha) = \left| \sum_{n=n_0}^{n_0+N-1} x(n)\ e^{\left(-j\alpha n^2\right)}\ e^{-\left(j\frac{2\pi}{N}kn\right)} \right|^2 \tag{9.3}$$

where α is proportional to the unknown slew rate of the linear chirp, k is the spectral location, N is the signal duration, and n_0 is the start time. If the start time is unknown, we can easily modify this approach by initializing the processing at every point of interest, similar to the window used in (9.2). The signal duration problem can be handled by using multiple duration transforms, each one identical in operation, differing only in the duration N. Of course, when working with energy one needs to account for the difference in integration (summation) length if an automated thresholding is employed.

9.3 CORRELATION

For wideband (i.e., transient) components, the cross-correlation is an ideal detection tool. The cross-correlation can be used to detect the presence of transient signals received at two spatially separated sensors (arrays). When the signals are lined up in time, a sharp peak occurs in the correlation domain indicating the presence of the transient and providing an estimate of the time difference of arrival (TDOA). We define the correlation function (based on the time average)

$$R_{XY}(\ell) = \frac{1}{N-|\ell|} \sum_{n=0}^{N-|\ell|-1} x(n)y^*(n+l) \quad \text{(unbiased estimate)}$$

$$R_{XY}(\ell) = \frac{1}{N} \sum_{n=0}^{N-|\ell|-1} x(n)y^*(n+l) \quad \text{(biased estimate)} \tag{9.4}$$

We realize that averaging of the lagged products corresponds to the projection of one vector onto a shifted version of the second vector (i.e., an inner product as discussed in earlier chapters). A typical application is the localization of an uncooperating target or emitter [4].

9.4 INSTANTANEOUS CORRELATION FUNCTION, WIGNER-VILLE DISTRIBUTION, SPECTRAL CORRELATION, AND THE AMBIGUITY FUNCTION

The publication of the papers by Claasen and Mecklenbräuker [5,6] renewed interest in time frequency distributions (TFDs). The basic idea is to take the Fourier transform of the instantaneous correlation function, where the instantaneous correlation function is approximated using the lagged product of the sequences under consideration.

9.4.1 Wigner-Ville Distribution (WVD)

Several textbooks [7–9] and papers [5,6,9–19] have been written describing aspects of the WVD. If we define $R(n,m)$ as $E\left[x(n+m/2)\, x^*(n-m/2)\right]$, and approximate it by its instantaneous value $x(n+m/2)\, x^*(n-m/2)$, then taking the 1-D Fourier transform over m leads to

$$WVD_X(n,k) = \sum_m R_X(n,m)\, e^{\left(-j\frac{2\pi}{N}km\right)}$$

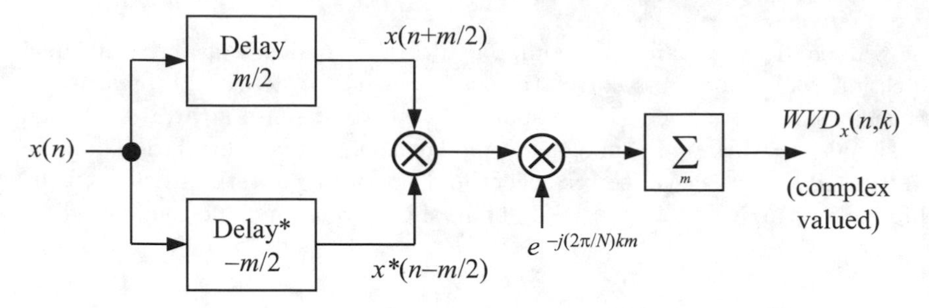

Figure 9.5: Block diagram of the WVD.

$$= \sum_{m} x\left(n + \frac{m}{2}\right) \, x^{*}\left(n - \frac{m}{2}\right) \, e^{\left(-j\frac{2\pi}{N}\, kn\right)} \tag{9.5}$$

where n is the time index, m is the delay, and k is the spectral location. The algorithm is illustrated with the block diagram (Figure 9.5), where one could easily replace the x-sequence with a y-sequence and create the cross Wigner-Ville distribution between the x and y sequences, denoted by $WVD_{XY}(n,k)$. Figure 9.6 shows the WVD and its three transform domains, which will be discussed in the following sections.

The product in the summand (i.e., $x(n+m/2)\, x^{*}(n-m/2)$, a non-linear operation) causes many artifacts in the time frequency domain. In particular, cross terms in time and in frequency do occur. Several approaches to minimize or remove these unwanted cross terms have been addressed [13,15,18,19]. In the time frequency (TF) plane, one can read off the onset, frequency, duration, bandwidth, spectral shape, and spectral dynamics of the signal of interest. The tiles of the TF representation can be averaged, in a magnitude sense, to enhance the detection of a waveform that follows the time-frequency positions governed by the averaging assignment of the TF tiles.

9.4.2 Spectral Correlation

As shown in Figure 9.6, the 2-D Fourier transform of the instantaneous correlation function leads to the spectral correlation function denoted by $C(\ell,k)$, as given by

$$R_X(\ell,k) = \sum_{n}\sum_{m} x\left(n + \frac{m}{2}\right) x^{*}\left(n - \frac{m}{2}\right) e^{+j(2\pi/N)\ell n} \; e^{-j(2\pi/N)km} \tag{9.6}$$

where ℓ and k are spectral locations. This result can also be obtained by taking a 1-D Fourier (inverse) transform over n, of the WVD as given by

$$C_X(\ell,k) = \sum_{n} WVD_X(n,k) \; e^{+j(2\pi/N)\ell n} \tag{9.7}$$

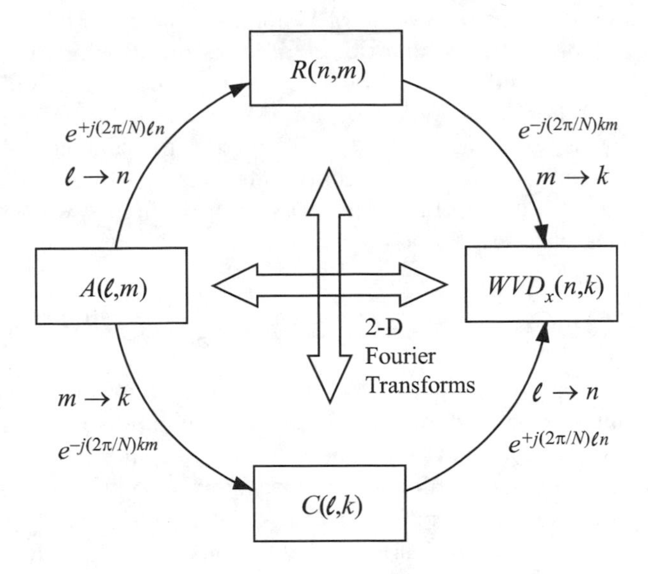

Figure 9.6: Transform domains of the WVD.

For our adopted sign convention, the 1-D forward Fourier transforms (i.e., the exponential kernel uses a negative sign) are indicated on Figure 9.6. The sign convention is arbitrary, but comes from making the narrowband ambiguity function definition agree with the generally accepted version.

9.4.3 Ambiguity Function

The ambiguity function (narrowband ambiguity function) is given by the 1-D Fourier transform of the instantaneous correlation function, by the 1-D Fourier transform of the spectral correlation function, or by the 2-D Fourier transform of the WVD. The most common form is given by

$$A_X(\ell, m) = \sum_n x(n + m/2)\, x^*(n - m/2)\, e^{-j(2\pi/N)\ell n} \tag{9.8}$$

where ℓ is the spectral location and m is the time (correlation shift) variable. This equation shows that it is the auto- (or cross-) correlation function of the sequence $x(n)$ with a frequency shifted version of $x(n)$.

Depending on the task and the scenario, one of the four representations can be exploited to obtain detection and parameter information directly from the surface or from the magnitude of the surface. Usually, the WVD is the preferred tool. As mentioned earlier, it suffers from the product operation of the two sequences involved. Many papers have dealt with the artifacts created by the non-linear operation. Most noteworthy are the smoothing using a Gaussian type window (Choi-Williams [18]) and the adaptive kernel techniques [15]. If the summation support is finite, which is usually the case, then the truncated version is called the pseudo Wigner-Ville distribution (PWVD). A host of TF distribution estimators is available, which can be obtained by using a particular kernel function in what is known as Cohen's class of joint time-frequency representations [7,10]. MATLAB based code and a user manual can be found at the website http://www.crttsnuniv-nantes.fr/~auger/tftb.html, free of charge.

9.5 CYCLO-STATIONARY PROCESSING

Cyclo-stationary processing performs a spectral correlation between the sidebands (or what is assumed to be the sideband components) relative to a given spectral signal location [21]. The resultant output surface obtained by cyclo-stationary processing will convey more information than a conventional power spectral density if the auto-correlation function of the signal of interest tends to be periodic. Different signal modulations result in different, unique outputs from a spectral correlation analyzer (SCA) [22], which can be used to

(a) Establish the presence of a signal.

(b) Identify the type of modulation.

(c) Extract some of the signal parameters.

Clearly, the first item is the detection part while the last two items address the estimation aspect. The time smoothed cyclic periodogram of the sequence $x(n)$ is defined as

$$S_X^{\alpha}(\ell, k)_{\Delta t} = \left\langle X_T\left(n, f + \frac{\alpha}{2}\right)\ X_T^*\left(n, f - \frac{\alpha}{2}\right)\right\rangle_{\Delta t} \tag{9.9}$$

where Δt is the data time span used in the process, $X_T(n, f \pm \alpha/2)$ are the complex envelopes of the narrowband bandpass sequence $x(n)$ filtered at $f \pm \alpha/2$, $*$ denotes complex conjugation, α is the spectral separation, and $\langle\ \rangle_{\Delta t}$ denotes time averaging over the data time span. The complex envelope is given by

$$X_T(n, f) = \sum_{m=-N'/2}^{N'/2-1} a(m)\ x(n-m)\ e^{-j2\pi f(n-m)T} \tag{9.10}$$

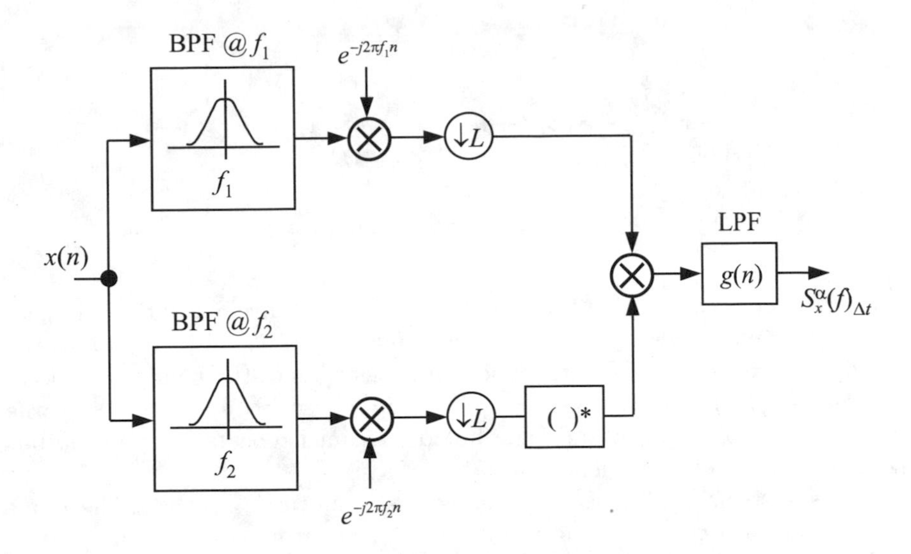

Figure 9.7: Smoothed cyclic periodogram.

where $a(n)$ is a data tapering window of length T ($T = N'\ T_s$), T_s is the sampling period, N' is the number of data samples in T seconds. By comparison, that standard time averaged periodogram is of the form

$$\begin{aligned} SP_X(n,f) &= \lim_{\alpha \to 0} S_X^\alpha(n,f)_{\Delta t} \\ &= S_X(f) \end{aligned} \tag{9.11}$$

Equation 9.11 shows that the classical PSD estimate resides at the $\alpha = 0$ locations of the surface.

If we are dealing with just one single, short data span (i.e., one segment), then we can remove the time dependency, in the argument list of (9.11). If there is more than one data segment, we retain the time dependency, symbolized with the symbol n. One way to create the smoothed cyclic periodogram is shown in Figure 9.7, where $\alpha = f_1 - f_2$, L is the decimation parameter (for no decimation, L will be one), f_1 is the center frequency of the top bandpass filter, f_2 is the center frequency of the bottom bandpass filter, and $g(n)$ allows the low pass filter to have desirable characteristics. One possible $g(n)$ sequence is that all $g(n)$ equal unity, i.e., a summer that sums up a finite number of samples, also called a boxcar integrator.

Figure 9.7 illustrates the general principle of spectral correlation very eloquently. Assuming for simplicity that $g(n)$ is a boxcar integrator, then we see that the algorithm is simply a correlator that correlates the output of the top channel with the conjugated output of the bottom channel. Focusing on

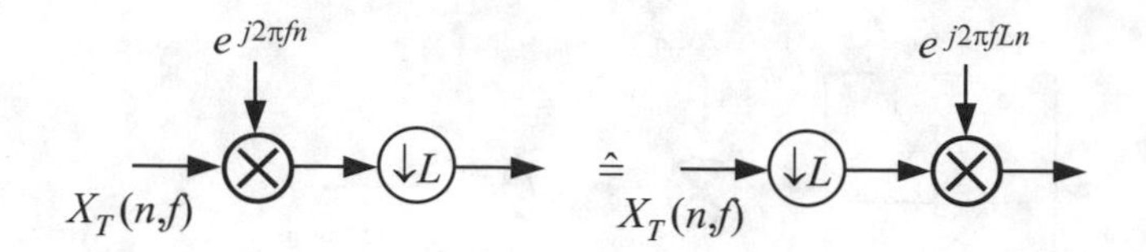

Figure 9.8: Rearranging hetrodyne and desample operations.

the top channel, we see that the signal is the complex envelope of the bandpass sequence centered at f_1, desampled by a factor L (i.e., the filter passes the bandpass process, the complex exponential hetrodynes the bandpass process to base band, and the desampler removes, if desired, redundancy due to the oversampling of the process). The bottom channel repeats the process, but selects as the center frequency f_2.

This system, in modern terminology, is a multirate system. The input is sampled at a rate f_s. The input to the correlator section is sampled at a rate f_s/L. The output is a three-dimensional surface indexed by f (i.e., $f = 1/2\ (f_1 + f_2)$, which is the center frequency), and α (i.e., $\alpha = f_1 - f_2$) which is the difference frequency. The difference frequency is the two-sided bandwidth relative to the center frequency f. If the proper f_0 and α_0 are selected we expect a large correlation peak at the coordinate (f_0, α_0). If that value, in magnitude, exceeds a given threshold, we declare a detection. The behavior of $|S_X^\alpha(f)|$ (i.e., peaks, valleys, and geometry) tells the investigator about the modulation type and modulation parameters. Increasing Δt, the data span, will increase detectability and improve the accuracy of the modulation parameters, provided the received signal maintains integrity of the parameters. Hence, a negative SNR (in dB) is not a problem if the number of samples can be made sufficiently large to make up for the low SNR. We also notice that (9.7) and the operations shown in Figure 9.7 can be related. That is, the smoothed cyclic periodogram can be interpreted as an evaluation of the spectral correlation function about a center frequency given by $(f_1+f_2)/2$ followed by an additional time average (i.e., LPF operation).

When we examine (9.9) and (9.10) and Figure 9.7, we notice the order of the hetrodyne and desample operations. This is also presented on the left-hand side of block diagram Figure 9.8.

The order of the resampler and hetrodyner can easily be reversed, as shown on the right-hand side of Figure 9.8. The new arrangement has the obvious advantage that by resampling first the number of data points has been reduced. Hence, the number of data points that need to be stored and operated on has been reduced while maintaining full accuracy. We can modify the algorithm of Figure 9.7 to take advantage of this idea. The result is shown in Figure 9.9.

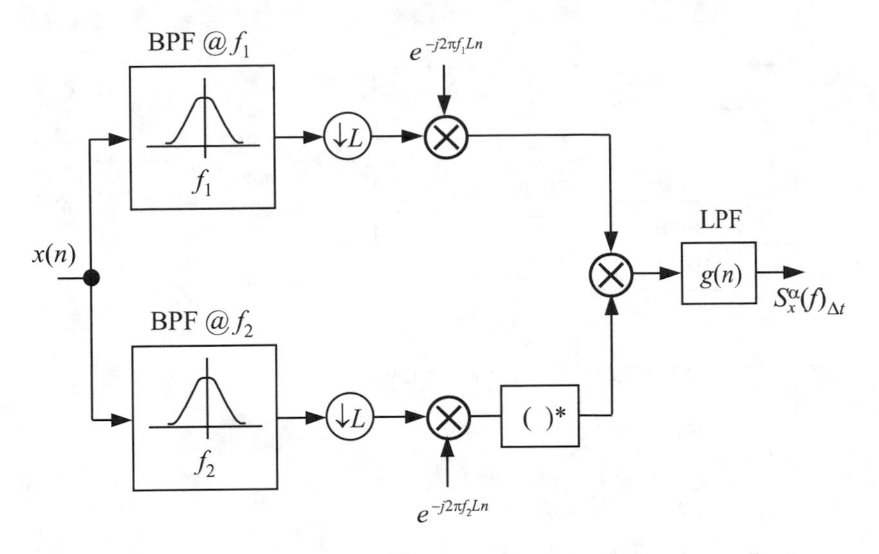

Figure 9.9: Smoothed cyclic periodogram.

We realize that the first two elements (blocks) in each leg (i.e., the band-pass filter and the down sampler) can be replaced with an FFT using the spectral bin corresponding to frequency f_1 and f_2, denoted by f_{ℓ_1} and f_{ℓ_2}, respectively. The desired down sampling is obtained with proper choice of the FFT size and the overlap factor. We can also replace the last processing components (i.e, the two hetrodyne, the product, and the low pass filtering operations) with an additional FFT of size P. This processing scheme is called the FFT accumulation method (FAM) [24] and is shown in Figure 9.10.

The variables are: i, the number of the ith FFT (or segment); N, the input FFT size governing the accuracy of choosing the particular center frequency f and the width of the spectral filter as well as the desampling rate; and P, the size of the output FFT governing the incremental resolution of the α parameter [25,26].

There is another processor implementation, the so-called strip spectral correlation algorithm (SSCA) as shown in Figure 9.11 [24]. The topic of cyclic detection, where the theoretical and measured spectral correlation density (SCD) are correlated, is addressed in Chapter 14, Section E of [27]. A possible detection statistic for continuous type variables is given in (9.12)

$$z = \int \hat{S}_X^\alpha(f) \, \frac{S_S^\alpha(f)^*}{S_N(f+\alpha/2)\, S_N(f-\alpha/2)} \, df \tag{9.12}$$

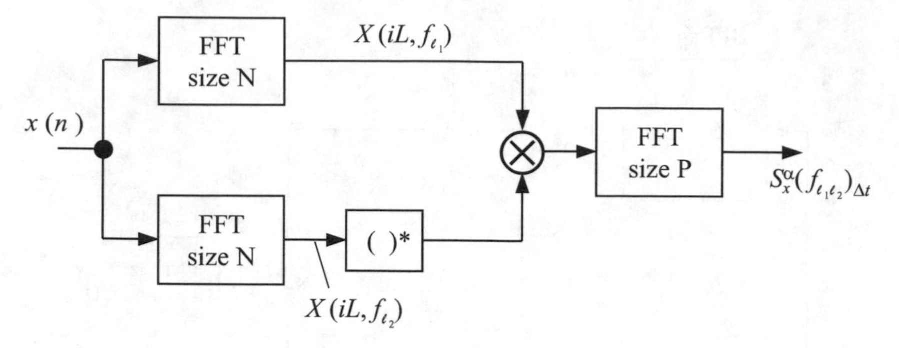

Figure 9.10: The FFT accumulation method (FAM).

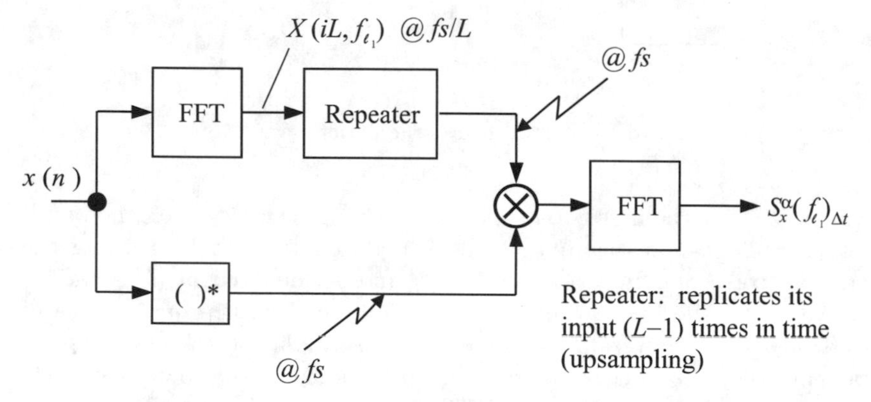

Figure 9.11: The strip spectral correlation algorithm (SSCA).

where $*$ denotes conjugation, $\hat{S}_X^{\alpha}(f)$ is a coarse estimate of $S_S^{\alpha}(f)$. We can interpret the operation in (9.12) as a correlation between a known cyclic spectra ($S_X^{\alpha}(f)$) and the data dependent estimate $\hat{S}_X^{\alpha}(f)$, which will be an optimal operation if the output noise can be thought of as white Gaussian noise. The denominator serves as a normalization. For additional details consult reference [27]. This does not include averaging over the α parameter. The variable z is, in general, complex valued, so that one works with the magnitude of z.

For more details, especially for plots of the magnitude of $S_X^{\alpha}(f)$ of various modulated digital communication signals, we refer the reader to [21,23–28]. Extensions of this work are very successful, when using higher order statistics [29,30]. We shall explore some general aspects of higher order statistics in the next section.

9.6 HIGHER ORDER MOMENTS AND POLY-SPECTRA

With ever faster computers, tasks that seemed to be overwhelming some time ago are becoming commonly solved problems. Of particular interest is the detection of transients and non-Gaussian type signals as well as the detection of signals in a non-stationary or non-Gaussian noise environment. One successfully employed technique relies on the use of higher order moments or higher order cumulants. For simplicity, we shall assume that the random sequences and their underlying random processes are real valued and stationary. Extensions to complex valued processes and sequences are available in the literature. As a starting point, we recommend the book by Nikias and Petropulu [31] and three tutorial papers [32,38,41]. From a probabilistic point of view, the first four moments are given by

$$m_1 = E\{x_1\} \tag{9.13}$$

$$m_2 = E\{x_1^2\} \tag{9.14}$$

$$m_3 = E\{x_1^3\} \tag{9.15}$$

$$m_4 = E\{x_1^4\} \tag{9.16}$$

The corresponding cumulants, using the sign convention [31,32] (i.e., plus signs in the expectations) are given by

$$c_1 = m_1 \text{ (mean)} \tag{9.17}$$

$$c_2 = m_2 - m_1^2 \text{ (variance)} \tag{9.18}$$

$$c_3 = m_3 - 3m_2m_1 + 2m_1^3 \tag{9.19}$$

$$c_4 = m_4 - 4m_3m_1 - 3m_2^2 + 12m_2m_1^2 - 6m_1^4 \tag{9.20}$$

If $\{x(m)\}$ is a real and stationary random sequence and its first n moments exist, then

$$\begin{aligned} &\text{mom}\{x(m), x(m+\tau_1), \ldots, x(m+\tau_{n-1})\} \\ &= E\{x(m), x(m+\tau_1), \ldots, x(m+\tau_{n-1})\} \\ &= m_n(x(m), x(m+\tau_1), \ldots, x(m+\tau_{n-1}) \end{aligned}$$

and will only depend on the time difference $\tau_1, \tau_2, \ldots, \tau_{n-1}$. For example, for orders $n = 1, 2, 3, 4$, the cumulants $c_n(\tau_1, \tau_2, \tau_3)$ of the zero mean sequence $\{x(m)\}$ are related to the moments $m_n(\tau_1, \tau_2, \tau_3)$ as follows

$$c_1 = m_1 = Ex(m) = 0$$

$$c_2(\tau_1) = m_2(\tau_1) - m_1^2(\tau_1) = m_X(\tau_1) = E\,x(m)\,x(m+\tau_1)$$

$$
\begin{aligned}
c_3(\tau_1, \tau_2) &= E\, x(m)\, x(m+\tau_1)\, x(m+\tau_2) \\
c_4(\tau_1, \tau_2, \tau_3) &= E\, x(m)\, x(m+\tau_1)\, x(m+\tau_2)\, x(m+\tau_3) \\
&\quad -E\, x(m)\, x(m+\tau_1)\, E\, x(m+\tau_2)\, x(m+\tau_3) \\
&\quad -E\, x(m)\, x(m+\tau_2)\, E\, x(m+\tau_1)\, x(m+\tau_3) \\
&\quad -E\, x(m)\, x(m+\tau_3)\, E\, x(m+\tau_1)\, x(m+\tau_2)
\end{aligned}
$$

The expressions become complicated as the order of the cumulant increases. Especially if we cannot claim that the first order moment is zero. A further complication arises if we allow the process to be complex valued. Then one needs to decide which terms in the expressions are to be conjugated (there may be several choices) [32,39].

Cumulants have some properties that make them more desirable than moments, as far as higher order statistics are concerned. Some of these properties are listed below [40,41]

(a) Each cumulant is independent of all lower order cumulants.

(b) For a Gaussian process (or sequence), all cumulants of order greater than two are equal to zero, that is, the Gaussian process is completely characterized by its first two moments. Hence, one can use cumulants to estimate the degree of non-Gaussianity of the process.

(c) Cumulants of the sum of two independent statistical processes equal the sum of their respective cumulants.

(d) Ideally, the cumulant of Gaussian noise will be zero, suggesting a processing approach wherein Gaussian noise will have little effect on the processors outcome.

From a data processing point of view, we cannot compute expectations and, assuming ergodicity of the moment (cumulant) under consideration, employ a time varying operation. For example

$$\hat{m}_j = 1/N \sum_{i=0}^{N-1} x^j(i)$$

where $x^j(i)$ is $x(i)$ raised to the jth power.

$$\hat{C}_2(\tau) = 1/N \sum_{i=0}^{N-1} x(i)\, x(i+\tau)$$

and

$$\hat{C}_3(\tau_1, \tau_2) = 1/N \sum_{i=0}^{N-1} x(i)\, x(i+\tau_1)\, x(i+\tau_2)$$

Complex valued data requires complex conjugation of some of the sequences [31,39]. Biased and unbiased estimates will use different normalizations.

9.6.1 Cumulant Spectrum

Suppose $x(n)\forall n$ is a real statistical random sequence with m^{th} order cumulant $c_m(\tau_1, \tau_2, \ldots, \tau_{m-1})$. Assuming that the cumulant sequence satisfies

$$\sum_{\tau_1=-\infty}^{\infty} \cdots \sum_{\tau_{m-1}=-\infty}^{\infty} |c_m(\tau_1, \tau_2, \ldots, \tau_{m-1})| < \infty$$

and

$$\sum_{\tau_1=-\infty}^{\infty} \cdots \sum_{\tau_{m-1}=-\infty}^{\infty} (1 + |\tau_j|)\, |c_m(\tau_1, \tau_2, \ldots, \tau_{m-1})| < \infty$$

for $j = 1, 2, \ldots, m-1$.

Then the mth order cumulant spectrum $C_m(\omega_1, \ldots, \omega_{m-1})$ of $\{x(n)\}$ exists, is continuous, and is defined by the $(m-1)$-dimensional Fourier transform of the mth order cumulant sequence. The second order cumulant spectrum is the power spectrum (power spectral density) given by

$$\begin{aligned} C_2(\omega) &= \sum_{\tau=-\infty}^{\infty} c_2(\tau)\, e^{-j\omega\tau} \\ &= \frac{1}{N} |X(\omega)|^2 \end{aligned} \tag{9.21}$$

for $-\pi \leq \omega \leq \pi$.

The second order cumulant sequence is the covariance sequence, while the second order moment sequence is the auto-correlation function. Of course, for a zero mean sequence, the auto-covariance and the auto-correlation sequence are identical. The Fourier transform relationship (9.21) is known as the Wiener-Khintchine relationship.

9.6.2 Bi-Spectrum

The bi-spectrum is defined as the 2-D Fourier transform of the third order cumulant sequence and is given by

$$\begin{aligned} C_3(\omega_1, \omega_2) &= \sum_{\tau_1=-\infty}^{\infty} \sum_{\tau_2=-\infty}^{\infty} c_3(\tau_1, \tau_2)\, e^{-j\omega_1\tau_1}\, e^{-j\omega_2\tau_2} \\ &= \frac{1}{N} X(\omega_1)\, X(\omega_2)\, X^*(\omega_1 + \omega_2) \end{aligned} \tag{9.22}$$

for $-\pi \leq \omega_1, \omega_2 \leq \pi$, and $|\omega_1 + \omega_2| \leq \pi$.

9.6.3 Tri-Spectrum

The tri-spectrum is given as the 3-D Fourier transform of the fourth order cumulant sequence and is given by

$$C_4(\omega_1, \omega_2, \omega_3) = \sum_{\tau_1=-\infty}^{\infty} \sum_{\tau_2=-\infty}^{\infty} \sum_{\tau_3=-\infty}^{\infty} c_4(\tau_1, \tau_2, \tau_3) \cdot e^{-j\omega_1\tau_1} e^{-j\omega_2\tau_2} e^{-j\omega_3\tau_3} \quad (9.23)$$

for $|\omega_1| \leq \pi$, $|\omega_2| \leq \pi$, $|\omega_3| \leq \pi$, and $|\omega_1 + \omega_2 + \omega_3| \leq \pi$.

9.6.4 Poly-Spectrum

In general, the poly-spectrum of order $L-1$ is given as the $(L-1)$-D Fourier transform of the Lth order cumulant sequence and is given by

$$C_L(\omega_1, \ldots, \omega_{L-1}) = \sum_{\tau_1=-\infty}^{\infty} \cdots \sum_{\tau_{L-1}=-\infty}^{\infty} c_L(\tau_1, \ldots, \tau_{L-1})\, e^{-j\sum_{i=1}^{L-1} \omega_i \tau_i} \quad (9.24)$$

Many papers have been published to show the properties of these spectral and their equivalent cumulant representations [32,38,41]. We want to point out that the cumulant sequences or the cumulant spectra can be used for detection and classification purposes. Some typical examples are

- Kletter and Messer use higher order spectral analysis (bi-spectrum) to detect a non-Gaussian signal embedded in Gaussian noise [33].
- Colonnese and Scarano use the third and fourth order statistics to detect transients in white Gaussian noise. The detectors are suboptimal and it is shown that the detector based on third order statistics outperforms the suboptimal detectors based on second and fourth order statistics [34].
- Wickert and Turcotte detect the symbol rate line of a digitally modulated carrier. Their results show that the tri-spectrum performs better than conventional detectors [35]. Robust beam forming is introduced by Nikias and Mendel [38].
- Sattar and Salomonsson study detection using filter banks and higher order statistics [36].
- Ferrari and Alengrin address the estimation of frequencies of a complex sinusoid using the fourth order statistics [37].

These examples are by no means exclusive, rather than that they are thought to serve as a starting point for the interested reader. Mathworks software package [42] maintains a higher order statistics toolbox that allows data processing using some standard algorithms (i.e., bi-spectrum).

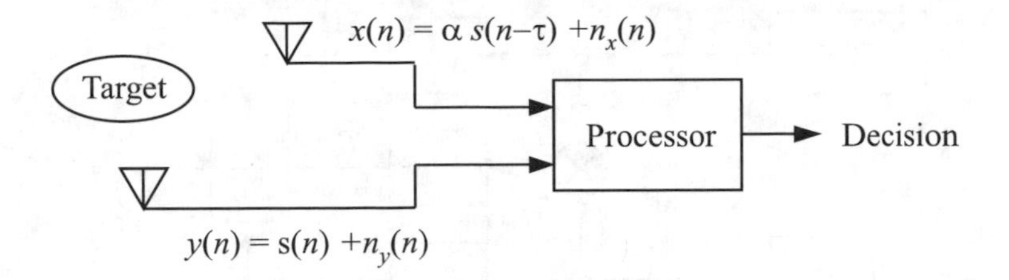

Figure 9.12: Receiver-target scenario.

9.7 COHERENCE PROCESSING

Coherence processing deals with the detection and localization of narrowband signal emitters using two or more receivers. Early applications are found in detection, classification, and localization of high value military targets (i.e., submarines, aircraft carriers, etc.) which emit an acoustic signal that may propagate over long distances in the ocean (i.e., see Figure 9.12). Actually, any narrowband or tonal component that propagates through the medium (that is the ocean) with little attenuation can serve as the detection and localization clue [43,44].

The output from the coherence processor, provided a particular threshold (i.e., CFAR philosophy) is exceeded, can fix the location by using the time difference of arrival (TDOA) and differential Doppler frequency estimate. One can use either information, or both, to help in the localization of the target. The generic continuous time algorithm is given by

$$\gamma_{XY}(f) = \frac{S_{XY}(f)}{\sqrt{S_X(f)\; S_Y(f)}} \tag{9.25}$$

where $S_{XY}(f)$ is the cross power spectral density, $S_X(f)$ and $S_Y(f)$ are the auto-power spectral densities. Since the cross PSD is complex valued, one typically works with the magnitude squared coherence (MSC) function as given for frequency f_k by

$$|\gamma_{XY}(f_k)|^2 = \frac{|S_{XY}(f_k)|^2}{S_X(f_k)\; S_Y(f_k)} \tag{9.26}$$

where $0 \leq |\gamma_{XY}(f_k)|^2 \leq 1$. This algorithm is shown in Figure 9.13.

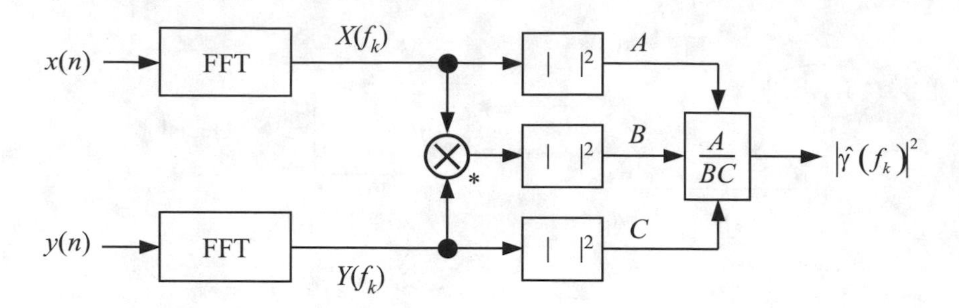

Figure 9.13: MSC estimator.

Usually, estimates are accomplished in discrete-time suggesting an estimate of the form

$$\hat{\gamma}_{XY}(k) = \frac{\sum_{i=1}^{M} X_i(k)\ Y_i^*(k)}{\sqrt{\sum_{i=1}^{M} |X_i(k)|^2 \sum_{i=1}^{M} |Y_i(k)|^2}} \tag{9.27}$$

for some given k, where $X_i(k)$ and $Y_i(k)$ are the Fourier transforms of the ith segments, k is the frequency of interest, M is the number of segments, and $*$ denotes complex conjugation. We can also interpret (9.27) as the average cross PSD evaluated at frequency k, averaged over M sequential segments and normalized by the square root of the product of the averaged auto-power spectral densities.

The MSC has several useful properties, some of which are listed below:

(1) If $x(n)$ and $y(n)$ are independent then $\gamma_{XY} = 0$.

(2) The MSC can be used to measure the linearity of a system, where $x(n)$ and $y(n)$ represent the input and output of the system, respectively.

(3) If $x(n)$ and $y(n)$ are the outputs of two parallel channels, which are both fed by a common signal and independent noise, then the MSC can be used to obtain an estimate of the input SNR to the two channels.

The PDF of the MSC estimate is given by [45–47]

$$\begin{aligned} p\left(|\hat{\gamma}|^2|N, |\gamma|^2\right) &= (N-1)\left(1-|\gamma|^N\right)\left(1-|\hat{\gamma}|^{N-2}\right)\left(1-|\gamma|^2|\hat{\gamma}|^2\right)^{1-2N} \\ &\quad \cdot F_{21}\left(1-N, 1-N; 1; |\gamma|^2\ |\hat{\gamma}|^2\right) \end{aligned} \tag{9.28}$$

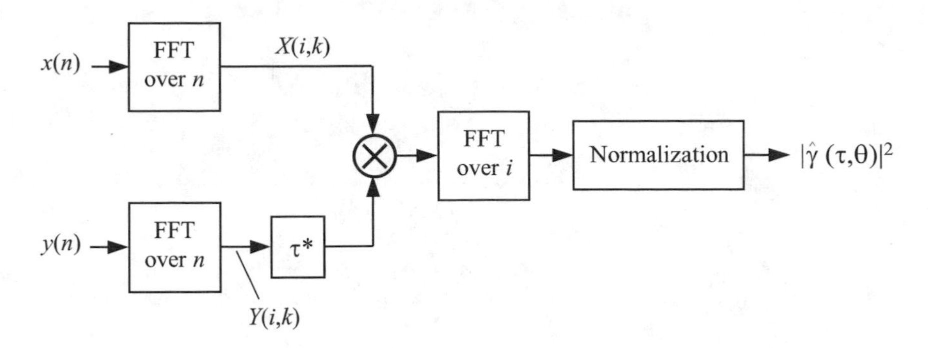

Figure 9.14: Doppler and delay compensated MSC estimator.

where

$$F_{21}\ (a, b; c; z) = \sum_{\ell=0}^{\infty} (a)_\ell\ (b)_\ell\ \frac{z^\ell}{(C)_\ell\ \ell!}$$

and

$$(a)_\ell = \frac{\Gamma(a+\ell)}{\Gamma(a)} \quad = \quad a(a+1)\ldots(a+\ell-1)$$

To allow for Doppler correction and time delay adjustments, the actual MSC algorithm is given by

$$|\hat{\gamma}_{XY}(\tau,\theta)|^2 = \frac{\left| \sum_i X_i\ (f_k)\ Y^*_{i+\tau}\ (f_k)\ e^{\ j\theta i} \right|^2}{\sum_i |X_i\ (f_k)|^2 \sum_i |Y_{i+\tau}\ (f_k)|^2} \tag{9.29}$$

This implementation is shown in Figure 9.14. A typical MSC type output is illustrated in Figure 9.15, where the MSC peaks at location $(\hat{\theta}, \hat{\tau})$.

9.8 WAVELET PROCESSING

We recall from the discussion in Chapter 3, see also Appendix E, that wavelets are bandpass filters whose quality factor (Q) is a constant, that is the ratio of center frequency to bandwidth is a given constant. We also recall that the filtering operation is a linear operation; hence, if signals are embedded in Gaussian noise, then the wavelet filter outputs still have the Gaussian characteristic. Depending on whether or not a signal component is present or not, the output of the wavelet filter (the bandpass portion is called a detail function) may or may not contain a desampled scaled version of the signal.

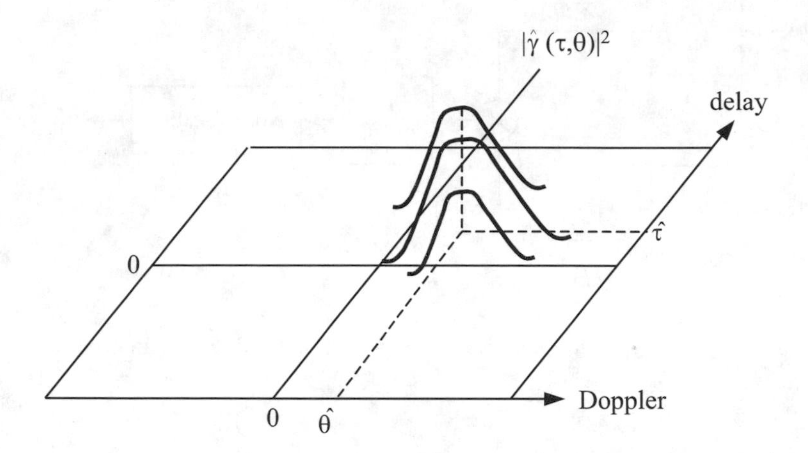

Figure 9.15: Typical MSC output.

If the signal is present, it will be embedded in Gaussian noise. Furthermore, if the input noise is white, the output noise tends to be white.

If the input noise is not white, the bandpass and decimating operations tend to make the output noise spectrally flat. A simple detection scheme is then just based on the detection of a signal embedded in white Gaussian noise. Since the wavelet output, properly desampled, is just a bandpass filtered version of its input that is being sampled at the Nyquist rate, detection of transients is fairly easily accomplished. This is in contrast with the classical narrowband (tonal) detector, which as we recall is achieved via an FFT operation followed by a magnitude (or magnitude squared operation). The periodogram is the optimal detector for a tonal that exists for the duration of the transform. But when the signal is a short duration event, i.e., a pulsed sinusoid or a transient, then the periodogram based detection scheme performs poorly since it averages in too much noise, does not take advantage of the signal bandwidth property, and cannot localize in time. Wavelet based processing is addressing these types of problems, since it allows a range of bandwidth and works in the neighborhood of the time in question. A typical detection setup is shown in Figure 9.16.

For some signals, in particular the ones that have jumps or discontinuities (in amplitude, phase and/or frequency) wavelet decomposition allows a sparse representation. We recall that the ideal scenario is the situation where the resultant is a delta function. Then all signal related information has been focused at one location in the detection space, while all interferences, i.e, noise, other signals, jamming waveforms, etc., are diffused over the detection

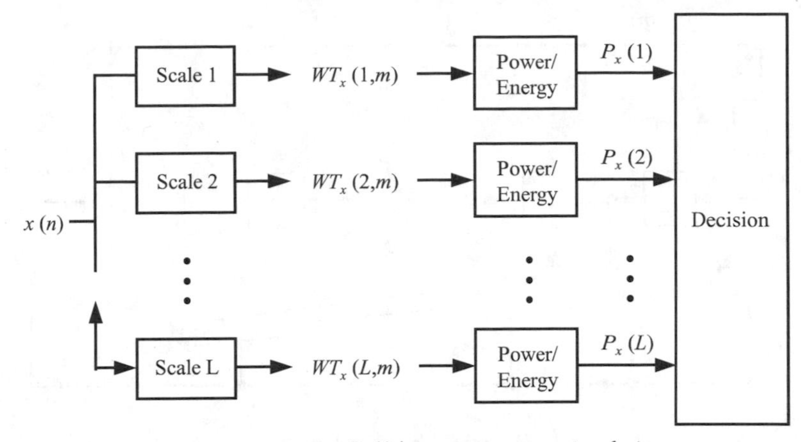

Figure 9.16: Wavelet detection.

space. This is the idea behind the inner product as was advocated in Chapters 4 and 6. Hence, a wavelet type decomposition can be the operation that allows good detection. Note, we left out the word optimal, since we did not derive a wavelet type structure from a Bayes point of view. Rather, we argue using a matched filter type argument, where we try to match the replica to the signal, provided the ambient noise is Gaussian. For introduction to wavelet transforms, we refer the reader to Chapter 3 and Appendix E, as well as to [49–56]. For some wavelet based detection application, we suggest [57–62]. The list is by no means complete, it is intended to serve as a starting point for further reading.

9.9 ADAPTIVE TECHNIQUES

The adaptive filter [63–65] can be used to enhance narrowband spectral line components which are embedded in white noise. Hence, this filter can be used as a detection tool. The adaptive filter can be set up to predict the predictable (i.e., correlated) part of the data under consideration. The so-called automatic line enhancer (ALE) constitutes such an implementation [66–68]. It can be used as a one-step linear predictor. A typical implementation is shown in Figure 9.17. The predicted output is subtracted from the actual and received data. The resulting error output is used in the filter update algorithm. This is similar to the auto-regressive modeling technique addressed in Chapter 7, Section 7.8.4. Rather than solving the matrix equation (7.126), the optimal weights are obtained by solving for them in an iterative fashion,

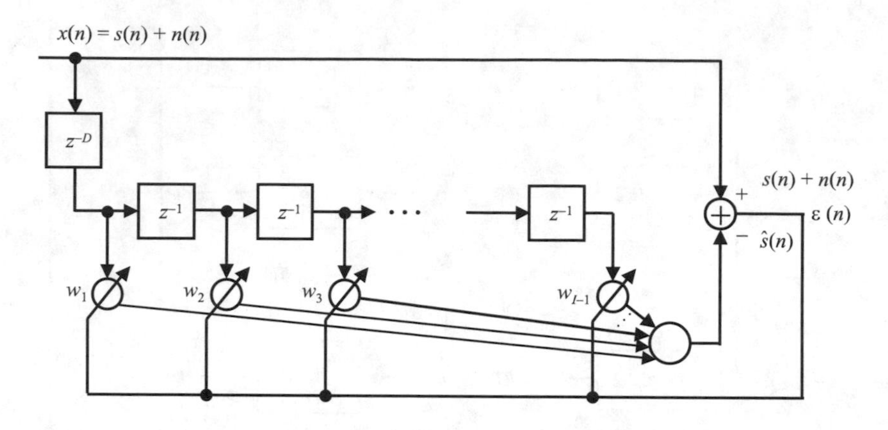

Figure 9.17: Block diagram of the adaptive line enhancer (ALE).

as given by

$$\mathbf{w}(n) = \mathbf{w}(n-1) + 2\,\mu\,\varepsilon(n)\,\mathbf{x}(n-D) \tag{9.30}$$

where D is the delay (typically one sample) between reference and delay channel, μ is a small positive constant, $\varepsilon(n)$ is the error at time n. The vectors $\mathbf{w}$ and $\mathbf{x}$ are given by $\mathbf{w}(n) = [w_1(n), w_2(n), \cdots, w_{I-1}(n)]^T$ (the filter weights at time n) and $\mathbf{x}(n-D) = [x(n-D), x(n-D-1), \cdots, x(n-D-I+1)]^T$ (the stored data in the delay channel), I is the number of multipliers in the FIR filter, and $\mathbf{w}(n-1)$ is the previous filter vector. This is called the least mean-square (LMS) algorithm [69], which is an implementation of the steepest descent method using an estimate or measurement of the gradient of the error surface.

This realization (adaptive filter) does not require knowledge of the correlation functions, but it needs a good choice of the step size parameter: too large a step size creates noise, while too small a step size does not allow quick convergence to the true values (i.e., it requires excessive time). The filter weights can be used to form an estimate of the power spectral density as given by (9.31), where the impulse response $h(i) = -w(i)$, for $i = 1, 2, \cdots, I-1$ and $h(0) = 1$

$$P_{\text{ALE}}(k) = \frac{\text{constant}}{\left|\displaystyle\sum_{i=0}^{N-1} h(i) e^{-j(2\pi/N)ki}\right|^2} \tag{9.31}$$

where $k = 0, 1, \cdots, N-1$ and $[\pi k/N]$ is the digital frequency. Notice that the vector $\mathbf{w}(n)$ is of length $I-1$ while N is typically larger than I. This is easily accomplished by zero padding the filter weight vector to the desired length. It is not unusual to append zeros so that the $\mathbf{h}$ vector is substantially longer than the **w** vector.

The filter is adaptive, hence it allows operation in a non-stationary environment or with dynamic spectral components, or it can accommodate both. Of course, this is all a function of how quickly the filter adjusts (i.e., learns), knowing that a quick response can also generate excessive noise.

9.10 SUMMARY

This chapter tries to trace out some of the late efforts associated with detection and detection related topics. With the progress in computational abilities, processing tasks once assumed to be too cumbersome have become a standard operation. In this spirit, some of the newer techniques have been identified in this chapter, providing some reference material that a motivated reader may easily be able to access. In Section 9.2, the periodogram, the spectrogram, and some averaging techniques are addressed. Section 9.3 briefly explores the correlation concept. The concept of instantaneous correlation function, Wigner-Ville distribution, spectral correlation, and ambiguity function are addressed in Section 9.4. Cyclo-stationary processing and some of its implementations are introduced in Section 9.5. Section 9.6 addresses higher order moments and cumulants as well as their Fourier transforms (i.e., higher order spectra). Coherence processing is discussed in Section 9.7, while Section 9.8 introduces wavelet processing in the detection context. Section 9.9 introduces the adapative filter in a detection related scenario and provides some references.

References

[1] Strum, R.D., and Kirk, D.E., *First Principles of Discrete Systems and Digital Signal Processing*, New York: Addison-Wesley Publishers, 1988.

[2] Ludeman, L.C., *Fundamentals of Digital Signal Processing*, New York: Harper & Row, 1986.

[3] Soliman, S.S., and Srinath, M.D., *Continuous and Discrete Signals and Systems*, Upper Saddle Cliffs, NJ: Prentice-Hall, 1998.

[4] Hippenstiel, R.D., and Aktas, U., "Time Difference of Arrival Estimation Using Wavelet Transforms," *42nd Mid-West Symp. on Circuits and Systems*, Las Cruces, NM, August 9–11, 1999.

[5] Claasen, T.A.C.M., and Mecklenbräuker, W.F.G., "The Wigner Distribution — A Tool for Time-Frequency Signal Analysis, Part I, Continuous Time Signals," *Philips J. of Research*, vol. 35, pp. 217–250, 1980.

[6] Claasen, T.A.C.M., and Mecklenbräuker, W.F.G., "The Wigner Distribution — A Tool for Time-Frequency Signal Analysis, Part II, Discrete-Time Signals," *Philips J. of Research*, vol. 35, pp. 276–300, 1980.

[7] Cohen, L., *Time-Frequency Analysis*, Englewood Cliffs, NJ: Prentice-Hall, 1995.

[8] Qian, S., and Chen, D., *Time-Frequency Analysis — Methods and Applications*, Upper Saddle Cliffs, NJ: Prentice-Hall, 1996.

[9] Flandrin, P., *Time-Frequency Time-Scale Analysis*, San Diego, CA: Academic Press, 1999.

[10] Cohen, L., "Time-Frequency Distributions — A Review," *Proc. of IEEE*, vol. 77, pp. 941–981, 1989.

[11] Amin, M.G., and Allen, F., "Separation of Signals and Cross Terms in Wigner Distribution Using Adaptive Filtering," *Proc. of 32nd Mid-West Symp. on Circuits and Systems*, vol. 2, pp. 853–856, 1989.

[12] Boashash, B., "Time-Frequency Signal Analysis," in *Advances in Spectrum Estimation and Array Processing*, S. Haykin, ed., Englewood Cliffs, NJ: Prentice-Hall, pp. 418–517, 1991.

[13] Hippenstiel, R.D., and Oliveira, P., "Spectral Estimation Using the Instantaneous Power Spectrum (IPS)," *IEEE Trans. Acoustics, Speech, and Signal Processing*, Oct. 1990.

[14] Marple, L., Brotherton, T., Barton, R., et al, "Travels Through the Time-Frequency Zone: Advanced Doppler Ultrasound Processing Techniques," *Proc. 27th Asilomar Conf. on Signals, Systems and Computers*, vol. 2, pp. 1469–1473, 1993.

[15] Baraniuk, R.G., and Jones, D.L., "A Signal-Dependent Time-Frequency Representation: Fast Algorithm for Optimal Kernel Design," *IEEE Trans. Signal Processing*, vol. 42, pp. 134–146, 1994.

[16] Flandrin, P., and Rioul, O., "Affine Smoothing of the Wigner-Ville Distribution," *Proc. IEEE ICASSP-90*, pp. 2455–2458, 1990.

[17] Boudreaux-Bartels, G.F., "Mixed Time-Frequency Signal Transformations," in *The Transforms and Applications Handbook*, A.D. Poularikas, ed., Boca Raton, FL: CRC Press, 1996.

[18] Choi, H.I., and Williams, W.G., "Improved Time-Frequency Representation of Multicomponent Signals Using Exponential Kernels," *IEEE Trans. Acoustics, Speech, and Signal Processing*, vol. 37, pp. 862–871, 1989.

[19] Atlas, L.E., Loughlin, P.J., and Pitton, J.W., "Signal Analysis with Cone Kernel Time-Frequency Representations and Their Applications," in *Time-Frequency Signal Analysis*, B. Boashash, ed., pp. 375–388, Longman Cheshire, 1992.

[20] Hlawatsch, F., and Boudreaux-Bartels, G.F., "Linear and Quadratic Time-Frequency Signal Representations," *IEEE SP Magazine*, pp. 21–67, April 1992.

[21] Gardner, W.A., "An Introduction to Cyclostationary Signals," in *Cyclostationarity in Communications and Signal Processing*, W. A. Gardner, ed., Chapter 1, New York: IEEE Press, 1994.

[22] Stephens, J.P., "Signal Analysis Using a Spectral Correlation Analyzer," *Rfdesign*, pp. 22–30, July 1999.

[23] Gardner, W.A., "Exploitation of Spectral Redundancy in Cyclostationary Signals," *IEEE SP Magazine*, pp. 14–36, April 1991.

[24] Roberts, R.S., Brown, W.A., and Loomis, Jr., H.H., "Computationally Efficient Algorithms for Cyclic Spectral Analysis," *IEEE SP Magazine*, pp. 38–49, April 1991.

[25] Gardner, W.A., *An Introduction to Random Processes with Applications to Signals and Systems*, 2nd ed., New York: McGraw-Hill, Inc., 1990.

[26] Gardner, W.A., "Signal Interception: A Unifying Theoretical Framework for Feature Detection," *IEEE Trans. on Communications*, vol. COM-36, no. 8, pp. 897–906, 1988.

[27] Gardner, W.A., *Statistical Spectral Analysis: A Non-probabilistic Theory*, Englewood Cliffs, N.J.: Prentice-Hall, 1988.

[28] Gardner, W.A., "Measurement of Spectral Correlation," *IEEE Trans. Acoustics, Speech, and Signal Processing*, vol. 34, no. 5, pp. 1111–1123, October 1986.

[29] Spooner, C.M., "Higher-Order Statistics for Nonlinear Processing of Cyclostationary Signals," in *Cyclostationarity in Communications and Signal Processing*, W. A. Gardner, ed., Chapter 2, New York: IEEE Press, 1994.

[30] Spooner, C.M., and Gardner, W.A., "Exploitation of Higher-Order Cyclostationarity for Weak-Signal Detection and Time-Delay Estimation," *Proc. of 6th Workshop on Statistical Signal and Array Processing*, Victoria, British Columbia, Canada, pp. 197–201, October 1992.

[31] Nikias, C.L., and Petropulu, A.P., *Higher Order Spectra Analysis: A Nonlinear Signal Processing Framework*, Englewood Cliffs, N.J.: Prentice-Hall, 1993.

[32] Nikias, C.L., and Raghuveer, M.R., "Bispectrum Estimation: A Digital Signal Processing Framework," *Proc. IEEE*, vol. 75, no. 7, pp. 869–891, July 1987.

[33] Kletter, D., and Messer, H., "Detection of a Non-Gaussian Signal in Gaussian Noise Using High-Order Spectral Analysis," *Workshop on Higher-Order Spectral Analysis*, pp. 95–99, 1989.

[34] Colonnese, S., and Scarano, G., "Transient Signal Detection Using Higher Order Moments," *IEEE Trans. on Signal Processing*, vol. 47, no. 2, pp. 515–520, February 1999.

[35] Wickert, M.A., and Turcotte, R.L., "Rate Line Detection Using Higher Order Spectra," *Military Communications Conf.*, MILCOM'92, pp. 1221–1224, 1992.

[36] Sattar, F., and Salomonsson, G., "On Detection Using Filter Banks and Higher Order Statistics," *IEEE Trans. on Aerospace and Electronic Systems*, vol. 36, no. 4, pp. 1179–1189, October 2000.

[37] Ferrari, A., and Alengrin, G., "Estimation of the Frequencies of a Complex Sinusoidal Noisy Signal Using Fourth Order Statistics," *IEEE*, pp. 3457–3460, 1991.

[38] Nikias, C.L., and Mendel, J.M., "Signal Processing with Higher Order Spectra," *IEEE Signal Processing Magazine*, vol. 10, no. 3, pp. 10–37, July 1993.

[39] Therrien, C.W., *Discrete Random Signals and Statistical Signal Processing*, Englewood Cliffs, NJ: Prentice-Hall, 1992.

[40] Nikias, C.L., "Higher Order Spectral Analysis," in *Advances in Spectrum Estimation and Array Processing*, S. Haykin, ed., Englewood Cliffs, NJ: Prentice-Hall, pp. 326–365, 1991.

[41] Mendel, J.M., "Tutorial on Higher-Order Statistics (Spectra) in Signal Processing and System Theory: Theoretical Results and Some Applications," *Proc. IEEE*, vol. 79, no. 3, pp. 278–305, March 1991.

[42] The Mathworks, Inc., "Higher Order Spectral Analysis Toolbox,", Version 5.2, Natick, MA, 1998.

[43] Bendat, J.S., and Piersol, A.G., *Random Data: Analysis and Measurement Procedures*, New York: John Wiley & Sons, 1971.

[44] Carter, G.C., "Estimation of the Magnitude-Squared Coherence Function (Spectrum)," in *Coherence Estimation*, New London, CT: Naval Underwater Systems Center.

[45] Nuttall, A.H., and Carter, G.C., "Bias of the Magnitude-Squared Coherence," *IEEE Trans. on Acoust., Speech, and Signal Processing*, vol. 24, pp. 582–583, 1976.

[46] Carter, G.C., and Nuttall, A.H., "Statistics of the Estimate of Coherence," *Proc. IEEE*, vol. 60, pp. 465–466, 1972.

[47] Mohnkern, G.L., "Maximum Likelihood Estimation of Magnitude-Squared Coherence," *IEEE Trans. on Acoust., Speech, and Signal Processing*, vol. 36, pp. 130–132, January 1988.

[48] Friedlander, B., and Porat, B., "Detection of Transient Signals by Gabor Representation," *IEEE Trans. on Acoust., Speech, and Signal Processing*, vol. 37, pp. 169–180, February 1989.

[49] Daubechies, I., "Orthonormal Bases of Compactly Supported Wavelets," *Communications on Pure and Applied Mathematics*, vol. 41, pp. 909–997, November 1988.

[50] Daubechies, I., "Ten Lectures on Wavelets," *SIAM*, Philadelphia, PA, 1992.

[51] Burrus, C.S., Gopinath, R.A., and Guo, H., *Introduction to Wavelets and Wavelet Transforms, A Primer*, Upper Saddle River, NJ: Prentice-Hall, 1998.

[52] Strang, G., and Nguyen, T., *Wavelets and Filter Banks*, Wellesley, MA: Wellesley-Cambridge Press, 1996.

[53] Vetterli, M., and Kovačević, J., *Wavelets and Subband Coding*, Englewood Cliffs, NJ: Prentice-Hall, 1995.

[54] Mallat, S.G., "A Theory for Multi-Resolution Signal Decomposition: the Wavelet Representation," *IEEE Trans. on Pattern Recognition and Machine Intelligence*, vol. 7, no. 11, pp. 674–693, July 1989.

[55] Vaidyanathan, P.P., *Multirate Systems and Filter Banks*, Englewood Cliffs, NJ: Prentice-Hall, 1992.

[56] Goswami, J.C., and Chan, A.K., *Fundamentals of Wavelets*, New York: John Wiley & Sons, 1999.

[57] Wang, Z., Willett, P., and Streit, R., "Wavelets in the Frequency Domain for Narrowband Process Detection," *Proc. IEEE Conf. on Acoustics, Speech and Signal Processing*, Salt Lake City, Utah, 2001.

[58] Erdol, N., Kyperountas, S., and Petljanski, B., "Time and Scale Evolutionary EVD and Detection," *Proc. IEEE Conf. on Acoustics, Speech and Signal Processing*, Salt Lake City, Utah, 2001.

[59] Erdol, N., "Multiscale Detection of Non-Stationary Signals," in *Wavelet Subband and Block Transforms in Communications and Multimedia*, A.N. Akansu, M.J. Medley, eds., pp. 183–205, New York: Kluwer, 1999.

[60] Frisch, M., and Messer, H., "The Use of the Wavelet Transform in the Detection of an Unknown Transient Signal," *IEEE Trans. on Information Theory*, vol. 38, no. 2, pp. 892–897, March 1992.

[61] Fargues, M., Overdyke, H., and Hippenstiel, R., "Wavelet-Based Detection of Frequency Hopping Signals," *31st Asilomar Conf. on Signals, Systems, and Computers*, Pacific Grove, CA, November 1997.

[62] Farrell, T.C., and Prescott, G., "A Low Probability of Intercept Signal Detection Receiver Using Quadratic Mirror Filter Bank Trees," *Proc. IEEE Conf. on Acoustics, Speech and Signal Processing*, Atlanta, Georgia, 1996.

[63] Haykin, S., *Adaptive Filter Theory*, 3rd ed., Upper Saddle River, NJ: Prentice-Hall, 1996.

[64] Alexander, S.T., *Adaptive Signal Processing: Theory and Applications*, New York: Springer-Verlag, 1986.

[65] Widrow, B., and Stearns, S.D., *Adaptive Signal Processing*, Englewood Cliffs, NJ: Prentice-Hall, 1985.

[66] Widrow, B., et al., "Adaptive Noise Cancelling: Principles and Applications," *Proc. IEEE*, Vol. 63, No. 12, December 1975, pp. 1692–1716.

[67] Widrow, B., et al., "Stationary and Non-Stationary Learning Characteristics of the LMS Adaptive Filter," *Proc. IEEE*, Vol. 64, No. 8, August 1976, pp. 1151–1162.

[68] Zeidler, J.P., "Performance Analysis of the LMS Adaptive Prediction Filter," *Proc. IEEE*, Vol. 784, No. 12, December 1990, pp. 1780–1806.

[69] Widrow, B., and Hoff, Jr., M., "Adaptive Switching Circuits," *IRE WESCON Conv. Rec.*, Pt. 4, 1960, pp. 96–104.

APPENDIX A

Probability, Random Processes, and Systems

This appendix summerizes some of the properties of probability, random processes, and systems. The nomenclature is

Density function: $f_X(x)$

Probability: $Pr\{X \leq x\}$

Distribution function: $F_X(x)$

Some properties of probability density functions (PDFs) using conventional dummy variables of integration (slight misuse in a mathematical sense) are:

$$f_X(x) \geq 0$$

$$\int_{-\infty}^{\infty} f_X\ (x)\ dx = 1$$

$$Pr\{X \leq x\} \ = \ F_X(x) = \int_{-\infty}^{x} f_X(x)\ dx$$

$$Pr\{X \leq x\,,\ Y \leq y\} = \int_{-\infty}^{x} \int_{-\infty}^{y} f_{XY}(x, y)\ dx\ \ dy$$

$$Pr\{X \leq x\} = \int_{-\infty}^{\infty} \int_{-\infty}^{x} f_{XY}(x, y) dx\ \ dy \quad \text{(marginal)}$$

$$f_X(x) = \int_{-\infty}^{\infty} f_{XY}(x, y)\ dy \quad \text{(marginal density)}$$

Bayes' rule:

$$f_{X|Y}(x|y) = \frac{f_{XY}(x, y)}{f_Y(y)} \quad \text{with} \quad f_Y(y) \neq 0$$

This is also true for multi-dimensional variables:

$$f(y_1, y_2, \cdots, y_N|x_1, x_2, \cdots, x_M) = \frac{f(y_1, y_2, \cdots, y_N, x_1, x_2, \cdots, x_M)}{f(x_1, x_2, \cdots, x_M)}$$

Some common densities are

Uniform PDF: for $0 \le a < b$

$$f(u) = \begin{cases} \dfrac{1}{b-a} \; ; & a < u \le b \\ \\ 0 \; ; & \text{else} \end{cases}$$

For example: If

$$u = \tan^{-1}\left(\frac{x}{y}\right) \quad \text{or} \quad u = \tan^{-1}\left(\frac{y}{x}\right)$$

where x, y i.i.d. and

$$\begin{aligned} x &\sim N(0, \sigma^2) \\ y &\sim N(0, \sigma^2) \end{aligned}$$

then

$$f(u) = \begin{cases} \dfrac{1}{2\pi} \; ; & -\pi < u \le \pi \\ \\ 0 \; ; & \text{else} \end{cases}$$

Exponential PDF: (χ^2 with 2 degrees of freedom)

$$f(x) = \begin{cases} ae^{-ax} & x \ge 0 \quad a > 0 \\ \\ 0 & x < 0 \end{cases}$$

Gaussian (Normal) PDF:

$$\begin{aligned} f(x) &= \frac{1}{\sqrt{2\pi\sigma^2}} e^{-(x-m)^2/2\sigma^2} \\ \text{where} \qquad m &= Ex \\ \text{and} \qquad \sigma^2 &= Ex^2 - E^2(x) \\ &= E\{x - E(x)\}^2 \end{aligned}$$

or

$$x \sim N(m, \sigma^2)$$

For N-real valued Gaussian random variables, the joint probability density function is given by

$$f_{\mathbf{X}}(x_1, x_2, \cdots, x_N) = \frac{1}{(2\pi)^{N/2} |\mathbf{\Lambda}_X|^{1/2}} \exp\left[-\frac{1}{2}(\mathbf{x} - \mathbf{m}_X)^T \mathbf{\Lambda}_X^{-1} (\mathbf{x} - \mathbf{m})\right]$$

where

$$\begin{aligned} \mathbf{x} &= (x_1, x_2, \cdots, x_N)^T \\ \mathbf{m}_X &= (m_1, m_2, \cdots, m_N)^T \end{aligned}$$

$$\mathbf{\Lambda}_X = (u_{ij})$$

where $(u_{ij}) = E(x_i - m_i)(x_j - m_j)$ and $|\mathbf{\Lambda}_X|$ is the determinant of the covariance matrix $\mathbf{\Lambda}_X$.

If the random variables are all uncorrelated, i.e.,

$$u_{ij} = \begin{cases} 0 & i \neq j \\ \sigma_i^2 & i = j \end{cases}$$

then $\mathbf{\Lambda}_X$ is a diagonal matrix having entries σ_i^2 along its diagonal.

Rayleigh PDF:

$$\text{Let} \qquad z = \sqrt{x^2 + y^2}$$

$$\text{where} \qquad x,\ y \sim N(0, \sigma^2) \quad \text{are i.i.d., and where } b = \sigma^2$$

$$\text{then} \qquad f(z) = \begin{cases} \dfrac{z}{b} e^{-z^2/2b}\ ; & z \geq 0 \qquad b > 0 \\ 0\ ; & z < 0 \end{cases}$$

Cauchy PDF:

$$\text{Let} \qquad c = \frac{x}{y}$$

$$\text{where} \qquad x,\ y \sim N(0, \sigma^2) \text{ are i.i.d.}$$

$$\text{then} \qquad f(c) = \frac{a}{\pi} \qquad \frac{1}{a^2 + c^2}\ ; \qquad a > 0$$

Chi-Squared PDF (N-degrees of freedom):

$$\text{Let} \qquad y = \sum_{i=1}^{N} x_i^2$$

$$\text{where} \qquad x_i \sim N(0, \sigma^2) \quad \text{are i.i.d.}$$

$$\text{then} \qquad f(y) = \frac{1}{2^{N/2}\Gamma(N/2)\sigma^N}\, y^{(N-2)/2} \exp - \frac{y}{2\sigma^2} \qquad U(y)$$

where $\Gamma(n+1) \triangleq n!$ and $U(\cdot)$ denotes the unit step.

Rician (non-central Rayleigh) PDF:

$$\text{Let} \qquad f(x,y) = \frac{1}{2\sigma^2\pi} \exp\left\{ - \left(\frac{(x - m_X)^2 + y^2}{2\sigma^2} \right) \right\}$$

$$\text{and} \qquad z = \sqrt{x^2 + y^2}$$

$$\text{then} \qquad f(z) = \begin{cases} \dfrac{z}{\sigma^2} \quad \exp -(z^2 + m_X^2)/2\sigma^2 \quad I_0\left(\dfrac{z\, m_X}{\sigma^2}\right) & ; \; z \geq 0 \\ 0 \qquad\qquad z < 0 & \end{cases}$$

$$\begin{aligned} \text{where} \qquad I_0(x) &= \text{modified Bessel function of order zero} \\ &= \sum_{n=0}^{\infty} \frac{x^{2n}}{2^{2n}(n!)^2} \end{aligned}$$

Log-Normal PDF:

$$\text{Let} \qquad y = \exp(x)$$

$$\text{where} \qquad x \sim N(m_X, \sigma_X^2)$$

$$\text{then} \qquad f(y) = \begin{cases} \dfrac{1}{\sqrt{2\pi}\sigma_X Y} \quad \exp - \left(\dfrac{(\ln y - m_X)^2}{2\sigma_x^2} \right) & ; \quad y \geq 0 \\ 0 & ; \quad y < 0 \end{cases}$$

Non-Central Chi-Squared PDF:

$$\text{Let} \qquad v = \sum_{i=1}^{N} (A + x_i)^2$$

$$\text{where} \qquad x_i \sim N(0, \sigma^2) \quad \text{are i.i.d. and } A \text{ is a fixed constant}$$

$$\text{then} \qquad f(v) = \frac{1}{2\sigma^2} \left(\frac{v}{\lambda}\right)^{(N-2)/4} \exp\left(-\frac{\lambda + v}{2\sigma^2}\right) I_{(N/2)-1}\left(\frac{(v\lambda)^{1/2}}{\sigma^2}\right) U(v)$$

$$\text{where} \qquad \lambda = A^2 N$$

A very useful rule, applicable when differentiation of an integral expression is required.

Leibnitz Rule:

$$\phi(x) = \int_{a(x)}^{b(x)} f(t, x) dt$$

$$\frac{d\phi}{dx} = \int_{a(x)}^{b(x)} \frac{\partial f(t, x)}{\partial x} dt + f(b(x), x) \frac{db(x)}{dx} - f(a(x), x) \frac{da(x)}{dx}$$

PROPERTIES OF DISTRIBUTION FUNCTIONS

$$\begin{aligned}
F(-\infty) &= 0 \qquad F(x) \triangleq Pr(X \le x) \\
F(\infty) &= 1 \\
F(x_1) &\le F(x_2) \quad \text{for } x_1 \le x_2 \quad \text{(non-decreasing function)} \\
F(x^+) &= F(x) \quad \text{(continuous from the right)}
\end{aligned}$$

If $x_1 < x_2$ then

$$F(X \in (x_1, x_2)) = Pr\{x_1 < X \le x_2\} = F(x_2) - F(x_1)$$

Note: for discrete random variables

$$F(x) = \sum_i Pr(X = x_i) \quad \text{for all} \quad i \ni x_i \le x$$

Density Function:

$$f(x) \triangleq \frac{dF(x)}{dx} \qquad f(x) \ge 0$$

$$F(x) = \int_{-\infty}^{x} f(x)dx$$

$$Pr\{X \in A\} = \int_A f(x)dx$$

Moments:

$$E(x^n) = \int_{-\infty}^{\infty} x^n f_X(x)dx$$

$$\text{mean} = Ex = \int_{-\infty}^{\infty} x f_X(x)dx$$

$$\text{variance} = E(x - m_X)^2 = E(x^2) - m_X^2$$

$$2^{\text{nd}} \text{ moment} = E(x^2)$$

If $y = g(x)$ then $E(y) = Eg(x) = \int g(x)\ f_X(x)\ dx.$

Covariance:

$$C_{XY} = cov(X, Y) = E\{(x - m_X)(y - m_Y)\} = R_{XY} - m_X m_Y$$

We realize that for zero mean random processes $C_{XY} = R_{XY}$.

$$\rho_{XY} = \frac{C_{XY}}{\sigma_X \sigma_Y} \quad \text{(normalized covariance coefficient)}$$

Uncorrelated: The random variables X and Y are uncorrelated *iff* (if and only if)

$$C_{XY} = cov(XY) = 0 = E\{(x - m_X)(y - my)\}$$

$$\Longrightarrow \quad Exy = ExEy$$

Orthogonal:

$$Exy = 0$$

Note: If zero mean and uncorrelated, then the components are orthogonal.

Independence:

$$f_{XY}(x, y) = f_X(x) f_Y(y)$$

$$\Longrightarrow \quad Exy = ExEy$$

Note: independence $\Longrightarrow$ uncorrelated

Uncorrelatedness does not imply independence. However, if the random variables are Gaussian and uncorrelated $\Rightarrow C_{XY} = 0$, then

$$\Longrightarrow \quad f_{XY}(x, y) = f_X(x) f_Y(y)$$

FUNCTIONS OF RANDOM VARIABLES

Given the random variable $\quad x$

and the transformation $\quad y = g(x)$

The Fundamental Theorem [2] Says:

$$f_Y(y) = \frac{f_X(x_1)}{|g'(x_1)|} + \frac{f_X(x_2)}{|g'(x_2)|} + \cdots + \frac{f_X(x_n)}{|g'(x_n)|} + \cdots$$

where

$$g'(x) = \frac{dg(x)}{dx} \qquad \text{(also called Jacobian)}$$

and x_i's are all the *real* valued roots of

$$\begin{aligned} y &= g(x_1) = \cdots \\ \cdots &= g(x_n) = \cdots \end{aligned}$$

Note:

1. The theorem does not apply when $g(\cdot)$ is constant over an interval.
2. If there are no real roots ($y = g(x)$) then the density $f_Y(y) = 0$.
3. Be careful to account for all roots.

Another way is through the probability distribution function:

$$\begin{aligned} Pr\{Y \leq y\} &= Pr\{g(X) \leq y\} \\ &= Pr\{X \leq g^{-1}(y)\} \\ &= F_X(g^{-1}(y)) \end{aligned}$$

Example: $y = ax + b$ for $a > 0$

$$\begin{aligned} F_Y(y) &= Pr\{Y \le y\} = Pr\{aX + b \le y\} \\ &= Pr\left\{X \le \frac{y-b}{a}\right\} = F_X\left(\frac{y-b}{a}\right) \end{aligned}$$

Example: $y = \alpha x \qquad \alpha > 0$

$$x \sim N(0,1) \quad \text{hence} \quad f_X(x) = \frac{1}{\sqrt{2\pi}} e^{-x^2/2}$$

via the fundamental theorem

$$x = y/\alpha \quad \text{hence} \quad \frac{dg(x)}{dx} = \frac{d(\alpha x)}{dx} = \alpha$$

$$f_Y(y) = f_X\left(\frac{y}{\alpha}\right)\left|\frac{1}{\alpha}\right| = \frac{1}{\sqrt{2\pi}\,\alpha} e^{-y^2/2\alpha^2}$$

$$\text{hence} \qquad Y \sim N(0, \alpha^2)$$

or: via the distribution function

$$\begin{aligned} Pr\{Y \le y\} &= Pr\left\{X \le \frac{y}{\alpha}\right\} \\ F_Y(y) - F_X\left(\frac{y}{\alpha}\right) &= \int_{-\infty}^{y/\alpha} \frac{1}{\sqrt{2\pi}} e^{-\xi^2/2} d\xi \\ f_Y(y) = \frac{d}{dy} F(y) &= \frac{d}{dy}\left(\frac{1}{\sqrt{2\pi}} \int_{-\infty}^{y/\alpha} e^{-\xi^2/2} d\xi\right) \\ &= \frac{1}{\sqrt{2\pi}} \frac{1}{\alpha} e^{-y^2/2\alpha^2} \end{aligned}$$

$\Uparrow$ Leibnitz rule

or: for Gaussian family (closed under linear transformation)

$$x \sim N(0,1) \qquad Ex = 0 \qquad Ex^2 = 1$$

$$Ey = \alpha Ex = 0$$

$$Ey^2 = \alpha^2 Ex^2 = \alpha^2 1 = \alpha^2$$

$$\Longrightarrow\ y \sim N(0, \alpha^2)$$

Example: Solve the fundamental theorem

$$y = x_1 + x_2\ ; \quad \text{where} \quad x_1 \quad \text{and} \quad x_2 \quad \text{are statistically independent}$$

The Auxiliary Variables Are:

$$y = x_1 + x_2 \qquad \text{let} \quad x_2 = x_2$$

$$y = g_1(x_1 x_2) \quad \text{hence} \quad x_2 = g_2(x_2)$$

$$|J| = \begin{vmatrix} \dfrac{\partial g_1}{\partial x_1} & \dfrac{\partial g_1}{\partial x_2} \\ \\ \dfrac{\partial g_2}{\partial x_1} & \dfrac{\partial g_2}{\partial x_2} \end{vmatrix} = \begin{vmatrix} 1 & 1 \\ \\ 0 & 1 \end{vmatrix} = 1$$

$$\begin{aligned} x_1 &= y - x_2 \\ x_2 &= x_2 \end{aligned}$$

$$f_{X_1X_2}(y_1, x_2) = f_{X_1X_2}(y - x_2, x_2)\frac{1}{|1|}$$

$$f_Y(y) = \int_{-\infty}^{\infty} f_{X_1X_2}(y - x_2, x_2)dx_2 \quad = \quad \int_{-\infty}^{\infty} f_{X_1}(y - x_2) f_{X_2}(x_2) dx_2$$

$\Uparrow$ statistically independent

or

$$\begin{aligned} F_Y(y) &= \int_{x_2=-\infty}^{\infty} \int_{x_1=-\infty}^{y-x_2} f_{X_1X_2}(x_1, x_2)dx_1 dx_2 \\ &= Pr\{Y \leq y\} \\ &= \int_{-\infty}^{\infty} f_{X_2}(x_2) \left(\int_{-\infty}^{y-x_2} f_{X_1}(x_1)dx_1 \right) dx_2 \\ f_Y(y) &= \frac{dF_Y(y)}{dy} \\ &= \int_{-\infty}^{\infty} f_{X_2}(x_2) f_{X_1}(y - x_2)dx_2 \end{aligned}$$

$\Uparrow$ Leibnitz rule

Characteristic Function:

$$C_X(j\omega) = \int_{-\infty}^{\infty} f(x)e^{j\omega x}dx$$

$$f(x) = \frac{1}{2\pi}\int_{-\infty}^{\infty} C_X(j\omega)e^{-j\omega x}d\omega$$

Example: Suppose the random variable is Gaussian, i.e., $x \sim N(0,1)$, then

$$C_X(j\omega) = \int_{-\infty}^{\infty} \frac{1}{\sqrt{2\pi}}e^{-x^2/2}e^{j\omega x}dx = e^{-\omega^2/2}$$

(i.e., linear transformation of a Gaussian is Gaussian)

General Moments:

$$Eg(x) = \int_{-\infty}^{\infty} g(x)f(x)dx$$

$$Ex^n = \int_{-\infty}^{\infty} x^n f(x)dx$$

$$Eg(x,y) = \int_{-\infty}^{\infty}\int_{-\infty}^{\infty} g(x,y)f(x,y)dx\,dy$$

$$Ex^n y^m = \int_{-\infty}^{\infty}\int_{-\infty}^{\infty} x^n y^m f(x,y)dx\,dy$$

$$ExyY = R_{XY} = \int_{-\infty}^{\infty}\int_{-\infty}^{\infty} xyf(x,y)dx\,dy$$

Note: if $Exy = Ex\,Ey$, then x and y are uncorrelated.

CONDITIONAL PROBABILITY

$$F_X(x|Y \in B) \triangleq Pr(X \le x|Y \in B) \quad \text{with} \quad (Pr\{(Y \in B)\} > 0)$$

$$f_X(x|Y \in B) \triangleq \frac{dF_X(x|Y \in B)}{dx}$$

$$f_X(x|Y \in B) \ge 0$$

$$\int_{-\infty}^{\infty} f_X(x|Y \in B)dx = 1$$

$$F_X(x|Y \in B) = \int_{-\infty}^{x} f_X(x|Y \in B)dx$$

$$Pr[X \in A|Y \in B)] = \int_A f_X(x|Y \in B)dx$$

$$F_{X|Y}(x|y) \triangleq F_X(x|Y = y) = \frac{\int_{-\infty}^{x} f_{XY}(x,y)dx}{f_Y(y)}$$

$$f_{X|Y}(x|y) \triangleq \frac{dF_{X|Y}(x,y)}{dx} = \frac{f_{XY}(x,y)}{f_Y(y)}$$

$$Pr[X \epsilon A|Y = y] = \int_A f_{X|Y}(x|y)dx$$

Bayes' Theorem:

$$Pr\{A|B\} = \frac{Pr(A,B)}{Pr(B)} = \frac{Pr(B|A)}{Pr(B)} Pr(A)$$

$$Pr(A|B,C) = \frac{Pr(A,B,C)}{Pr(B,C)}$$

Note: Bayes' theorem holds also for densities.

$$f(x) = \int_{-\infty}^{\infty} f(x|y) f(y) dy$$

where

$$f(x|y) = \frac{f(x,y)}{f(y)}$$

so

$$\int f(x,y)dy = \text{marginal of } f(x)$$

Typically used when an uncertainty is present. We condition it on the uncertainty (need the density function of the uncertainty) and average it out.

For example,

$$f(x|z) = \int_{-\infty}^{\infty} f(x|y,z) f(y|z) dy$$

and

$$\begin{aligned} f(x) &= \int_{-\infty}^{\infty} f(x|z) f(z) dz \\ &= \int_{-\infty}^{\infty} f(x|y,z) f(y|z) f(z) dy dz \end{aligned}$$

RANDOM PROCESSES

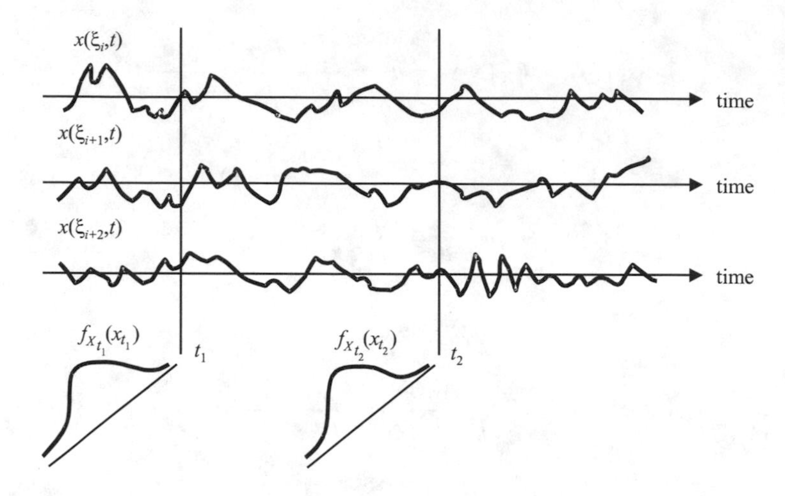

1st Order Density:

$$f_{X_{t_1}}(x_{t_1})\ ,\quad f_{X_{t_2}}(x_{t_2}),\cdots \quad \text{(any } t_i \text{ combination)}$$

2nd Order Density:

$$f_{X_{t_1}X_{t_2}}(t_1, t_2),\cdots \quad \text{(any } t_i, t_j \text{ combination)}$$

3rd Order Density:

$$f_{X_{t_1}X_{t_2}X_{t_3}}(x_{t_1}, x_{t_2}, x_{t_3}),\cdots \quad \text{(any } t_i, t_j, t_k \text{ combination)}$$
$$\vdots$$

For a complete description, we need all moments (joint moments) of all orders for all possible times. However, most analysis is done using the first two moments (requires knowledge of at most the joint density). Sometimes we need to know up to the 4th order moment.

ENSEMBLE AVERAGES

$$R_{X_{t_1}X_{t_2}}(t_1,t_2) = R_X(t_1,t_2) = E\{x_1x_2\} = \int_{-\infty}^{\infty}\int_{-\infty}^{\infty} x_1x_2 f_{X_1X_2}(x_1,x_2)dx_1dx_2$$

(We use the notation t_i and i interchangeably.)

for $t_1 = t$ and $t_2 = t - \tau$ we have

$$R_X(t, t-\tau) = \int\int x_t x_{t-\tau} f_{X_t X_{t-\tau}}(x_t, x_{t-\tau}) dx_t dx_{t-\tau}$$

for *wide sense stationary* (w.s.s.) process

$$f_{X_t X_{t-\tau}}(x_t, x_{t-\tau}) = f_{X_{t+T} X_{t+T-\tau}}(x_{t+T} - x_{t+T-\tau})$$

$\Longrightarrow$ 1st moment is constant:

$$Ex(t) = m$$

2nd moment is a function of time difference:
(for general complex valued processes)

$$Ex(t)x^*(t-\tau) = R_X(\tau)$$

so

$$R_X(t, t-\tau) = R_X(\tau)$$

$$\rho_X(t_1, t_2) \triangleq \frac{E[(x_1 - m_1)(x_2 - m_2)]}{\sigma_1 \sigma_2} = \frac{R_X(t_1, t_2) - m_1 m_2}{\sigma_1 \sigma_2}$$

where

$$\begin{aligned} m_i &= Ex_i \qquad i = 1, 2 \\ \sigma_i^2 &= E(x_i - m_i)^2 \end{aligned}$$

for w.s.s. process (m_i = constant = m)

$$\rho_X(\tau) = \frac{R_X(\tau) - m^2}{\sigma^2}$$

Cross-Correlation Function:

$$\begin{aligned} R_{XY}(t_1, t_2) &= E\{x_1 x_2^*\} \\ R_{XY}(\tau) &= E\{x_t x_{t-\tau}^*\} \end{aligned}$$

Properties:

$$\begin{aligned} R_X(\tau) &= R_X^*(-\tau) \qquad \text{for complex process} \\ R_X(\tau) &= R_X(-\tau) \qquad \text{for real process} \end{aligned}$$

$$
\begin{aligned}
R_{XY}(\tau) &= R^*_{YX}(-\tau) \\
R_X(0) &= \sigma_X^2 \\
|R_X(\tau)| &\le R_X(0) = \sigma_X^2 \\
|R_{XY}(\tau)| &\le \frac{1}{2}[R_{XX}(0) + R_{YY}(0)] \\
|R_{XY}(\tau)| &\le \sqrt{R_X(0)R_Y(0)}
\end{aligned}
$$

Time Averages: (ergodic processes: times averages = ensemble averages)

$$
\begin{aligned}
\langle x(t) \rangle &\triangleq \lim_{T\to\infty} \frac{1}{2T} \int_{-T}^{T} x(t)dt = \text{mean} \\
R_X(\tau) &= \lim_{T\to\infty} \frac{1}{2T} \int_{-T}^{T} x(t)x^*(t-\tau)dt \\
R_{XY}(\tau) &= \lim_{T\to\infty} \frac{1}{2T} \int_{-T}^{T} x(t)y^*(t-\tau)dt
\end{aligned}
$$

POWER SPECTRAL DENSITY (PSD)

$$
\begin{aligned}
S_X(\omega) &= \int_{-\infty}^{\infty} R_X(\tau)e^{-j\omega\tau}dt = F\{R_X(\tau)\} \\
R_X(\tau) &= \frac{1}{2\pi} \int_{-\infty}^{\infty} S_X(\omega)e^{j\omega\tau}d\omega \\
\text{Note}: \quad R_X(0) &= \frac{1}{2\pi} \int_{-\infty}^{\infty} S_X(\omega)d\omega = E(x_t^2)
\end{aligned}
$$

Properties:

1. $S_X(\omega) \ge 0$
2. $S_X(\omega) = S_X(-\omega)$ for real x_t
3. $S_X(\omega)$ is real for real x_t

Note:

$$
S_X^*(\omega) = \int R_X^*(\tau)e^{j\omega t}dt
$$

$$= \int R_X(\tau)e^{-j\omega t}dt = S_X(\omega) = \text{is always a real valued function}$$

$$\Longrightarrow \text{imag } S_X(\omega) = 0$$

$$\Longrightarrow \text{an even function (for real process)}$$

since

$$R_X(\tau) = \text{real} \Longrightarrow S_X^*(\omega) = \int R_X(\omega)e^{+j(-\omega)\tau}dt = S_X(-\omega)$$

Cross-Spectra/Correlation:

$$\begin{aligned} S_{XY}(\omega) &= \int R_{XY}(\tau)e^{-j\omega\tau}dt \\ R_{XY}(\tau) &= \frac{1}{2\pi}\int S_{XY}(\omega)e^{j\omega\tau}d\omega \end{aligned}$$

$$\begin{aligned} S_{XY}^*(\omega) &= S_{YX}(\omega) \qquad \text{for complex processes} \\ S_{XY}(\omega) &= S_{XY}(-\omega) \qquad \text{for real processes} \end{aligned}$$

LINEARITY: (linear filter with input $f(t)$)

$$Tf(t) = y(t) \qquad f(t) \longrightarrow \boxed{T} \longrightarrow y(t)$$

$$\begin{aligned} T\{\alpha f_1(t) + \beta f_2(t)\} &= \alpha T\{f_1(t)\} + \beta T\{f_2(t)\} \\ &= \alpha\, y_1(t) + \beta\, y_2(t) \end{aligned}$$

$$y(t) = \int_{-\infty}^{\infty} h(\tau)f(t-\tau)dt = h(t) * f(t)$$

where $h(t)$ is the impulse response

or

$$F\{y(t)\} = F\{h(t)\}F\{f(t)\}$$

$$Y(\omega) = H(\omega)F(\omega)$$

BIBO STABILITY

If

$$|f(t)| = M < \infty$$

then

$$|y(t)| = \left| \int_{-\infty}^{\infty} f(t)x(t-\tau)dt \right| \leq \int_{-\infty}^{\infty} M|h(\tau)|dt \leq M \int_{-\infty}^{\infty} |h(\tau)|dt < \infty$$

In general we'll work with *stable time invariant* systems.

Random Inputs:

Let $x(t)$ be a realization (sample function) of the random process $x(t, \omega)$

$$x(t) \longrightarrow \boxed{h(t)} \longrightarrow y(t) = \int_{-\infty}^{\infty} h(\tau)x(t-\tau)dt$$

$$\begin{aligned} m_Y = \bar{y} = Ey(t) &= \int_0^{\infty} h(\tau)m_X dt \\ &= m_X \int_0^{\infty} h(\tau)dt \end{aligned}$$

$$\boxed{m_Y = m_X H(0)}$$

$$\begin{aligned} E(y_t y_{t-\tau}) &= \int_0^{\infty}\int_0^{\infty} h(\xi)h(\gamma)E\{x_{t-\xi}x_{t-\tau-\gamma}\}d\xi\, d\tau \\ &\downarrow \quad \text{w.s.s. input} \\ &= \int_0^{\infty}\int_0^{\infty} h(\xi)h(\gamma)R_X(\tau+\gamma-\xi)d\xi\, d\gamma \\ &= R_Y(\tau) = R_X(\tau) * h^*(-\tau) * h(\tau) \\ S_Y(\omega) &= F\{R_Y(\tau)\} \\ &= \int_{-\infty}^{\infty} R_Y(\tau)e^{-j\omega\tau}d\tau \\ S_Y(\omega) &= \int_0^{\infty} h(\xi)e^{-j\omega\xi}d\xi \int_0^{\infty} h(\gamma)e^{j\omega\gamma}d\gamma \\ &\quad \cdot \int_{-\infty}^{\infty} R_X(\tau+\gamma-\xi)e^{-j\omega(\tau+\gamma-\xi}d\xi\, d\gamma\, dt \\ &= |H(j\omega)|^2 S_X(\omega) \Longleftrightarrow R_Y(\tau) = R_X * h(\tau) * h^*(-\tau) \end{aligned}$$

Given the parallel networks:

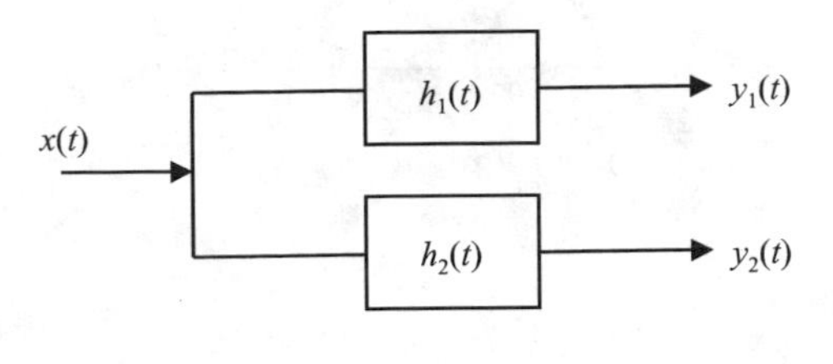

$$R_{Y_1Y_2}(\tau) = \int_0^\infty \int_0^\infty h_1(u)h_2(v)R_x(\tau + v - u)du\,dv$$
$$= \frac{1}{2\pi}\int_{-\infty}^\infty H_1(j\omega)H_2^*(j\omega)S_X(\omega)e^{j\omega\tau}d\omega$$

$$S_{Y_1Y_2}(\omega) = H_1(j\omega)H_2^*(j\omega)S_X(\omega)$$

White Noise:

$$S_N(\omega) = \frac{N_0}{2} \quad \text{for all } \omega$$

$S_N(\omega)$

$N_0/2$

ω

0

$$R_N(\tau) = \frac{N_0}{2}\delta(\tau)$$

$R_N(\tau)$

$N_0/2$

τ

0

Bandlimited White Noise (lowpass process):

$$S_N(\omega) = \begin{cases} \frac{A\pi}{W} & -W < \omega < W \\ 0 & \text{else} \end{cases}$$

$S_N(\omega)$

$A\,\pi/W$

ω

$-W$ 0 W

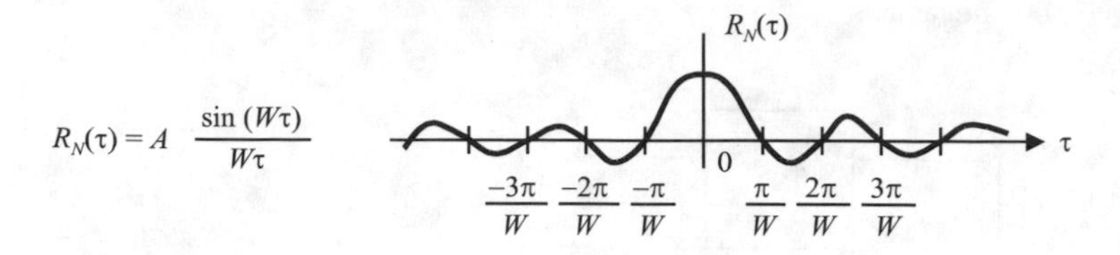

$$S_N(\omega) = \begin{cases} \dfrac{A\pi}{W} & \omega_0 - W < \omega < \omega_0 + W \\ \dfrac{A\pi}{W} & -\omega_0 - W < \omega < -\omega_0 + W \\ 0 & \text{else} \end{cases}$$

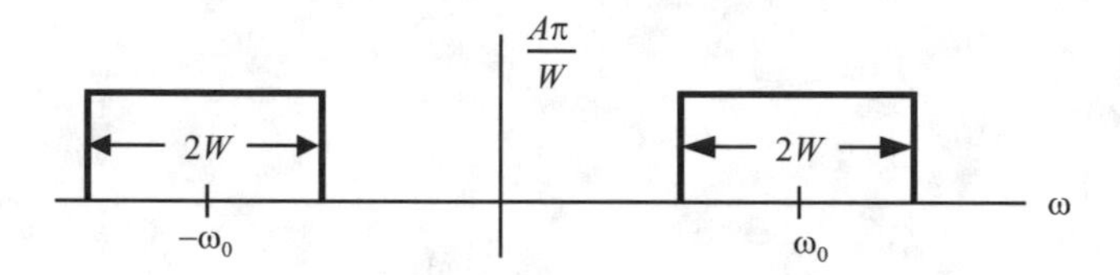

$$R_N(\tau) = A\frac{\sin(W\tau/2)}{(W\tau/2)}\cos(\omega_0\tau)$$

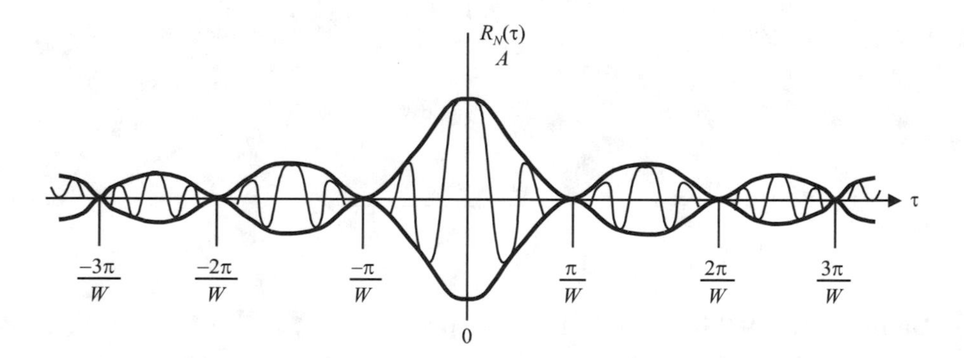

References

[1] Papoulis, A., *Probability, Random Variables, and Stochastic Processes*, New York: McGraw-Hill, 1965.

[2] Leon-Garcia, A., *Probability and Random Processes for Electrical Engineering*, 2nd ed., Addison-Wesley Publ. Co., 1994.

[3] Helstrom, C.W., *Probability and Stochastic Processes for Engineers*, 2nd ed., New York: MacMillan, Inc., 1991.

[4] Thomas, J.B., *Applied Probability and Random Processes*, New York: John Wiley & Sons, 1971.

[5] Mohanty, N., *Random Signals Estimation and Identification: Analysis and Applications*, New York: Van Nostrand Reinhold Co. Inc., 1986.

[6] Shanmugan, K.S., and Breipohl, A.M., *Random Signals: Detection, Estimation and Data Analysis*, New York: John Wiley & Sons, 1988.

[7] Whalen, A.D., *Detection of Signals in Noise*, New York: Academic Press, 1971.

[8] Maisel, L., *Probability, Statistics, and Random Processes*, New York: Simon & Schuster, 1971.

APPENDIX B

Signals and Transforms

Given a signal $f(t)$ we say it has a Fourier transform [1–3] if it is absolutely integrable

$$\int_{-\infty}^{\infty} |f(t)| dt < \infty$$

or if it is square integrable

$$\int_{-\infty}^{\infty} |f(t)|^2 dt < \infty$$

then

$$F(j\omega) = \int_{-\infty}^{\infty} f(t) e^{-j\omega t} dt$$

Many times we suppress the j dependency, so $F(j\omega)$ is denoted by $F(\omega)$. The inverse is given by

$$f(t) = \frac{1}{2\pi} \int_{-\infty}^{\infty} F(\omega) e^{+j\omega t} d\omega$$

For periodic signals (Fourier series expansion):

$$\begin{aligned}
f(t) &= f(t+T); \qquad \forall\, t \text{ and } \qquad T = \frac{2\pi}{\omega_0} \\
f(t) &= \sum_{k=-\infty}^{\infty} c_k e^{jk\omega_0 t} \\
\text{where } \; c_k &= \frac{1}{T} \int_{-T/2}^{T/2} f(t) e^{-j(k\omega_0 t)} dt
\end{aligned}$$

Discrete-Time Fourier Transform:

$$\begin{aligned}
F(t_k) &= \frac{1}{2\pi} \int_{-\pi}^{\pi} F(e^{j\omega}) e^{j\omega t_k} d\omega \\
F(e^{j\omega}) &= \sum_{k=-\infty}^{\infty} f(t_k) e^{-j\omega t_k}
\end{aligned}$$

Discrete Fourier Transform (DFT):

$$F(k) = \sum_{n=0}^{N-1} f(n)e^{-j(2\pi/N)kn}$$

$$f(n) = \frac{1}{N}\sum_{k=0}^{N-1} F(k)e^{j(2\pi/N)kn}$$

For Fourier reference related material consult [1,2,3] and review Chapter 3.

Hilbert Transform: [4,5]

$$\hat{x}(t) = \mathcal{H}\{x(t)\} = x(t) * \frac{1}{\pi t}$$

$$= \frac{1}{\pi}P\int_{-\infty}^{\infty} \frac{x(\tau)}{t-\tau}d\tau$$

where P denotes the principle value:

$$P\int_{-\infty}^{\infty} \equiv \int_{-\infty}^{\epsilon}\int_{\epsilon}^{\infty}$$

$$\lim_{\epsilon\to 0} \quad \text{(simultaneously)}$$

$$F\{x(t)\} = X(\omega)$$

$$F\left\{\frac{1}{\pi t}\right\} = \frac{1}{j}sgn(\omega) = -j\ sgn(\omega) = \begin{cases} -j\ ;\ \omega \geq 0 \\ j\ ;\ \omega < 0 \end{cases}$$

$$F\{\hat{x}(t)\} = -jX(\omega)sgn(\omega) = \begin{cases} -jX(\omega)\ ;\ \omega \geq 0 \\ jX(\omega)\ ;\ \omega < 0 \end{cases}$$

$\Longrightarrow$ Hilbert transform flips the sign of the Fourier transform of the positive part of the frequency components and multiplies all spectral components by $j = \sqrt{-1}$.

Example:

$$x(t) = \cos\omega_0 t = \frac{e^{j\omega_0 t}}{2} + \frac{e^{-j\omega_0 t}}{2}$$

$$\Big\downarrow F$$

$$
\begin{aligned}
F\{\cos\omega_0 t\} &= \pi[\delta(\omega-\omega_0)+\delta(\omega+\omega_0)] \\
&\Big\downarrow \mathcal{H} \\
\mathcal{H}\{F\{\cos\omega_0 t\}\} &= +j\pi[-\delta(\omega-\omega_0)+\delta(\omega+\omega_0)] \\
&\Big\downarrow F^{-1} \\
F^{-1}\{\mathcal{H}\{F\{\cos\omega_0 t\}\}\} &= \mathcal{H}\{x(t)\} \\
\mathcal{H}\{\cos\omega_0 t\} &= \hat{x}(t) \\
&= \frac{1}{2}j\left(-e^{j\omega_0 t}+e^{-j\omega_0 t}\right) \\
&= \frac{1}{2j}\left(e^{+j\omega_0 t}-e^{-j\omega_0 t}\right) = \sin\omega_0 t
\end{aligned}
$$

Example: If $a(t)$ is bandlimited (i.e., $\mathcal{F}\{a(t)\} = A(\omega) = 0$ for $|\omega| > W$), then

$$
\begin{aligned}
\mathcal{H}\{a(t)\}\cos\omega t\} &= a(t)\sin\omega_0 t \\
\mathcal{H}\{a(t)\sin\omega t\} &= -a(t)\cos\omega_0 t
\end{aligned}
$$

Properties:

1. $x(t)$ and $\hat{x}(t)$ are orthogonal
2. $\hat{x}(t) = \mathcal{H}\{x(t)\}$
3. $\mathcal{H}\{\hat{x}(t)\} = -x(t)$
4. Hilbert transform behaves as a 90° phase shifter for sinusoidal signals
5. If
$$y(t) = h(t) * x(t)$$
then
$$\hat{y}(t) = h(t) * \hat{x}(t) = x(t) * \hat{h}(t)$$
6. If $a(t)$ is bandlimited then
$$
\begin{aligned}
\mathcal{H}\{a(t)\cos\omega_0 t\} &= a(t)\sin\omega_0 t \\
\mathcal{H}\{a(t)\sin\omega_0 t\} &= -a(t)\cos\omega_0 t
\end{aligned}
$$

7. Time averages:

$$
\begin{aligned}
R_{XX}(\tau) &= R_{\hat{X}\hat{X}}(\tau) \\
R_{\hat{X}X}(\tau) &= \lim_{T\to\infty} \frac{1}{2T}\int_{-T}^{T} \hat{x}(t)x(t-\tau)dt \\
&= \hat{R}_{XX}(\tau) \\
R_{X\hat{X}}(\tau) &= -\hat{R}_{XX}(\tau) \\
R_{\hat{X}X}(\tau) &= -R_{\hat{X}X}(-\tau) \\
R_{X\hat{\mathbf{X}}}(-\tau) &= -R_{X\hat{X}}(\tau) \\
R_{X\hat{X}}(0) &= 0
\end{aligned}
$$

Expectations:

$$
\begin{aligned}
EX_t\hat{X}_{t-\tau} &= R_{X\hat{X}}(\tau) \\
R_{X\hat{X}}(\tau) &= -\hat{R}_{XX}(\tau) \\
R_{\hat{X}X}(\tau) &= \hat{R}_{XX}(\tau) \\
R_{\hat{X}\hat{X}}(\tau) &= R_{XX}(\tau) \\
S_{XX}(\omega) &= S_{\hat{X}\hat{X}}(\omega)
\end{aligned}
$$

Pre-Envelopes, Analytic Signal and Complex Envelopes

We define the pre-envelope of $x(t) = x_p(t)$

$$x_p(t) = x(t) + j\hat{x}(t)$$

where $\hat{x}(t)$ = Hilbert transform of $x(t)$

$$
\begin{aligned}
x(t) &= \text{real}\ \{x_p(t)\} \\
\hat{x}(t) &= \mathcal{H}\{x(t)\} = \text{imag}\ \{x_p(t)\} \\
X(\omega) &= F\{x(t)\} \\
X_p(\omega) &= F\{x_p(t)\} \\
&= X(\omega) + j[-j\ sgn(\omega)]X(\omega) \\
&= \begin{cases} 2X(\omega) & \omega > 0 \\ X(0) & \omega = 0 \\ 0 & \omega < 0 \end{cases}
\end{aligned}
$$

The two sketches drawn below illustrate the idea behind the concept of the pre-envelope. It shows, in the Fourier domain, how the bandpass and DC components are manipulated.

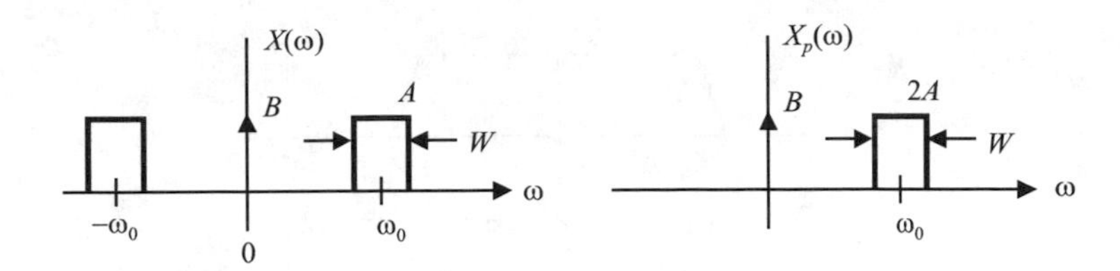

We can get $x_p(t)$ two different ways:

1. get $\mathcal{H}\{x(t)\} = \hat{x}(t)$ and use an earlier equation

$$x_p(t) = x(t) + j\hat{x}(t)$$

   or

2. get $X(\omega) = F\{x(t)\}$ let

$$X_p(\omega) = \begin{cases} 2X(\omega) & \omega > 0 \\ X(0) & \omega = 0 \\ 0 & \omega < 0 \end{cases}$$

   and inverse transform

$$x_p(t) = \frac{2}{2\pi} \int_0^{\infty} X(\omega) e^{j\omega t} d\omega$$

We also call the pre-envelope the analytic signal.

If the carrier frequency is removed, then the analytic signal becomes the complex envelope denoted by $\tilde{x}(t)$. This is shown with the next example.

Example: Bandpass signal

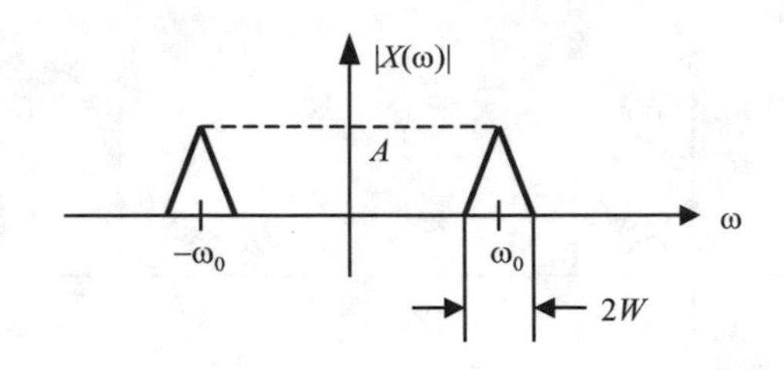

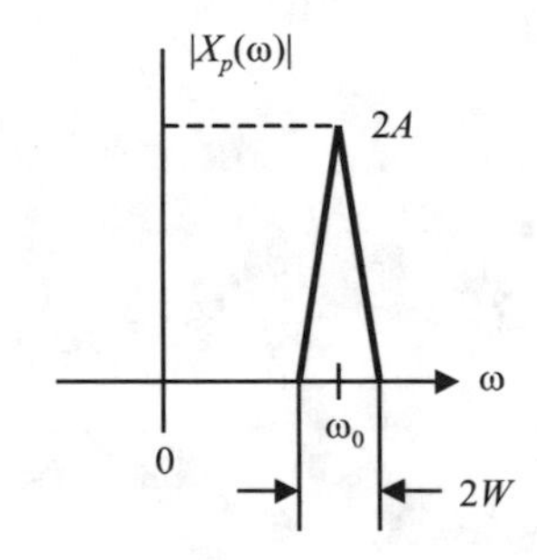

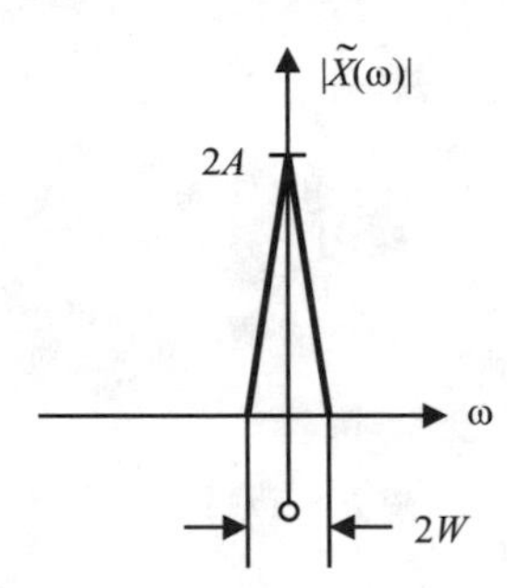

We see that

$$\tilde{X}(\omega) = \text{shifted version of } X_p(\omega)$$

or

$$\tilde{x}(t) = x_p(t)e^{-j\omega_0 t}$$

Example:

$$f(t) \quad \text{a narrowband signal}$$

$$\begin{aligned}
f(t) &= x(t)\cos\omega_0 t - y(t)\sin\omega_0 t \\
\hat{f}(t) &= x(t)\sin\omega_0 t + y(t)\cos\omega_0 t \\
f_p(t) &= x(t) + j\hat{x}(t) \\
&= x(t)\cos\omega_0 t - y(t)\sin\omega_0 t \\
&\quad +j(x(t)\sin\omega_0 t + y(t)\cos\omega_0 t) \\
&= (x(t) + jy(t))(\cos\omega_0 t + j\sin\omega_0 t) \\
&= (x(t) + jy(t))e^{j\omega_0 t} \\
\tilde{f}(t) &= f_p(t)e^{-j\omega_0 t} = x(t) + jy(t)
\end{aligned}$$

The envelope is defined to be the magnitude of the complex envelope

$$f(t) = |\tilde{f}(t)| = \sqrt{x^2(t) + y^2(t)}$$

Example: Bandlimited white noise

$$n(t) = x(t)\cos\omega_0 t - y(t)\sin\omega_0 t$$

$$\begin{array}{lcl} n(t) \text{ is w.s.s.} & \longrightarrow & x(t) \text{ and } y(t) \text{ are w.s.s.} \\ \text{with } R_{XX}(\tau) = R_{YY}(\tau) & & \text{and } R_X(0) = R_Y(0) = \sigma^2 \\ S_X(\omega) = S_Y(\omega) & & \text{(narrowband process)} \end{array}$$

$$\hat{n}(t) = x(t)\sin\omega_0 t + y(t)\cos\omega_0 t$$

so

$$\begin{aligned} x(t) &= n(t)\cos\omega_0 t + \hat{n}(t)\sin\omega_0 t \\ y(t) &= \hat{n}\cos\omega_0 t - n(t)\sin\omega_0 t \end{aligned}$$

If $n(t)$ is Gaussian then $\hat{n}(t)$ is Gaussian, therefore $x(t)$ and $y(t)$ are Gaussian.

Note:

$$n(t) = \underbrace{x(t)\cos\omega_0 t}_{\text{in-phase component}} - \underbrace{y(t)\sin\omega_0 t}_{\text{quadrature phase component}}$$

Let

$$\begin{aligned} e(t) &= \sqrt{x^2(t) + y(t)} && \text{(Rayleigh PDF)} \\ \phi(t) &= \tan^{-1}(y(t)/x(t)) && \text{(uniform PDF)} \end{aligned}$$

then

$$n(t) = e(t)\cos(\omega_0 t + \phi(t))$$

For more detail on signals, consult [4,5].

DFT Interpretation:

Suppose we have a complex valued time signal (i.e., signal with a one-sided narrowband spectrum), say $x(n) = a(n)e^{j\omega_0 n}$ where $a(n)$ is low pass narrowband process.

$$\begin{aligned}
X_n(e^{j\omega}) &= \sum_{m=n}^{n+N-1} x(m)e^{-j\omega m} \ ; \quad \text{where } n \text{ is the time index} \\
&\quad \text{and the DFT output rate equals the input rate.} \\
&= \sum_{m=n}^{n+N-1} a(m)e^{-jm(\omega-\omega_0)} = A_n(\omega - \omega_0) \\
&\quad \text{If } a(n) \text{ does not vary much over the } N \text{ data points,} \\
&\quad \text{then the transforms can be approximated by} \\
&\cong a(n) \sum_{m=n}^{n+N-1} e^{-jm(\omega-\omega_0)} \\
&= \begin{cases} Na(n) & \omega = \omega_0 \\ 0 & \text{else} \end{cases} \\
x_p(n) &= x(n) \quad \text{(since signal is already one-sided)} \\
\tilde{x}(n) &= x_p(n)e^{-j\omega_0 n} \\
&= x(n)e^{-j\omega_0 n} = a(n)e^{j\omega_0 n}e^{-j\omega_0 n} \\
&= a(n)
\end{aligned}$$

Example: One-sided narrowband spectral data (residing in one spectral bin). For the DFT, the output at the particular bin corresponds to a desampled complex envelope. If the maximum overlap (change one point per transform) is used, the output rate equals the input.

Note: The complex time data out of that spectral bin corresponds to the complex envelope. If real valued narrowband data is inputed (resides at a given spectral $\pm k$ location) then the output at spectral location corresponds to the complex envelope (scaled by a factor of $N/2$ and resampled).

References

[1] Oppenheim, A.V., Willsky, A.S., and Young, I.T., *Signals and Systems*, Englewood Cliffs, NJ: Prentice-Hall, 1983.

[2] Ludeman, L.C., *Fundamentals of Digital Signal Processing*, New York: Wiley, Harper & Row, Publishers, 1986.

[3] Oppenheim, A.V., and Schafer, R.W., *Digital Signal Processing*, Englewood Cliffs, NJ: Prentice-Hall, 1975.

[4] Mohanty, N., *Random Signals Estimation and Identification: Analysis and Applications*, New York: Van Nostrand Reinhold Co. Inc., 1986.

[5] Shanmugan, K.S., and Breipohl, A.M., *Random Signals: Detection, Estimation and Data Analysis*, New York: John Wiley & Sons, 1988.

APPENDIX C

Mathematical Structures

Group: A group consists of

1. A set K.
2. A rule (or operation) which associates with each pair element x, y in K an element xy in $K \ni$.
   (a) $x(yz) = (xy)z, \quad \forall\, x, y$ and z in K
   (b) $\exists$ an element e in $K \ni ex = xe = x \;\forall\, x$ in K
   (c) to each element x in K there corresponds an element x^{-1} in $K \ni$ $xx^{-1} = x^{-1}x = e$

Commutative Group: $xy = yx \;\; x, y \in K$

Ring: A ring is a set K, together with two operations $(x, y) \longrightarrow x + y$ and $(x, y) \longrightarrow xy$ satisfying

1. K is a commutative group under the operation $(x, y) \longrightarrow x + y$ (i.e., K is a commutative group under addition).
2. $(xy)z = x(yz)$ (multiplication is associative).
3. $\left.\begin{array}{rcl} x(y+z) & = & xy + xz \\ (y+z)x & = & yx + zx \end{array}\right\}$ (distribution law holds).

If $xy = yx$ for all x and y in K, the ring K is **commutative**. If there is an element 1 in $x \ni x = x1 = x$ for each x, K is said to be a ring with identity 1 (it is called the identity for K).

Field: Let K be the set of real or complex numbers.

Field axioms:

1. Addition is commutative

$$x + y = y + x \quad \text{for all } x, y \text{ in } K$$

2. Addition is associative

$$(x + y) + z = x + (y + z) \quad \text{for all } x, y, \text{ and } z \text{ in } K$$

3. There is a unique element 0 (zero) in $K \ni x + 0 = x$ for all x in K.

4. To each x in K there corresponds a unique element $-x$ in $K \ni x + (-x) = 0$ (i.e., additive inverse).

5. Multiplication is commutative

$$\forall\, x, y \in K$$

$$xy = yx$$

6. Multiplication is associative

$$(xy)z = x(yz) \;\forall\, x, y, z \in K$$

7. There is a unique element 1 (one) in $K \ni x1 = x \;\forall\, x$ in K.

8. To each non-zero x in K there corresponds a unique element x^{-1} in $K \ni xx^{-1} = 1$ (i.e., multiplicative inverse).

9. Multiplication distribution over addition $x(y+z) = xy+xz, \quad \forall\, x, y, z \in K$.

Order Axioms:

1. If $x \in y$ in K then only one is true

$$x < y\,, \qquad y < x \quad \text{or} \quad x = y$$

2. If $x < y$ if and only if $0 < y - x$

3. If $0 < x$ and $0 < y$ then $0 < (x + y)$ and $0 < xy$

Completeness Axioms: If S and T are the non-empty subsets of $K \ni$

(a) $K = S \cup T$.

(b) $s < t$ for every s in T and every $t < T$ then either there exists a largest number in the set S or there exists a smallest number in the set T.

Vector Space or Linear Space: A vector space consists of

1. Field K of scalars

2. A set V of objects (called vectors)

3. A rule (operation) called vector addition such that all scalars $c \in K$

(a) addition is commutative

(b) addition is associative

(c) $\exists$ unique vector $0 \ni x + 0 = x$, for all x $\in V$

(d) $\exists$ unique vector $-x \ni x + (-x) = 0$, for all x $\in V$

4. A rule (operation) called scalar multiplication

   (a) $1x = x$, for all x $\in V$

   (b) $(c_1 c_2)x = c_1(c_2 x)$, for all x $\in V$

   (c) $c(x + y) = cx + cy$, for all x,y $\in V$

   (d) $(c_1 + c_2)x = c_2 x + c_2 x$, for all x $\in V$

(i.e., vector space = field + a set of vectors and two operations with certain special properties).

Algebra: A (linear) algebra over the field K is a vector space $\mathcal{A}$ over K with an additional operation of vectors which associates with each pair of vectors x, y in $\mathcal{A}$ a vector xy in $\mathcal{A}$ called the product of x and y in such a way that

1. Multiplication is associative $x(yz) = (xy)z$

2. Multiplication is distributive with respect to addition

$$x(y + z) = xy + xz \quad \text{and} \quad (x + y)z = xz + yz$$

3. For each scalar c in K

$$c(xy) = (cx)y = x(cy)$$

If there is an element 1 in $\mathcal{A} \ni 1x = x1 = x$ for each x in $\mathcal{A}$, then $\mathcal{A}$ is a linear algebra with identity over K. Algebra $\mathcal{A}$ is called commutative if $xy = yx$ for all x, y in $\mathcal{A}$.

Normed (Linear) Space $\mathcal{L}$: Normed vector space

Definition: $\mathcal{L}$ a linear space with a norm $|| \ ||$ defined on it with these properties

(a) $||x|| \geq 0$

(b) $||x + y|| \leq ||x|| + ||y||$

(c) $||Cx|| = |C| \ ||x||$

and if $||x|| = 0$ then $x = 0$

If the space is complete and has a norm $\longrightarrow$ we call it a Banach space.

Inner Product Space: An inner product space is a linear space with an inner product on $\mathcal{L}$.

Inner product:

(a) $\langle x, x \rangle \geq 0$ if $\langle x, x \rangle = 0$ then $x = 0$

(b) $\langle y, x \rangle = \langle x, y \rangle^*$

(c) $\langle cx + y, z \rangle = c\langle x, z \rangle + \langle y, z \rangle$

A finite-dimensional normed linear space is called a *Euclidian space.* It is a normed linear space which consists of linear space R^n together with the norm

$$
\begin{aligned}
|\mathbf{x}| &= (|x_1|^2 + |x_2|^2 + \cdots + |x_n|^2)^{1/2} \\
&= \text{length}
\end{aligned}
$$

A complete inner product space is called a *Hilbert space.*

For additional reference consult [1,2].

References

[1] Hoffman, K., and Kunze, R., *Linear Algebra*, Englewood Cliffs, NJ: Prentice-Hall, 1961.

[2] Hoffman, K., *Analysis in Euclidcan Space*, Englewood Cliffs, NJ: Prentice-Hall, 1975.

APPENDIX D

Some Mathematical Expressions and Moments of Probability Density Function

$$I_0(x) = \frac{1}{2\pi}\int_0^{2\pi} e^{x\sin\theta} d\theta \qquad [1]$$

$$I_0 = \text{modified Bessel function } (1^{\text{st}} \text{ kind}) \text{ of order zero}$$

$$I_0(x) = \sum_{n=0}^{\infty} \frac{x^{2n}}{2^{2n}(n!)^2} = 1 + \frac{x^2}{2^2} + \frac{x^4}{2^2 \cdot 4^2} + \frac{x^6}{2^2 \cdot 4^2 \cdot 6^2} + \cdots$$

$$I_n(x) = j^{-n} J_n(jx) = e^{-j(n\pi/2)} J_n(jx)$$

$$I_0(x) = J_0(jx)$$

$$J_0(x) = \frac{1}{\pi}\int_0^{\pi} \cos(x\sin\theta) d\theta = 1 - \frac{x^2}{2^2} + \frac{x^4}{2^2 \cdot 4^2} - \frac{x^6}{2^2 \cdot 4^2 \cdot 6^2} + \cdots \qquad [1]$$

$$= \frac{1}{2\pi}\int_0^{2\pi} e^{jx\sin\theta} d\theta$$

$$I_0(x) = \frac{1}{2\pi}\int_0^{2\pi} \exp(x\cos(\theta+\phi)) d\theta$$

$$\boxed{\text{for small } x \quad I_0 \approx 1 + \frac{x^2}{4}} \Longleftarrow$$

↙ asymptotic expansion (large arguments)

$$I_0(x) = \frac{\exp x}{\sqrt{2\pi x}}\left(1 + \frac{1}{8x} + \frac{1^2 \cdot 3^2}{2(8x)^2} + \cdots\right)$$

$$\boxed{\text{for large } x \quad I_0(x) \approx \frac{\exp(x)}{\sqrt{2\pi}\sqrt{x}}} \Longleftarrow$$

erf (x)

Define

$$erf(x) = \frac{1}{\sqrt{2\pi}} \int_{-\infty}^{x} e^{-y^2/2} dy$$

$$erfc(x) = \frac{1}{\sqrt{2\pi}} \int_{x}^{\infty} e^{-y^2/2} dy$$

$$= 1 - erf(x) = Q(x)$$

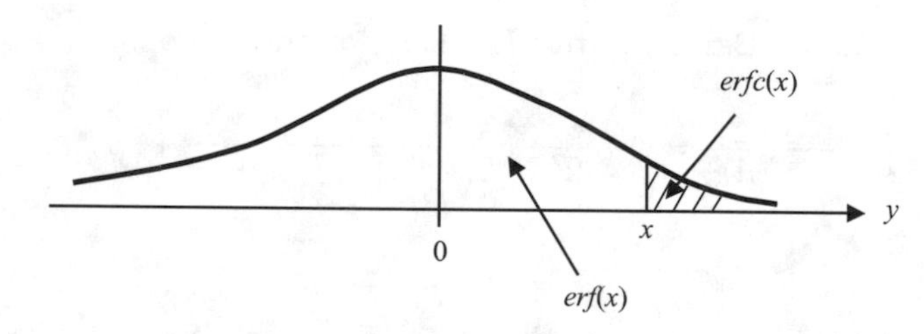

Note 1:

$x_1 < 0 \quad \Rightarrow$ We see

↙ a negative number

$erfc(x_1) = erf(|x_1|)$ ← a positive number

$erf(x_1) = erfc(|x_1|)$

$\text{erf }(x_1) = 1 - erf(-x_1)$

Note 2: Beware of *definition* (table) of the error function.

$erfc(x) \cong \left(\sum_{i=1}^{5} a_i t^i\right) e^{-x^2/2}$	for $x > 0$ accuracy : error $< 7.5 \cdot 10^{-8}$

where

$a_1 = 0.127414796$

$a_2 = -0.142248368 \qquad t = 1/(1 + ex)$

$a_3 = 0.710706871$

$a_4 = -0.726576013 \qquad e = 0.2316419$

$a_5 = 0.530702714$

See Abramowitz and Stegun [2].

We can also solve for x iteratively, when $0 < P_{FA} < 1/2$

$$x = erfc^{-1}(PFA)$$

Guess a reasonable x_0

$$
\begin{aligned}
x_1 &= \left[2\ln\left(\frac{R(t_0)}{P_{FA}}\right)\right]^{1/2} && \text{where} \quad t_0 = \frac{1}{1+ex_0} \\
x_2 &= \left[2\ln\left(\frac{R(t_1)}{P_{FA}}\right)\right]^{1/2} && t_1 = \frac{1}{1+ex_1} \\
&\vdots
\end{aligned}
$$

$$
\text{repeat until} \quad x_i - x_{i-1} < \in
$$

where $R(t)$ is from last page

$$
R(t) = \sum_{i=1}^{5} a_i t^i
$$

Note: For $N \sim (m, \sigma^2)$ over $(-\infty, x))$

$$
\boxed{
\begin{aligned}
&\frac{1}{\sqrt{2\pi}\,\sigma}\int_{-\infty}^{x} e^{-(y-m)^2/2\sigma^2}\,dy = erf\left(\frac{x-m}{\sigma}\right) \\
&= \frac{1}{\sqrt{2\pi}}\int_{-\infty}^{x-m/\sigma} e^{-t^2/2}dt \\
&= 1 - Q\left(\frac{x-m}{\sigma}\right)
\end{aligned}
}
$$

where

$$
\begin{aligned}
Q\left(\frac{(x-m)}{\sigma}\right) &= \int_{(x-m)/\sigma}^{\infty} (2\pi)^{-1/2} \exp - \left(\frac{x^2}{2}\right) dx \\
&= erfc\left(\frac{(x-m)}{\sigma}\right)
\end{aligned}
$$

Complementary Error Function Table

$$ERFC(x) = \frac{1}{\sqrt{2\pi}} \int_x^{\infty} e^{-u^2/2} du$$

The values are computed using the MATLAB 4 or 5 version (The Mathworks, Inc.) as given below.

$$ERFC(x) = \frac{1}{2} erfc_{\mathrm{MATLAB}} \left(\frac{x}{\sqrt{2}} \right)$$

x	0	0.0100	0.0200	0.0300	0.0400	0.0500	0.0600	0.0700	0.0800	0.0900
0	0.5000	0.4960	0.4920	0.4880	0.4840	0.4801	0.4761	0.4721	0.4681	0.4641
0.1000	0.4602	0.4562	0.4522	0.4483	0.4443	0.4404	0.4364	0.4325	0.4286	0.4247
0.2000	0.4207	0.4168	0.4129	0.4090	0.4052	0.4013	0.3974	0.3936	0.3897	0.3858
0.3000	0.3821	0.3783	0.3745	0.3707	0.3669	0.3632	0.3594	0.3557	0.3520	0.3483
0.4000	0.3446	0.3409	0.3372	0.3336	0.3300	0.3264	0.3228	0.3192	0.3156	0.3121
0.5000	0.3085	0.3050	0.3015	0.2981	0.2946	0.2912	0.2877	0.2843	0.2810	0.2776
0.6000	0.2743	0.2708	0.2676	0.2643	0.2611	0.2578	0.2546	0.2514	0.2483	0.2451
0.7000	0.2420	0.2389	0.2358	0.2327	0.2296	0.2266	0.2236	0.2206	0.2177	0.2148
0.8000	0.2119	0.2090	0.2061	0.2033	0.2005	0.1977	0.1949	0.1922	0.1894	0.1867
0.9000	0.1841	0.1814	0.1788	0.1762	0.1736	0.1711	0.1685	0.1660	0.1635	0.1611
1.0000	0.1587	0.1562	0.1539	0.1515	0.1492	0.1469	0.1446	0.1423	0.1401	0.1379
1.1000	0.1357	0.1335	0.1314	0.1292	0.1271	0.1251	0.1230	0.1210	0.1190	0.1170
1.2000	0.1151	0.1131	0.1112	0.1093	0.1075	0.1056	0.1038	0.1020	0.1003	0.0985
1.3000	0.0968	0.0951	0.0934	0.0918	0.0901	0.0885	0.0869	0.0853	0.0838	0.0823
1.4000	0.0808	0.0793	0.0778	0.0764	0.0749	0.0735	0.0721	0.0708	0.0694	0.0681
1.5000	0.0668	0.0655	0.0643	0.0630	0.0618	0.0606	0.0594	0.0582	0.0571	0.0559
1.6000	0.0548	0.0537	0.0526	0.0516	0.0505	0.0495	0.0485	0.0475	0.0465	0.0455
1.7000	0.0446	0.0436	0.0427	0.0418	0.0409	0.0401	0.0392	0.0384	0.0375	0.0367
1.8000	0.0359	0.0351	0.0344	0.0336	0.0329	0.0322	0.0314	0.0307	0.0301	0.0294
1.9000	0.0287	0.0281	0.0274	0.0268	0.0262	0.0256	0.0250	0.0244	0.0239	0.0233
2.0000	0.0228	0.0222	0.0217	0.0212	0.0207	0.0202	0.0197	0.0192	0.0188	0.0183
2.1000	0.0179	0.0174	0.0170	0.0166	0.0162	0.0158	0.0154	0.0150	0.0146	0.0143
2.2000	0.0139	0.0136	0.0132	0.0129	0.0125	0.0122	0.0119	0.0116	0.0113	0.0110
2.3000	0.0107	0.0104	0.0102	0.0099	0.0096	0.0094	0.0091	0.0088	0.0087	0.0084
2.4000	0.0082	0.0080	0.0078	0.0075	0.0073	0.0071	0.0069	0.0068	0.0066	0.0064
2.5000	0.0062	0.0060	0.0058	0.0057	0.0055	0.0054	0.0052	0.0051	0.0049	0.0048
2.6000	0.0047	0.0045	0.0044	0.0043	0.0041	0.0040	0.0039	0.0038	0.0037	0.0036
2.7000	0.0035	0.0034	0.0033	0.0032	0.0031	0.0030	0.0029	0.0028	0.0027	0.0026
2.8000	0.0026	0.0025	0.0024	0.0023	0.0023	0.0022	0.0021	0.0021	0.0020	0.0019
2.9000	0.0019	0.0018	0.0018	0.0017	0.0016	0.0016	0.0015	0.0015	0.0014	0.0014

Some Handy Expressions

1. $$\sum_{i=0}^{N-1}(a+id)=\frac{1}{2}N(a+e)$$

   where $e=a+(N-1)d$

   $$1+2+3+\cdots+N=\frac{1}{2}N(N+1)$$

   $$1+3+5\cdots+(2N_1)=N^2$$

2. $$\sum_{i=0}^{N-1}ar^i=\frac{a(1-r^N)}{1-r} \qquad r\neq 1$$

   if $-1<r<1$ then $\displaystyle\sum_{i=0}^{\infty}ar^i=\frac{a}{1-r}$

3. $$\ln(1+x)=x-\frac{x^2}{2}+\frac{x^3}{3}-\frac{x^4}{4}+\cdots\;; \qquad -1<x\leq 1$$

4. $$I_0(x)=1+\frac{x^2}{2^2}+\frac{x^4}{2^2\cdot 4^2}+\frac{x^6}{2^2\cdot 4^2\cdot 6^2}+\cdots$$

   $$\cong 1+\frac{x^2}{4} \qquad \text{for small } x$$

5. $$I_0(x)=\frac{\exp(x)}{\sqrt{2\pi x}}\left\{1+\frac{1}{8x}+\frac{1^2\cdot 3^2}{2(8x)^2}+\cdots\right.$$

   $$\sim\frac{\exp(x)}{\sqrt{2\pi x}} \qquad \text{for large } x$$

Density	Functional Form	Characteristic Function	Mean	σ^2
Poisson	$P_r(x) = \frac{a^x e^{-a}}{x!}$; $x = 0, 1, \cdots$ $a > 0$	$\phi_X(\omega) = e^{a(e^{j\omega}-1)}$	a	a
Chi-square	$f_X(x) = \frac{1}{\left(\frac{n}{2}-1\right)!2^{n/2}} x^{n/2-1} e^{-x/2}$ for $x > 0$	$\phi_X(\omega) = \frac{1}{(1-2j\omega)^{n/2}}$	n	$2n$
Exponential	$f_X(x) = ae^{-ax}$ for $x > 0$	$\phi_X(\omega) = \frac{a}{(a - j\omega)}$	$\frac{1}{a}$	$\frac{1}{a^2}$
Gaussian	$f_X(x) = \frac{1}{\sqrt{2\pi\sigma^2}} e^{-(x-\bar{x})^2/(2\sigma^2)}$	$\phi_X(\omega) = e^{j\omega\bar{x}} e^{-\omega^2\sigma^2/2}$	$\bar{x}$	σ^2
Rayleigh	$f_X(x) = \frac{x}{a^2} e^{-x^2/2a^2}$ for $x > 0$		$a\sqrt{\frac{\pi}{2}}$	$\left(2 - \frac{\pi}{2}\right) a^2$
Uniform	$f_X(x) = \frac{1}{b-a}$ for $a < x < b$ $-\infty < a < b < \infty$	$\phi_X(\omega) = \frac{e^{j\omega b} - e^{j\omega a}}{j\omega(b-a)}$	$\frac{a+b}{2}$	$\frac{(b-a)^2}{12}$
Maxwell for $x > 0$	$f_X(x) = \sqrt{\frac{2}{\pi}} a^3 x^2 e^{-a^2x^2/2}$ $a > 0$		$\sqrt{\frac{8}{\pi}} \frac{1}{a}$	$\left(3 - \frac{8}{\pi}\right) \frac{1}{a^2}$

Density	Functional Form	Characteristic Function	Mean	σ^2
Erlang	$f_X(x)$ $= \dfrac{a^n x^{n-1} e^{-ax}}{(n-1)!}$ for $x > 0$ $n = 1, 2, \cdots$ $a > 0$	$\phi_X(\omega)$ $= \dfrac{a^n}{(a - j\omega)^n}$	$\dfrac{n}{a}$	$\dfrac{n}{a^2}$
Log normal	$f_X(x)$ $= \dfrac{\bar{e}^{-\{\ln(x-a)-b\}^2/2\sigma^2}}{\sqrt{2\pi\sigma^2}\,(x-a)}$ $\sigma > 0,\ x > a$ $-\infty < a/b < 0$		$a + e^{b+\sigma^2/2}$	$e^{2b+\sigma^2}(e^{\sigma^2} - 1)$
Weibull	$f_X(x)$ $= abx^{b-1}e^{-ax^b}$ for $x > 0$ $a, b > 0$		$\left(\dfrac{1}{a}\right)^{1/b}$ $\Gamma\left(1 + \dfrac{1}{b}\right)$	$\left(\dfrac{1}{a}\right)^{2/b}$ $\left\{\Gamma\left(1 + \dfrac{2}{b}\right)\right.$ $\left.\left[\Gamma\left(1 + \dfrac{1}{b}\right)\right]^2\right\}$
Cauchy	$f_X(x)$ $= \dfrac{a/\pi}{a^2 + (x-b)^2}$; $a > 0$ for $-\infty < x < \infty$	$\phi_X(\omega)$ $= e^{jb\omega - a\lvert\omega\rvert}$		

References

[1] Spiegel, M.R., *Schaum's Outline Series Theory and Problems: Mathematical Handbook of Formlulas and Tables*, New York: McGraw-Hill Book Co., 1968.

[2] Abramowitz, M., and Stegun, I.A., *Handbook of Mathematical Functions*, New York: Dover Publications, 1970, p. 932.

APPENDIX E

Wavelet Transforms

The purpose of this appendix is to provide a quick introduction into wavelet transform processing. The wavelet transform (WT) takes a one-dimensional function and decomposes it into a two-dimensional expression. For simplicity, we will use a time function or, if appropriate, a time sequence of one dimension. The nature of the variables of the transformation (continuous or discrete-time and continuous or discrete co-domain parameters) determine the name of the transformation. This is similar to conventional Fourier transform terminology [1], as is briefly indicated next.

Fourier Transform (FT)

1. When both domains are continuous and non-periodic, the transform is called the Fourier transform (FT).
2. If the time domain is discrete, the transform is called the discrete-time Fourier transform (DTFT). The frequency domain representation will be periodic.
3. If the function is continuous and periodic in time, the transform is called the Fourier series (FS). The frequency domain representation will be discrete.
4. If the function is discrete in time and finite in duration, the transform is called the discrete Fourier transform (DFT). The frequency representation will be discrete and periodic. Its fast cousin is the well-known fast Fourier transform (FFT). Some readers will recognize that the DFT is just the discrete-time Fourier series.

Wavelet Transform (WT)

The wavelet transform (WT) has a one-dimensional time domain and a two-dimensional transform domain. The two transform dimensions are the shift (also called delay) and scaling dimensions. The delay shifts the basis function along the segment of data under consideration, while the scale sets up a particular spectral filter frequency and bandwidth. The WT can be characterized as in the FT case.

1. If the input is continuous and outputs that are continuous in both dimensions (delay and scale), the transform is called the continuous wavelet transform (CWT).

2. If the input is continuous and the outputs are discrete (shift and scale are evaluated on a grid), the transformation is called the discrete wavelet transform (DWT).
3. If the time signal is discrete and the transformed dimensions are discrete, the transform is called the discrete-time wavelet transform (DTWT). The DTWT is the one that is usually performed in WT applications. Using the so-called Mallat algorithm leads to a fast implementation, sometimes called the fast wavelet transform.
4. If the input time signal is discrete and the transformed dimensions are continuous, the transform is called the discrete-time continuous wavelet transform (DTCWT). The DTCWT is mentioned for completeness sake. The author is not aware of any applications of this transform.

We will examine the CTWT and the DTWT in some detail. The discussion and notation will follow the ones advocated in [2]. This appendix is written so that it can be followed with little difficulty. The only oddity is our convention to label (order) scales in the same spectral sense as the frequencies. From a vector space point of view and from a simpler labeling ability, our ordering is meaningful. We note scale labeling is totally arbitrary. Common sense suggests that the labeling goes in the opposite direction that frequency labeling goes, which indeed most articles and books adhere to. We will use the words shift and delay interchangeably. A wavelet, denoted by $\psi(t)$, is the term coined to describe an oscillatory function that exists over a relatively short period of time and has no DC component. We start the introduction by considering continuous time domain functions and the CTWT.

Continuous Time Wavelet Transform (CTWT)

We assume that the time function of interest is band limited and that we can represent any band limited function by a weighted combination of basis functions. We assume that these basis functions are such that any possible signal function, band limited as discussed, can be modeled (expressed) as a linear combination of these basis functions. It may help to think in terms of a particular basis function set, such as the FT, that is any band limited function can be expressed as a linear combination of complex exponentials, i.e.,

$$f(t) = \int_{-\infty}^{\infty} F(f)\, e^{\,(j\ \pi\ f\ t)}\, df$$

where $F(f)$ is the weight and $\{e^{(j2\pi t)}\}$ are the basis functions.

In the wavelet transform representation, one is not limited to sinusoids, rather than that, any band pass function can be used. This means the wavelet (i.e., band pass) functions do not allow a DC component, and must be oscillatory and of finite duration (i.e., compact support). This means that only bandpass phenomena can be represented with wavelets. One extreme case is the set of the Haar basis functions, also called Daubechies wavelets of order 2. They consist of functions that are positive for the first and negative for the second half of the time support. The magnitude is the same for both halves of the support. Another interpretation of the Haar function is that of a hard clipped sine wave. We note that the time support is substantially shorter than the data length. To span another bandpass region, the wavelet function has to be dilated or expanded. The original wavelet function is also called the mother wavelet. If the wavelet (basis) function is dilated (compressed) the frequency range is decreased (increased). Suppose that we have a prototype that is a real valued wavelet function (mother wavelet) denoted by $\psi(t)$, we define the wavelet transform as

$$\begin{aligned} W_f(a,b) &= WT\{f(t)\} \\ &= \langle \psi_{ab}(t), f(t) \rangle \\ &= \int_{-\infty}^{\infty} \psi_{ab}(t)\ f(t)\ dt \end{aligned}$$

The inverse exists if

$$C_\phi = \int_0^\infty \frac{|\psi(\Omega)|}{\Omega}\ d\Omega \ < \infty$$

and

$$\Psi(0) = \int_{-\infty}^{\infty} \psi(t)\ dt = 0$$

then the original time function is given by

$$f(t) = \frac{1}{C_\psi} \int_{-\infty}^{\infty} \int_0^\infty WF_f(a,b)\ \psi_{ab}(t)\ da \frac{db}{a^2}$$

Discrete Wavelet Transform (DWT)

The CWT is very redundant and not very practical. To eliminate these problems, the shift (delay) and scale parameters are allowed to take on discrete values only (i.e., sample the CWT at discrete points in its transform domain). Given any spectral region $(-f_B, f_B)$, we can partition the spectral region into two equal sized parts (i.e., a low pass and a high pass region). Hence we can talk about two different sets of basis functions that span these two spectral regions. The low pass region is spanned by the scaling (i.e., low pass) functions while the band pass region is spanned by the wavelet functions. Let us take the example of band limited spectral region as plotted in the drawing on the next page and interpret the different spectral regions.

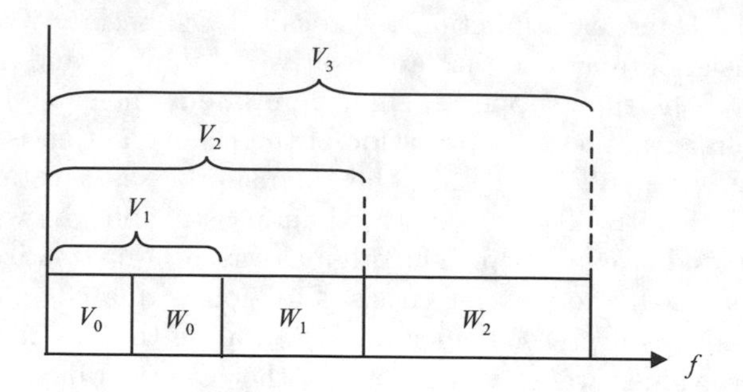

The regions W_0, W_1, and W_2 are band pass regions, while V_0, V_1, V_2, and V_3 are low pass regions. The direct sum of $V_0 \oplus W_0 = V_1$ and $V_1 \oplus W_1 = V_2$, etc. Suppose we have a prototype real valued function $\psi(t)$, which is a bandpass (wavelet) function and a low pass (scaling) function $\phi(t)$.

Define $\phi_k(t) = \phi(t-k)$, for k $\in$ Z, and $\phi \in \mathrm{L}^2$ so that

$$V_0 = \text{span}\,\{\phi_k(t)$$

The subspaces, as indicated in the above drawing, are arranged as

$$\{0\} \ldots \subset V_{-2} \subset V_{-1} \subset V_0 \subset V_1 \subset V_2 \ldots L^2$$

where $V_j \oplus W_j = V_{j+1}$.

Suppose that $f(t) \in V_j$ then $f(2t) \in V_{j+1}$. The inner product of the scaling and wavelet function is given by

$$\langle \phi_{jk}(t), \psi_{jl}(t) \rangle = \int \phi_{jk}(t)\, \psi_{jl}(t)\; dt = 0$$

which says that for all scales labeled j and for all delays, the scaling and wavelet functions are orthogonal. Also

$$\int \phi(x)\; \phi(x-m)\; dx = K\delta(m)$$

which says that the scaling function (in a given scale) is delta correlated. The energy of the scaling function is denoted by K. If K equals one, then the basis functions are normalized.

In general, any function $f(t)$ can be written as

$$f(t) = \sum_{k=-\infty}^{\infty} c(k)\; \phi(t-k) + \sum_{m=0}^{\infty} \sum_{k=-\infty}^{\infty} d(m,k)\; \psi_{m,k}(t)$$

where $\phi_{mk}(t) = 2^{m/2}\; \phi(2^{m/2}\; t - k)$.

Since $\phi(2t-n)$ spans $V_1 = V_0 \oplus W_0$, $\phi(t-n)$ spans V_0, and $\psi(t-n)$ spans W_0 a weighted sum will span V_0, as given by

$$\phi(t) = \sum_n h(n)\sqrt{2}\ \phi(2t-k) \quad \text{the scaling equation}$$

and a weighted sum will span W_0 as given by

$$\psi(t) = \sum_n h_1(n)\sqrt{2}\ \phi(2t-k) \quad \text{the wavelet equation}$$

As it turns out, $h_1(n)$ is such that $h_1(n) = (-1)^n\ h(l-n)$. It is a time reversed, odd ordered locations sign reversed, copy of the FIR filter denoted by $h(n)$.

Discrete-Time Wavelet Transform (DTWT)

With some work, one can obtain a recursive relationship between $c_m(k)$ and $d(m,k)$, allowing addressing discrete-time problems by replacing c_{m+1} with $f(n)$, the low pass signal with the highest sampling rate (i.e., original sampled time data):

$$\begin{aligned} c_m(k) &= \sum_l h(l-2k)\ c_{m+1}(l) \\ d(m,k) = d_m(k) &= \sum_l h_1(l-2k)\ c_{m+1}(l) \end{aligned}$$

This arrangement is shown, suppressing the time dependency, in the following drawings.

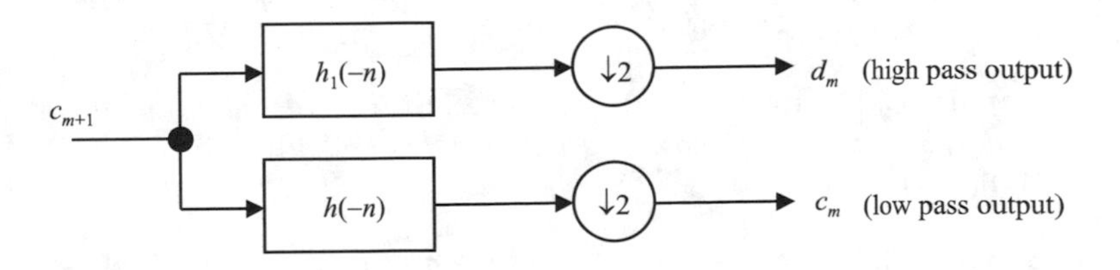

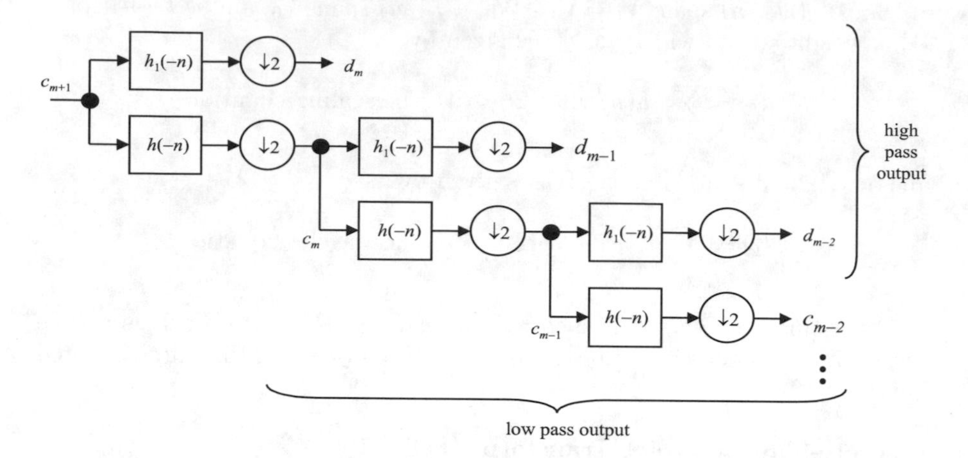

In this decomposition (Mallat's algorithm), d_m represents the detail (high pass), while c_m stands for the low pass output, at scale m. The decomposition can be stopped anywhere in the scale space. If it is allowed to continue until it runs out of data, then at the bandpass filter and the associated low pass filter at the last scale will output only one data point. Typically, the data length is chosen to be $N = 2^r$, where r is an integer. Then the number of stages (i.e., scales) is $ln\ 2^r = r$, if indeed a total decomposition is desired. The process can easily be inverted to recover the original signal $f(t)$.

Chapter 3 and Appendix E constitute an attempt to provide a simplified introduction to wavelet processing. For more detail, see references [2,3].

References

[1] Oppenheim, A.V., Willsky, A.S., and Nawab, S.H., *Signals and Systems*, 2nd ed., Upper Saddle River, NJ: Prentice-Hall, 1997.

[2] Burrus, C.S., Gopinath, R.A., and Guo, H., *Introduction to Wavelets and Wavelet Transforms: A Primer*, Upper Saddle River, NJ: Prentice-Hall, 1998.

[3] Strang, G., and Nguyen, T., *Wavelets and Filter Banks*, Wellesley, MA: Wellesley-Cambridge Press, 1996.

Index

I

K

L

M

N

O

P

Q

R

S

T

U

W

Z